Making Sense of Quantum Mechanics

Jean Bricmont

Making Sense of Quantum Mechanics

 Springer

Jean Bricmont
IRMP
UCLouvain
Louvain-la-Neuve
Belgium

ISBN 978-3-319-25887-4   ISBN 978-3-319-25889-8   (eBook)
DOI 10.1007/978-3-319-25889-8

Library of Congress Control Number: 2015956126

Printed on acid-free paper

This Springer imprint is published by SpringerNature
The registered company is Springer International Publishing AG Switzerland

Preface

This book is both apparently ambitious and modest in its aims. Ambitious, as it attempts to achieve something that has been declared impossible by some of the greatest physicists since the 1920s: making sense of what quantum mechanics really means. But modest, because that goal was actually already attained many years ago in the work of Louis de Broglie, David Bohm, and John Bell. I will simply try to explain what they achieved.

This book is written especially for all those students who feel that they have not understood the subject of quantum mechanics, not because they fail to master the mathematics or because they cannot do the exercises, but because they do not see what the theory means.

However, no prior knowledge of quantum mechanics is required. Most of the technical parts have been put in appendices, which can be skipped if one is willing to take certain results for granted.

Hopefully this book should also interest philosophers and historians of science, in particular Chaps. 1, 3, 7, and 8.

The analysis presented here has benefited from such a large number of discussions, seminars, and exchanges with so many people that thanking them all by name would scarcely be possible.

However, I must stress that I learned most of what I know about the subject through discussions with Detlef Dürr, Tim Maudlin, Nino Zanghì, and especially with Sheldon Goldstein.

Many readers of parts of this book have made useful comments and corrections and I wish to thank them: Xavier Bekaert, Serge Dendas, Lajos Diósi, Michel Ghins, Michel Hellas, Dominique Lambert, Vincent Mathieu, Alexis Merlaud, Amaury Mouchet, and Alan Sokal. My special thanks to Ward Struyve for helping me with his great scientific expertise on the topic of this book. I thank also Stephen Lyle for his careful reading of the manuscript and many clarifying exchanges. Needless to say, all remaining errors are mine.

I wish to thank Cathy Brichard warmly for her indispensable secretarial help. And finally, heartfelt thanks to my editor Angela Lahee, without whom this book would not have seen the light of day, for her encouragement and her patience.

Contents

Chapter 1
Physicists in Wonderland

1.1 What This Book Is About

According to the French newspaper *Le Monde*, a famous English rugby player, Jonny Wilkinson, claims to have been "saved from depression" by studying quantum physics [308]. The player even held a public conference with two well known French physicists, Jean Iliopoulos and Étienne Klein, attended by 500 people, and the conference was published (in French) under the title "Quantum Rugby" [519]. Interviewed by *Le Monde*, Étienne Klein says that the player did not really know quantum physics, and mostly relied on metaphors, linked to his interest in Buddhism.

If what Wilkinson understood of quantum mechanics is uncertain, one can be reasonably sure that nobody would claim to have been saved from a depression by studying any physical theory other than quantum theory.

Since its beginnings in 1900, the quantum theory[1] has led to the most spectacularly well confirmed predictions ever made in science (some experimental results agree with the theoretical predictions up to one part in a billion), and it underpins all modern electronics and telecommunications. It explains the stability of atoms and of stars, and lies at the foundation of the whole of particle physics, but also solid state physics, chemistry, and thus, in principle, biology. It is truly our most fundamental theory of the world. Yet, to quote the famous American physicist Richard Feynman,[2] "nobody understands quantum mechanics" [185].

---

[1] Since this book deals mostly with non-relativistic quantum physics, we will use the expressions "quantum mechanics", "quantum physics", or "quantum theory" interchangeably.

[2] Feynman was comparing the situation in quantum mechanics with the one in the theory of relativity [185]: "There was a time when the newspapers said that only twelve men understood the theory of relativity. I do not believe there ever was such a time. There might have been a time when only one man did, because he was the only guy who caught on, before he wrote his paper. But after people read the paper a lot of people understood the theory of relativity in some way or other, certainly more than twelve. On the other hand, I think I can safely say that nobody understands

© Springer International Publishing Switzerland 2016
J. Bricmont, *Making Sense of Quantum Mechanics*,
DOI 10.1007/978-3-319-25889-8_1

In a nutshell, the goal of this book is to explain, in the simplest possible terms, what is so bizarre about quantum mechanics and, nevertheless, to try to show that one can to some extent understand its mysteries in rational terms.[3] This is not a book that will teach quantum mechanics in its technical aspects (there are plenty of good books doing that[4]). It will deal only with the conceptual problems associated with quantum mechanics.

"In the simplest possible terms" means with a minimum of mathematics, but a minimum that is not zero: this is not a "popular" book. However, the level of mathematics required is only what is typically taught in first or second year courses of mathematics for scientists and engineers: elementary linear algebra, including vector spaces and matrices, complex numbers, Fourier transforms, basic differential equations and some classical mechanics; but even that will be largely discussed in the appendices. Some technical aspects will be put in the footnotes (hoping that the experts will not be too irritated by the simplifications introduced in the text), where one will also find references to the more advanced literature.

In this introduction, we will discuss the mysteries of quantum mechanics, or at least the way they are often presented to the public, in non-technical terms. It will give a feel for what is so strange in quantum mechanics.

Let us start with excerpts from an article on the "queerness of quanta" in *The Economist* [157] (quoted by Jeremy Bernstein in [56, p. 6]):

1. There are no such things as 'things.' Objects are ghostly, with no definite properties (such as position or mass) until they are measured. The properties exist in a twilight state of 'superposition' until then.

2. All particles are waves, and waves are particles, appearing as one sort or another depending on what sort of measurement is being performed.

3. A particle moving between two points travels all possible paths between them simultaneously.

4. Particles that are millions of miles apart can affect each other instantaneously.

The queerness of quanta [157]

These sentences express the two main "mysteries" of quantum mechanics, as they are usually presented, and which can be formulated as follows:

(Footnote 2 continued)

quantum mechanics." Many people, including some famous physicists, claim that the difficulty in understanding quantum mechanics is similar to that in understanding relativity, but this is just not so.

[3]There is an enormous amount of pseudo-scientific literature claiming to base itself on quantum mechanics. But we will not be concerned with that; given the way respectable scientists talk about quantum mechanics, as we will see in this book, its exploitation by the pseudo-sciences, while perfectly unfounded, may not be so surprising.

[4]In his critique of the standard discussions of the conceptual problems of quantum mechanics [46], Bell mentions three good books: those by Dirac [137], Landau and Lifshitz [302], and Gottfried [236], as well as an article by van Kampen [494]. These are classics and so is the one by Bohm [61]. One might add to that list the more recent one by Shankar [447]. However, Bell shows in [46] that even the good books do not deal in a satisfactory way with the conceptual problems.

a. Quantum objects can possess mutually exclusive properties, like traveling along different paths simultaneously. Further, quantum objects obtain definite properties when they are "measured", and *only* then. Moreover, those properties depend on which measurements we choose to make.

b. Particles can interact instantaneously even when they are arbitrarily far apart.

The last statement may sound violently counterintuitive, but it is not in principle impossible, and in fact it is the only one of the two that is essentially true. However, people who have "learned" from the special theory of relativity that "nothing goes faster than light" may wonder how interactions can be both instantaneous and take place between objects that are arbitrarily far apart. "Learned" here is put in scare quotes, because, although the statement "nothing goes faster than light" is frequently repeated, the actual implications of the theory of relativity are quite subtle, as we will discuss later.[5]

The statement about particles "traveling along different paths simultaneously", on the other hand, is obviously self-contradictory: by definition, a "particle" is something localized in space (as opposed to a wave, for example); to say that it follows two (or more) different paths at the same time makes no sense. How are we supposed to understand that? As a metaphor? But a metaphor of what? Is the particle divided into tiny parts, each of which follows a different path? Are we merely saying that we have no way of knowing which path is being taken (which is a meaning sometimes given to that statement)? In that case, the situation would be understandable, and not terribly surprising (why would gross creatures such as ourselves be able to follow the trajectories of tiny particles?), but admitting one's ignorance through a self-contradictory statement about the world is surely a rather strange way to express oneself.

The statement that "quantum objects obtain definite properties only when they are measured" may be the most fundamental and the most problematic. In almost all the talk about quantum mechanics, we find words such as "observer", "observation", and "measurement" playing a central role.[6] It is not just because observations are needed to verify or confirm the theory—that is true of all scientific theories. One would never hear a biologist speak of "observations" the way quantum physicists do, although biology is also an empirical science and thus is also "based on observations". For example, if biologists speak of dinosaurs, they speak of animals that lived in the past, not just of bones of dinosaurs, even though the bones are the only thing that we directly observe. Biology claims to study the properties of living beings, even when they are not observed, but the usual formulation of quantum mechanics speaks of systems having definite properties *only* when they are observed.

One of the main critics of the dominant discourse about quantum mechanics, the Irish physicist John Bell, raised the following objection:

---

[5]In Chap. 4 and in Sect. 5.2.2.

[6]The physical reasons for this emphasis on measurements or observations will be explained in Chap. 2.

The problem of measurement and the observer is the problem of where the measurement begins and ends, and where the observer begins and ends. Consider my spectacles, for example: if I take them off now, how far away must I put them before they are part of the object rather than part of the observer? There are problems like this all the way from the retina through the optic nerve to the brain and so on. I think, that—when you analyze this language that the physicists have fallen into, that physics is about the results of observations—you find that on analysis it evaporates, and nothing very clear is being said.

John S. Bell [107, p. 48].

There is another obvious objection, also raised by Bell [46, p. 34]: "What exactly qualifies some physical systems to play this role of 'measurer'?" And what happened before humans were around? Or maybe before modern laboratories were created? And what about the vast parts of the universe where there are no observers? Do the laws of physics cease to apply there or in the past? How can a physical theory, which is the most fundamental of all, which deals with atoms and elementary particles, and applies in principle to the entire universe, require for its very formulation something so contingent as certain manipulations ("measurements") done during the last 100 years by a few members of a particular species, *Homo sapiens*, living on a particular planet somewhere in the cosmos?

Since the seventeenth century, science has decentered human beings; for instance, by realizing that the Earth is not at the center of the universe and also by showing that humans are not the object of a special act of creation, but rather the result of a long and contingent evolution. Quantum mechanics seems to have put humans back at the center of the picture: it is sometimes claimed that it abolishes the distinction between subject and object or that it gives a special role, in the formulation of our most fundamental physical theory, to human consciousness. If that were so, one might wonder how humans got to be there in the first place: if it is through evolution, then how is that supposed to work? Biology is based on chemistry, whose mechanisms are explained through quantum mechanics. But what role did the human subject have during this whole process, before the appearance of *Homo sapiens*?

To understand where all these strange-looking ideas came from, we must go back to the beginning of quantum mechanics.

1.2 Back to Copenhagen

One may object that *The Economist* is not a scientific journal and that what is quoted here is just due to the desire of popularizers to make spectacular statements. But one finds similar statements coming from the founding fathers of quantum mechanics, especially those associated with the "Copenhagen interpretation" of quantum mechanics, including Niels Bohr, Max Born, Werner Heisenberg, Pascual Jordan, Wolfgang Pauli, and later John von Neumann.[7] The name "Copenhagen" comes from

[7]One should add to this list of founding fathers precursors like Max Planck and Albert Einstein, but also Louis de Broglie, Paul Dirac, and Erwin Schrödinger. However, Dirac was rather neutral on

the city where the Danish physicist Bohr lived and worked. However, it is far from clear that there is a unified doctrine, or even a well defined one, that can be systematically associated with the expression "Copenhagen interpretation" of quantum mechanics, since, for example, Heisenberg and Bohr had divergent views on many topics (see, e.g., [52, 275]). But there is a sort of vulgate in the popular literature, in philosophical reflections on quantum mechanics, and also in the teaching of quantum mechanics, whenever reference is made to "Copenhagen", that stresses the central role of observations in the very formulation of the theory.

Let us consider some of those associated with "Copenhagen", whether rightly or wrongly, and see what they said. Heisenberg,[8] for instance, wrote:

> [...] the idea of an objective real world whose smallest parts exist objectively in the same sense as stones or trees exist, independently of whether or not we observe them [...] is impossible [...]

> Werner Heisenberg [259, p. 129]

And again:

> We can no longer speak of the behavior of the particle independently of the process of observation. As a final consequence, the natural laws formulated mathematically in quantum theory no longer deal with the elementary particles themselves but with our knowledge of them. Nor is it any longer possible to ask whether or not these particles exist in space and time objectively [...].

> [...] Science no longer confronts nature as an objective observer, but sees itself as an actor in this interplay between man and nature. The scientific method of analysing, explaining, and classifying has become conscious of its limitations [...] method and object can no longer be separated.

> Werner Heisenberg [260, pp. 15, 29]

Here we encounter words such as "impossible", "no longer", etc., which are rather common in the "Copenhagen" rhetoric and which the historian of quantum mechanics Mara Beller calls the "rhetoric of inevitability" [52, Chap. 9], or what one might call the quantum mechanical version of Margaret Thatcher's TINA ("there is no alternative").

But how do they know that something is impossible? The fact that a theory is successful and that it does not permit us to answer certain questions or to think in certain ways does not, by any means, prove that one can never answer such questions or think otherwise in the future. To prove that something is impossible, one has to give arguments beyond the mere limitations of present-day science. Such arguments are sometimes given but, as we will see, they do not even begin to prove what is claimed.

(Footnote 7 continued)
these conceptual issues, de Broglie changed his views more than once, and Einstein and Schrödinger were strongly opposed to the Copenhagen interpretation.

[8]This is quoted and discussed by Sheldon Goldstein in [221].

Turning to Niels Bohr, Aage Petersen, who was his assistant for many years, characterized his views as follows[9]:

> When asked whether the algorithm of quantum mechanics[10] could be considered as somehow mirroring an underlying quantum world, Bohr would answer: "There is no quantum world. There is only an abstract physical description. It is wrong to think that the task of physics is to find out how nature is. Physics concerns what we can say about nature."
>
> Aage Petersen [395, p. 12]

Claims that "quantum theory no longer deals with the elementary particles themselves but with our knowledge of them" or that "physics concerns what we can say about Nature" were often heard in physics courses when I was a student, but it didn't make any sense to me then (or now). Indeed, if we say something about Nature or if we have knowledge of elementary particles, then we know something about the world, not just about our knowledge.

Besides, nothing in physics discusses the biological, psychological, or sociological factors that are usually associated with the acquisition of knowledge. So why this emphasis on knowledge? Only because quantum mechanics refers to some abstract "observer" which is a *deus ex machina* that gives definite properties to objects, but without explaining how this happens.

Going even further, Pascual Jordan, who was a very important contributor in the early days of quantum mechanics,[11] wrote:

> In a measurement of position, "the electron is forced to a decision. We compel it *to assume a definite position*; previously, it was, in general, neither here nor there; it had not yet made its decision for a definite position [...] If, in another experiment, the *velocity* of the electron is measured, this means: the electron is compelled to decide itself for some exactly defined value of the velocity; and we observe *which* value it has chosen. In such a decision, the decision made in the preceding experiment is completely obliterated."
>
> Pascual Jordan [285], quoted and translated by M. Jammer [281, p. 161] (original italics)

The defenders of the Copenhagen school sometimes present themselves as "hard-nosed scientists", whose views were driven by facts, while their opponents such as Einstein and Schrödinger were unable to accept the deep lessons of quantum mechanics because of their ideological and philosophical prejudices. For example, Max Born wrote that Einstein "could no longer take in certain new ideas in physics which contradicted his own firmly held philosophical convictions" [79, p. 72]. And Werner Heisenberg wrote:

---

[9]See [396] for a detailed presentation of Bohr's philosophy.

[10]This algorithm will be explained in Chap. 2. (Note by J.B.).

[11]As Norton Wise has shown [520], Jordan had rather strange views on biology (vitalism), parapsychology, and psychoanalysis and he was a committed member of the National Socialist party, mixing up his views on quantum mechanics with his politics. The subject-centered aspect of quantum mechanics was good news for him, since it put one more nail in the coffin of the Enlightenment. After the war, Jordan reincarnated himself as a democratic cold warrior, arguing for the nuclear armament of Germany, and denouncing the "naïve illusions" of pacifist-minded people such as Max Born. Concerning Jordan and his relationship with Bohr, see Heilbron [255].

Most scientists are willing to accept new empirical data and to recognize new results, provided they fit into their philosophical framework. But in the course of scientific progress it can happen that a new range of empirical data can be completely understood only when the enormous effort is made to enlarge this framework and to change the very structure of the thought process. In the case of quantum mechanics, Einstein was apparently no longer willing to take this step, or perhaps no longer able to do so.

Werner Heisenberg [79, p. X]

But the following quotes from Bohr, Pauli, and Born show that some of the founding fathers were not hostile to linking quantum mechanics and non-scientific speculations (to put it mildly)[12]:

[...] this domain [psychology] [...] is distinguished by reciprocal relationships which depend on the unity of our consciousness and which exhibit a striking similarity with the physical consequences of the quantum of action. We are thinking here of well-known characteristics of emotion and volition which are quite incapable of being represented by visualizable pictures. In particular, the apparent contrast between the conscious onward flow of associative thinking and the preservation of the unity of the personality exhibit a suggestive analogy with the relation between the wave description of the motions of material particles, governed by the superposition principle,[13] and their indestructible individuality.

Niels Bohr [71, p. 99]

[...] science and religion *must* have something to do with each other. (I do *not* mean "religion within physics", nor do I mean "physics inside religion", since either one would certainly be "one-sided", but rather I mean the placing of both of them within a whole.) I would like to make an attempt to give a name to that which the new idea of reality brings to my mind: the idea of reality of the symbol. [...] It contains something of the old concept of God as well as the old concept of matter (an example from physics: the atom. The primary qualities of filling space have been lost. If it were not a symbol how could it be "both wave and particle"?). The symbol is symmetrical with respect to "this side' and 'beyond" [...]. The symbol is like a god that exerts an influence on man but which also demands from man that he have a back effect on him (the God symbol).

Wolfgang Pauli [375, pp. 193–194] (italics in the original)

This comes from a private letter, but the reader can find several favorable references to Jungian psychoanalysis in Chaps. 17 and 21 of Pauli's *Writings on Physics and Philosophy* [382].[14]

The thesis 'light consists of particles' and the antithesis 'light consists of waves' fought with one another until they were united in the synthesis of quantum mechanics. [...] Only why not apply it to the thesis Liberalism (or Capitalism), the antithesis Communism, and expect a synthesis, instead of a complete and permanent victory for the antithesis? There

---

[12]The three quotes here come from Mara Beller's article [51], which we will discuss in Chap. 8. The quote from Pauli comes from a *private* letter.

[13]This principle will be explained in Chap. 2. (Note by J.B.).

[14]See also [306, 16] for Pauli's views on religion and "deep psychology".

seems to be some inconsistency. But the idea of complementarity[15] goes deeper. In fact, this thesis and antithesis represent two psychological motives and economic forces, both justified in themselves, but, in their extremes, mutually exclusive. [...] there must exist a relation between the latitudes of freedom Δf and of regulation Δr, of the type $\Delta f \Delta r \sim p$ which allows a reasonable compromise. But what is the 'political constant' p? I must leave this to a future quantum theory of human affairs.

Max Born [78, pp. 107–108]

Finally, according to the Belgian physicist Léon Rosenfeld, a close friend of Bohr's and one of the most vehement defenders of the Copenhagen interpretation,[16] Bohr seems sometimes to have suffered from delusions of grandeur:

On one of those unforgettable strolls during which Bohr would so openly disclose his inner-most thoughts, we came to consider that what many people nowadays sought in religion was a guidance and consolation that science could not offer. Thereupon Bohr declared, with intense conviction, that he saw the day when complementarity would be taught in the schools and become part of general education; and better than any religion, he added, a sense of complementarity would afford people the guidance they needed.

Léon Rosenfeld [422]. Reprinted in [424, p. 535]

1.3   What Do Physicists Say Now?

The reader might think that these quotes are old, going back to the very beginning of quantum mechanics, and that the situation has been clarified since then. As we will see in Chaps. 4 and 5, the situation was actually clarified, first in 1952 through the work of David Bohm, and then in 1964 through that of John Bell, but few physicists have paid much attention to those contributions. So what did the generations following the founding fathers say, even long after 1964?

John Archibald Wheeler, who studied with Bohr and is well known for his contri-butions both to nuclear physics and to cosmology, is famous for saying [509, p. 192]: "No elementary phenomenon is a phenomenon until it is a registered (observed) phe-nomenon." Wheeler linked that idea to what he called "the participatory principle":

According to it we could not even imagine a universe that did not somewhere and for some stretch of time contain observers because the very building materials of the universe are these acts of observer-participancy. You wouldn't have the stuff out of which to build the universe otherwise.

John Archibald Wheeler [508]

Wheeler emphasized that this means that *past* events did not really occur until they are recorded *now*:

---

[15] A basic concept of Bohr, which will be discussed in Chap. 2, particularly in Appendix 2.C. (Note by J.B.).

[16] As we will see in Chap. 7.

[…] we have to say that we ourselves have an undeniable part in shaping what we have always called the past. The past is not really the past until it has been registered. Or put another way, the past has no meaning or existence unless it exists as a record in the present.

John Archibald Wheeler [107, pp. 67–68]

Wheeler even wondered: "Are billions upon billions of acts of observer-participancy the foundation of everything?" Without answering this question affirmatively, he observed [509, p. 199]: "The very fact that we can ask such a strange question shows how uncertain we are about the deeper foundations of the quantum and its ultimate foundation."

Eugene Wigner, co-recipient of the 1963 Nobel Prize in physics for his contributions to quantum and nuclear physics was quite explicit about the role of consciousness in physics, supposedly revealed by the quantum revolution:

[…] It will remain remarkable, in whatever way our future concepts may develop, that the very study of the external world led to the scientific conclusion that the content of the consciousness is an ultimate reality.

[…] The preceding argument[17] for the difference in the roles of inanimate observation tools and observers with a consciousness—hence for a violation of physical laws where consciousness plays a role—is entirely cogent so long as one accepts the tenets of orthodox quantum mechanics in all their consequences.

Eugene Wigner [514]; reprinted in [510, pp. 169,178]

Wigner added that the "weakness" of this argument came from the "ephemeral nature of physical theories" and because it relied on "the tenets of orthodox quantum mechanics in all their consequences". But he was absolutely convinced[18] that it was [510, p. 169] "not possible to formulate the laws of quantum mechanics in a fully consistent way without reference to the consciousness."

Rudolf Peierls, who studied with Heisenberg and Pauli, and was a major theoretical physicist, both in quantum and statistical physics, wrote in 1979:

In recent years the debate on these ideas has reopened, and there are some who question what they call "the Copenhagen interpretation of quantum mechanics—as if there existed more than one possible interpretation of the theory.

Rudolf Peierls [388, p. 26]

Rudolf Peierls also declared in an interview published in 1993:

---

[17]The argument is based on the reduction or collapse of the quantum state, which will be defined in Sect. 2.3. (Note by J.B.).

[18]Here, Wigner refers in a footnote to some of the statements by Heisenberg quoted in Sect. 1.2. According to Wigner, Heisenberg had "expressed this [idea] most poignantly". He also refers to London and Bauer [312] who wrote in 1939 a detailed theory of measurement in quantum mechanics, which stressed "the essential role played by the consciousness of the observer" [510, p. 251]. To be fair to Wigner, one must add that his ideas on the role of consciousness in quantum mechanics changed over time (see Esfeld [177]).

You see, the quantum mechanical description is in terms of knowledge. And knowledge requires *somebody* who knows.

Rudolf Peierls [107, p. 74] (italics in the original)

Not surprisingly, when asked about the role of consciousness "in the nature of reality", Peierls answered [107, p. 74]: "I don't know what reality is."

In 1979, a well-known French theoretical physicist and philosopher of science, Bernard d'Espagnat, wrote an article in *Scientific American* with the following title[19]:

The doctrine that the world is made up of objects whose existence is independent of human consciousness turns out to be in conflict with quantum mechanics and with facts established by experiment.

Bernard d'Espagnat [124, p. 158]

D'Espagnat published another article, in the *Guardian* on 20 March 2009, with the title [125]: "Quantum weirdness: What we call 'reality' is just a state of mind."

The Cornell university physicist David Mermin, well known for his work in statistical and condensed matter physics and who also worked a lot on foundations of quantum mechanics, wrote in 1981:

We now know that the moon is demonstrably not there when nobody looks.[20]

David Mermin [331, p. 397]

In 2005, Anton Zeilinger, who has performed in Vienna some of the most remarkable quantum experiments, wrote in *Nature*:

So, what is the message of the quantum? [...] I suggest that [...] the distinction between reality and our knowledge of reality, between reality and information, cannot be made.

Anton Zeilinger [526, p. 743]

Finally, in the age of Twitter, Sean Carroll, theoretical physicist at Caltech, cosmologist and author of several popular books, considers that the best answer to "how to

---

[19]In 2009, D'Espagnat won the Templeton Prize, which rewards a person who "has made an exceptional contribution to affirming life's spiritual dimension, whether through insight, discovery, or practical works".

[20]This refers to the following remark by Abraham Pais about his conversations with Einstein who, as we will see in the following section, was irritated by all the talk about "observations" [369, p. 907]: "We often discussed his notions on objective reality. I recall that during one walk Einstein suddenly stopped, turned to me and asked whether I really believed that the moon exists only when I look at it." Pais adds: "The rest of the walk was devoted to a discussion of what a physicist should mean by the term 'to exist'." Of course, Mermin may not have meant literally what he said about the *moon*. But what he really meant is not obvious. We will come back to that quote in Sect. 3.3. (Note by J.B.).

summarize quantum mechanics in five words?" comes from the physicist and science writer Aatish Bhatia (@aatishb):

> Quantum mechanics in 5 words. Don't look: wave. Look: particle.

Sean Carroll [92, p. 35]

Of course, the statements quoted here do not reflect the views of most physicists (indifference to such questions or some form of "pragmatism" being the view of the majority), but their authors are certainly not marginal either and the statements should be sufficiently surprising to make the reader wonder what is going on.

However, there have also been views explicitly opposed to those mentioned here.

1.4 But There has Never Been a Consensus

From the early days of quantum mechanics, people like de Broglie, Schrödinger, and especially Einstein objected to the Copenhagen doctrine, but the dominant discourse misunderstood or ignored those objections. We will see that in detail in Chap. 7, but it is interesting to note the strength with which the opposition was expressed, at least in private letters. For example, in 1928, Einstein wrote to Schrödinger:

> The Heisenberg–Bohr tranquilizing philosophy—or religion?—is so delicately contrived that, for the time being, it provides a gentle pillow for the true believer from which he cannot very easily be aroused.

Albert Einstein [163]

In another letter to Schrödinger, Einstein referred to Bohr as the "Talmudic philosopher" for whom "reality is a frightening creature of the naive mind" [165]. Einstein also referred to Bohr as [167] "the mystic, who forbids, as being unscientific, an enquiry about something that exists independently of whether or not it is observed, i.e., the question as to whether the cat is alive[21] at a particular instant before an observation is made (Bohr)." Schrödinger was equally critical[22]:

> Bohr's [...] approach to atomic problems [...] is really remarkable. He is completely convinced that any understanding in the usual sense of the word is impossible. Therefore the conversation is almost immediately driven into philosophical questions, and soon you no longer know whether you really take the position he is attacking, or whether you really must attack the position he is defending.

Erwin Schrödinger [438]

---

[21] Here Einstein is referring to a famous thought experiment due to Schrödinger in which, if one follows the standard rules of quantum mechanics, a cat could be both alive and dead at the same time, before one looks at it. See Sects. 2.5 and 7.3 for further discussion of this argument. (Note by J.B.).

[22] Some of the quotes below come from Goldstein [221] and are discussed there.

And again:

> With very few exceptions (such as Einstein and Laue), all the rest of the theoretical physicists
> were unadulterated asses and I was the only sane person left.
>
> […]
>
> If I were not thoroughly convinced that the man [Bohr] is honest and really believes in the
> relevance of his—I do not say theory but—sounding word,[23] I would call it intellectually
> wicked.

Erwin Schrödinger [444]

Schrödinger put his finger on one of the main problems of the "Copenhagen" view, namely, the attempt to find an idealistic philosophical solution to the conceptual problems of quantum mechanics, when he wrote[24]:

> […] the reigning doctrine rescues itself or us by having recourse to epistemology. We are
> told that no distinction is to be made between the state of a natural object and what I know
> about it, or perhaps better, what I can know about it if I go to some trouble. Actually—so
> they say—there is intrinsically only awareness, observation, measurement.

Erwin Schrödinger [441], reprinted in [510, p. 157]

Schrödinger did not even try to hide his feelings when he wrote to "the other side", for example, to Max Born:

> Maxel, you know I love you and nothing can change that. But I do need to give you once
> a thorough head washing. So stand still. The impudence with which you assert time and
> again that the Copenhagen interpretation is practically universally accepted, assert it without
> reservation, even before an audience of the laity—who are completely at your mercy—it's
> at the limit of the estimable […]. Have you no anxiety about the verdict of history? Are you
> so convinced that the human race will succumb before long to your folly?

Erwin Schrödinger [445]

In a more constructive mode, Einstein nicely summarized his position in 1949:

> I am, in fact, firmly convinced that the essentially statistical character of contemporary quan-
> tum theory is solely to be ascribed to the fact that this (theory) operates with an incomplete
> description of physical systems […].[25]

Albert Einstein [170, p. 666]

---

[23] Schrödinger was referring to the word "complementarity", which was the foundation of Bohr's approach and will be discussed in Appendix 2.C. (Note by J.B.).

[24] This was written after he introduced his famous "cat" in [441], which is supposed to be "both alive and dead". The idealism, implicit in the view that Schrödinger rejects, will be criticized in Chap. 3.

[25] He added [170, p. 672]: "In a complete physical description, the statistical quantum theory would […] take an approximately analogous position to the statistical mechanics within the framework of classical mechanics." This refers to an idea, developed at the end of the nineteenth century, according to which the laws of thermodynamics could be derived from an application of statistical reasoning to the motion of atoms, the latter giving a more complete description of matter than the one given by thermodynamics or fluid mechanics.

The root of the difference between Einstein and the Copenhagen school of thought was exactly about this issue of completeness: Einstein thought that the existing quantum mechanics was incomplete, i.e., that a more detailed description of the microscopic world was possible and that such a description would eliminate the need to refer to an observer.[26] We will see in Chap. 5 that Einstein's hope was not only reasonable but was even realized during his lifetime, although in ways that he did not like (for other reasons, that we will discuss in Sect. 7.6.2).

But even putting that aside, which position is the more radical? Einstein's hope expressed here or Heisenberg's view (shared by many physicists) that "we can no longer speak of the behavior of the particle independently of the process of observation", where "no longer" means that we will never be able to do so? Yet, most contemporaries of Einstein thought that he was the unreasonable fellow, who had become too old to appreciate the depth of the quantum revolution.

After World War II, the critique of the mainstream view was taken up for the main part by David Bohm and John Bell. We will discuss their objections in detail later (in Chaps. 4 and 5), but one can get their flavor by reading the following answer given by Bell in an interview with the BBC:

> One wants to be able to take a realistic view of the world, to talk about the world as if it is really there, even when it is not being observed. I certainly believe in a world that was here before me, and will be here after me, and I believe that you are part of it! And I believe that most physicists take this point of view when they are being pushed into a corner by philosophers.

> John Bell [107, p. 50]

Bell even met Bohr once. As he recalled later, in an interview, with the magazine *Omni*:

> I went up in a hotel lift with him. I didn't have the nerve to say. 'I think your Copenhagen interpretation is lousy'. Besides the lift ride wasn't very long. Now, if the lift had gotten stuck between floors, that would have made my day! In which way, I don't know.

> John Bell [45, p. 85]

Note that the interviewers wrote [45, p. 86]: "We first asked Bell over the telephone whether he himself felt he had demonstrated that 'reality doesn't exist'. He responded by warning us that he is an impatient, irascible sort who tolerates no nonsense."

Murray Gell-Mann, Nobel prizewinner and discoverer of quarks, said about finding "an adequate philosophical presentation" of quantum mechanics:

> Bohr brainwashed a whole generation of physicists into thinking that the job was done 50 years ago.

> Murray Gell-Mann [200, p. 29]

---

[26]He probably also thought that this description would render the theory deterministic, but it is doubtful that he was mainly concerned with determinism. We will discuss that in Sect. 7.1.

But the most emphatic reaction to the alleged need to put ourselves, the "observers", at the center of things is probably due to Bertrand Russell, who disliked nothing more than anthropocentrism and subjectivism, and who wrote the following "pessimistic meditation" about "Modern Physics":

> […] Formerly, the cruelty, the meanness, the dusty fretful passion of human life seemed to me a little thing, set, like some resolved discord in music, amid the splendour of the stars and the stately procession of geological ages. What if the universe was to end in universal death? It was none the less unruffled and magnificent. But now all this has shrunk to be no more than my own reflection in the windows of the soul through which I look out upon the night of nothingness. The revolutions of nebulae, the birth and death of stars, are no more than convenient fictions in the trivial work of linking together my own sensations, and perhaps those of other men not much better than myself. No dungeon was ever constructed so dark and narrow as that in which the shadow physics of our time imprisons us, for every prisoner has believed that outside his walls a free world existed; but now the prison has become the whole universe. There is darkness without, and when I die there will be darkness within. There is no splendour, no vastness, anywhere; only triviality for a moment, and then nothing. Why live in such a world? Why even die?

Bertrand Russell [430, p. 374]

All this may look strange, but there is still a natural question to discuss.

1.5 Why Bother?

Most physicists are rather indifferent to the sort of issues discussed in this book, regarding them as "metaphysical". A good example of such a reaction is due to Pauli:

> As O. Stern said recently, one should no more rack one's brain about the problem of whether something one cannot know anything about exists all the same, than about the ancient question of how many angels are able to sit on the point of a needle. But it seems to me that Einstein's questions are ultimately always of this kind.

Wolfang Pauli [79, p. 223]

Similarly, another founding father of quantum mechanics, the great British physicist Paul Dirac distinguished between two kinds of difficulties with quantum theory[27]:

> The difficulties in quantum theory are of two kinds. I might call them Class One difficulties and Class Two difficulties. Class One difficulties are the difficulties I have already mentioned: How can one form a consistent picture behind the rules for the present quantum theory? These Class One difficulties do not really worry the physicist. If the physicist knows how to calculate results and compare them with experiment, he is quite happy if the results agree

---

[27]The Class Two difficulties, which he discusses in the rest of this article, are those related to the mathematical formulation of quantum field theories.

with his experiments, and that is all he needs. It is only the philosopher, wanting to have a satisfying description of nature, who is bothered by Class One difficulties.

Paul Dirac [138]

Those physicists are both right and wrong. They are right in the sense that ordinary quantum mechanics works perfectly FAPP, to use an acronym introduced by John Bell, meaning "for all practical purposes" [46]. The theory, specially in quantum electrodynamics, makes spectacularly precise predictions confirmed by experiment, and nobody needs to understand quantum mechanics beyond what is in textbooks in order to make computers and telecommunications work. Nothing that is written in this book puts into question those facts. As an algorithm (described in Chap. 2) allowing us to predict results of experiments, and to use various powerful technologies, quantum mechanics is perfect.

But it is precisely *because* it works so well that trying to understand why it works makes sense. Obviously if quantum mechanics worked half of the time, so to speak, there would be no reason to try to understand it in depth. Many models in physics are known to be applicable within certain limits and, once we know that, there are no further questions to be raised about those models. But quantum mechanics works on all known scales and is not contradicted by any experiment whatsoever.[28] Isn't it worthwhile to ask *why* it works so well?

To that question, there are typically two different types of answers, one "official" and one "implicit". The official reply is that the goal of physics is to predict results of experiments, or to account for what we see, or perceive, and nothing else (as proposed by Dirac in the above quote). But this is inverting the means and the goal. Experiments are needed to test our theories in order to avoid falling into idle speculation or "metaphysics", but our theories are about the world, not about the experiments themselves. As Bell said:

But experiment is a tool. The aim remains: to understand the world. To restrict quantum mechanics to be exclusively about piddling laboratory operations is to betray the great enterprise. A serious formulation [of quantum mechanics] will not exclude the big world outside the laboratory.

John Bell [46, p. 34]

Of course, it may be that it is simply impossible to understand the quantum world and that we have to content ourselves with predicting results of experiments. After all, who are we but somewhat evolved creatures and why should we expect to be able to understand how the world is? Isn't the fact that quantum mechanics looks weird to us simply a consequence of the limitations of our minds? That may be the case, but one needs some argument to show that and not simply rely on the "rhetoric of inevitability".

Besides, there is a serious issue of consistency raised by the notion that the only goal of physics is to predict results of experiments. If indeed that was all there is to

---

[28]Putting aside the problem of quantum gravity, which is indeed a difficult and unsolved problem, but it cannot be considered as a refutation of quantum mechanics.

physics, why do experiments in the first place? The need to finance costly experiments is "sold" to politicians and the public by saying that we are discovering the fundamental laws of Nature. But, if "it is wrong to think that the task of physics is to find out how Nature is" (Bohr according to Petersen), or if, in quantum mechanics, we "no longer deal with the elementary particles themselves but with our knowledge of them" (Heisenberg), then how can we claim that we are trying to find the fundamental laws of Nature? What would the funders say if they read those statements? Wouldn't they at least be puzzled and ask for some clarification? Isn't it therefore simply a matter of intellectual honesty to ask ourselves how we would clarify those statements?

On the other hand, most physicists probably do not really believe the official answer. They do give some meaning to quantum mechanics beyond our "observations". Physicists do speak of particles going this way or that way, having a certain polarization or a "spin" or speed, or some other properties, even when those particles are not observed. From that point of view, which I call the "implicit" one, there is no problem about quantum mechanics and the centrality of "observations". However, the main defect of that "solution" is that it is never spelled out clearly: what exactly can we say about the world out there?

Moreover, and that will be one of the main points emphasized in this book, the sort of thing that people have in the back of their minds when they give a meaning to quantum mechanics outside of "measurements" is sometimes inconsistent and sometimes even in contradiction with the consequences of quantum mechanics.[29] This lack of clarity leads to a general uncertainty about what one is "allowed" to say about the world (or about what is "speakable and unspeakable" to use John Bell's expression [49]) and that in turn induces many physicists to fall back on the standard talk about measurements being all there is to physics, which they feel is safe, even if it "betrays the great enterprise".

What we need is a theory which tells a story about what is going on in the world, even when we do not "observe" it, and which makes the same predictions as ordinary quantum mechanics whenever we do make "observations" or experiments. If such a theory existed, then all the confusing talk about the centrality of observations would disappear and we could analyze that theory in order to see how it helps us to understand the quantum world.

Amazingly, such a theory actually does exist, and has even existed, in some preliminary form, since the beginning of quantum mechanics, i.e., since 1927; it was proposed by Louis de Broglie at that time and developed by David Bohm in 1952. One of the main goals of this book is to make this theory better known.

Since this theory implies no practical change to what we do as physicists in our daily lives when we use quantum mechanics, it should be good news to all those who do not want to be bothered with "foundational" issues. It "simply" clarifies what quantum mechanics is all about and allows us to get rid of entire libraries of confused talk about the centrality of human observations in science. Of course, whether that "simply" is important or not is a matter of taste.

---

[29]See Sect. 2.5, and especially the theorem at the end of that section.

1.6 Outline of the Book

In this book, we will defend several theses. The first one is that there are genuine conceptual problems within the usual quantum formalism: the latter does not allow us to speak of the world beyond what happens in our laboratories and that is obviously unsatisfactory. Next, we will argue that there is no philosophical solution to this problem, contrary to the impression that one sometimes gets when one reads the proponents of the Copenhagen school.

Moreover, we will explain that there exists a way to complete the quantum formalism so that this problem is eliminated, and the completion is, in some sense, simpler to understand than the usual formalism. Finally, the refusal by a large part of the physics community to face the difficulties intrinsic to the quantum formalism has led it to ignore or misunderstand what is probably the main novel feature of quantum mechanics, namely the existence of nonlocal actions, or "particles interacting instantaneously even when they are arbitrarily far apart."

Now, in more detail. The second chapter is devoted to the first mystery of quantum mechanics: interference phenomena and the superposition principle, which lead to statements such as "quantum objects can possess mutually exclusive properties, like traveling along different paths simultaneously" and "quantum objects obtain definite properties when they are 'measured', but only then, and those properties depend on which measurements we choose to make". We will have to distinguish carefully between the phenomena to be explained, the quantum formalism that allows us to predict them, and the commentaries or mental pictures which accompany the formalism and which lead to the statements just mentioned.

This will do justice to the way physicists often speak about quantum phenomena. The phenomena are strange and all the talk about observations affecting reality is not based on pure prejudice, although it is not inevitable either.

In Sect. 2.5, we will define four possible reactions or attitudes with respect to that "first mystery" and to how one thinks of the quantum formalism. In the rest of the book, we will try to connect the various questions that we deal with to those four basic positions.

The third chapter can be skipped by those who are not interested in "philosophy". But for those with a philosophical background, the statements associated with "Copenhagen" should remind them of the writings of Bishop Berkeley, Immanuel Kant, Ernst Mach,[30] or sometimes the logical positivists, even if the connection is

---

[30]Ernst Mach was an Austrian physicist and philosopher, active at the end of the 19th century and the beginning of the 20th, whose views were somewhat similar to those of the Copenhagen school, long before the advent of quantum mechanics. For example, in 1897, he wrote:

Bodies do not produce sensations, but complexes of elements (complexes of sensations) make up bodies. If, to the physicist, bodies appear the real, abiding existences, whilst the "elements" are regarded merely as their evanescent, transitory appearance, the physicist forgets, in the assumption of such a view, that all bodies are but thought-symbols for complexes of elements (complexes of sensations).

Ernst Mach [314, p. 29]

not straightforward. In this chapter, we will try to clarify notions such as "realism" or "determinism" which have the disadvantage of often being ill-defined and used in ways that confuse rather than clarify the discussion.

For example, when the people who interviewed Bell for the magazine *Omni* [45] asked him whether he felt that he had shown that "reality does not exist", what could they possibly have meant? After all, the telephone they used, Bell himself, and his answers were all part of reality. And if Bell had thought that reality did not exist, he would presumably have been a solipsist (meaning someone who thinks that only his own mind exists and that everything else is some sort of dream going on in his mind) and he would therefore have thought that the interviewers themselves did not exist (outside of Bell's mind). No wonder Bell replied that he "tolerates no nonsense".

The fourth chapter deals with Bell's result and the problem of nonlocal action, which is the second mystery of quantum mechanics. There, we will explain the meaning of the statement that "particles can interact instantaneously even when they are arbitrarily far apart" and discuss the extent to which it is true.

The fifth chapter, which is the heart of this book, is about the de Broglie–Bohm theory (nowadays also called Bohmian mechanics), first introduced by Louis de Broglie before 1927, and then quickly abandoned by him; it was rediscovered and developed by David Bohm in 1952, popularized by John Bell, and further developed and defended by, among other people, David Albert, Chris Dewdney, Detlef Dürr, Sheldon Goldstein, Basil Hiley, Peter Holland, Anastasios Kyprianidis, Tim Maudlin, Nelson Pinto-Neto, Ward Struyve, Stefan Teufel, Roderich Tumulka, Antony Valentini, Jean-Pierre Vigier, Nino Zanghì and their collaborators. In a nutshell, the de Broglie–Bohm theory is a theory of matter in motion, just like any other physical theory; but the motion is quite strange, as one would expect, given all the strange phenomena that led to the discovery of quantum mechanics in the first place. However, the strangeness does not come from putting the observer at the center of everything.

Indeed, the main virtue of the de Broglie–Bohm theory is that it is a clear theory about what is going on in the world, whether we look at it or not. So the vagueness and subjectivity of the notion of "observer" or of "measurement" simply disappear in this theory. Of course, the theory does make empirical predictions, and the latter are the same as those of ordinary quantum mechanics, but the de Broglie–Bohm theory and ordinary quantum mechanics are *not* the same theory, because the de Broglie–Bohm theory is a theory about microscopic reality, while ordinary quantum mechanics is not: it is an algorithm for very accurately predicting results of experiments, an algorithm that is, in fact, a consequence of the de Broglie–Bohm theory, as we will see in Chap. 5.

Using the de Broglie–Bohm theory, one can easily explain why all the arguments that are supposed to prove that such a theory is impossible are false. While deterministic, the de Broglie–Bohm theory also accounts naturally for the apparent indeter-

---

(Footnote 30 continued)
Mach always rejected the existence of atoms. His philosophy influenced the school of logical positivism, which itself had an influence on the orthodox view of quantum mechanics. We will discuss logical positivism and its influence in physics in Sect. 7.7 and in Chap. 8.

minism of quantum phenomena. Finally, it explains the "active role" of measuring devices (the apparent effects of observations on reality), so strongly emphasized by the Copenhagen school, but by making it a consequence of the theory and not of some a priori philosophical doctrine. It also explains the nonlocal actions inherent in quantum phenomena.

What more could we ask for? As John Bell explained, after recalling the arguments claiming to show that a theory such as de Broglie–Bohm theory is impossible:

> But in 1952 I saw the impossible done. It was in papers by David Bohm. Bohm showed explicitly how parameters could indeed be introduced, into nonrelativistic wave mechanics, with the help of which the indeterministic description could be transformed into a deterministic one. More importantly, in my opinion, the subjectivity of the orthodox version, the necessary reference to the 'observer', could be eliminated. [...] Should it not be taught, not as the only way, but as an antidote to the prevailing complacency? To show us that vagueness, subjectivity, and indeterminism, are not forced on us by experimental facts, but by deliberate theoretical choice?
>
> John Bell [49, p. 160]

In the sixth chapter, we will consider the main theories, other than the de Broglie–Bohm theory, that have been proposed in order to solve the conceptual problems of quantum mechanics. Their clarity and consistency will be compared with those of the de Broglie–Bohm theory.

In the seventh chapter, we will address various historical misunderstandings regarding Einstein, de Broglie, Schrödinger, Bohm, and Bell. All these authors, whether in their critique of the usual interpretation of quantum mechanics or in their attempt to complete it, have been ignored or misunderstood by the majority of physicists of their time and often also of today.

The eighth chapter will outline some conjectural thoughts about the general cultural impact of the various interpretations and misinterpretations of quantum mechanics.

Many of the ideas defended here are heterodox and may even seem shocking. However, the intention of this book is not to give final answers to the conceptual problems of quantum mechanics, but rather to open the reader's mind to the possibility that answers can be given beyond what is taught in standard quantum mechanics courses. The student I once was, who could not understand sentences such as "physics does not deal with Nature, but with our knowledge of it", would have been delighted to read such a book.

It should nevertheless be emphasized that this book is written in the same spirit as the following statement, where one could replace "philosophy" by "the conceptual problems of quantum mechanics":

> Philosophy is to be studied, not for the sake of any definite answers to its questions since no definite answers can, as a rule, be known to be true, but rather for the sake of the questions themselves; because these questions enlarge our conception of what is possible, enrich our intellectual imagination and diminish the dogmatic assurance which closes the mind against speculation [...]
>
> Bertrand Russell [425, pp. 249–250]

Chapter 2
The First Mystery: Interference and Superpositions

2.1 The Spin

We will start with the simplest quantum mechanical situation, the one concerning the "spin" of a particle.[1] Despite its simplicity, it will allow us to explain one of the basic "mysteries" of quantum mechanics. Some particles, electrons for example, possess a property called "spin", which is a quantity that can be measured in different directions and takes, in each direction, only two values, denoted up ↑ and down ↓. We will consider here only two directions in which the spin can be measured, denoted 1 and 2, so that we can have four possibilities: spins that are up 1 ↑ or down 1 ↓ in direction 1 and up 2 ↑ or down 2 ↓ in direction 2. One should not confuse the directions 1 or 2 in which the spin is measured and the values up or down that can be the result of those measurements in each direction.

There is no need for the moment to try to understand what the property of "spin" means. We will start from a completely "phenomenological" attitude about the spin, namely, we will simply describe what happens in experiments that are "measuring" the spin, without at first trying to explain how these experiments work.

We put scare quotes here around the word "measuring" because, as we will see in Sect. 2.5 (and this will be one of the most important themes of this book), the notion that there is an intrinsic property of a particle corresponding to its spin in a given direction and that is being measured when one "measures its spin" is untenable. We will not put quotation marks everywhere, but it should be remembered that when we use the word "measurement" we do not want to suggest that some intrinsic property of a particle is being discovered.

The whole discussion using the spin may seem rather abstract, but it is easy to analyze mathematically, as we will see in Sect. 2.3. There is another, similar example,

---

[1]The first four sections of this chapter draw heavily on David Albert's book *Quantum Mechanics and Experience*. We emphasize that the "experiments" here are meant to illustrate the theory rather than real experiments. The latter are generally carried out with photons, whose polarization plays a role similar to the spin here. But all the experiments described below correspond to what quantum mechanics predicts.

© Springer International Publishing Switzerland 2016
J. Bricmont, *Making Sense of Quantum Mechanics*,
DOI 10.1007/978-3-319-25889-8_2

Fig. 2.1 Measuring the spin

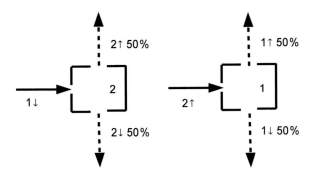

the double-slit experiment, which may seem more familiar, but is less easy to analyze and which will be discussed in Appendix 2.E.

Let us see what happens in the experiments described in Fig. 2.1, where many particles are sent, so that we get statistical results. Note also that here, as in every experiment described in this book, particles are sent (in principle at least) one at a time, so that there are no possible interactions between different particles that could account for their strange behavior. So in Fig. 2.1, we have two devices that "measure the spin" of the particle in two different directions (they are unrelated to each other). The reader who wants a more realistic view of these experiments can look at Fig. 5.4.

If we send a particle through one such device, the particle comes out through one of two holes, depending on the value of its spin, as shown in Fig. 2.1. We also suppose that we can select particles having a given spin value, up or down, in either of the two directions 1 or 2 (we will explain below how to do that). By "having a given spin value", we mean that if one measures, say, in direction 1 the spin of a particle that is up in that direction 1, we will always get up.

First, we send particles that are down in direction 1 into a device that measures the spin in direction 2 (Fig. 2.1 left); if we repeat that operation many times, we will get 50 % 2 ↑, 50 % 2 ↓. If we had started with particles that were up in direction 1, we would have gotten the same result. Likewise, if we send particles that are up in direction 2 into a device that measures the spin in direction 1 (Fig. 2.1 right), we get 50 % 1 ↑, 50 % 1 ↓ and we would get the same results starting with particles that are down in direction 2. So far, so good: there is no particular mystery here!

This explains also how one can select particles having a given spin value, up or down, in either of the two directions 1 or 2: just take the particle exiting, say, the device that measures the spin in direction 2 through the 2 ↑ hole and we will have selected particles that have spin up in direction 2 (i.e. if we then send those particles through a device that measures the spin in direction 2, they will always exit through the 2 ↑ hole). And likewise for the other possibilities.

Now, we may ask: can we select particles that are, say, down in direction 1 and up in direction 2? We might think that one way to do this, at least naively, is to send particles that are up in direction 2 in a device that measures the spin in direction

Fig. 2.2 Trying to measure
the spin in both directions
simultaneously

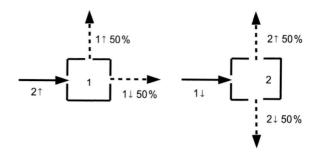

1 and select those that are down in that direction. That way, the particle should be down in direction 1 and up in direction 2 (see Fig. 2.2 left).

But if we want to check that we really have particles that are down in direction 1 and up in direction 2, we might measure the spin in direction 2 of those selected particles (see Fig. 2.2 right).[2] However, we find that the result is again 50 % 2 ↑, 50 % 2 ↓. This is our first surprise.

The same result would occur if we tried to have particles that are, say, down in direction 2 and up in direction 1, or with any of the four possible combinations. It seems that, by measuring in direction 1, the spin of a particle that is up in direction 2, we "erase" the fact that it is up in direction 2. Indeed, the results of a later measurement of the spin in direction 2 of the particles which are up or down in direction 1 are just what they would be for such particles, independently of the fact that they had a spin up in direction 2, before the measurement in direction 1.

This is a simple example of the Heisenberg uncertainty relations or of what Bohr called "complementarity"[3]: we cannot measure simultaneously the spin in two different directions, because they require different devices and, applying one device in one spin direction and then a second one in another direction, destroys the result of the first device. So, one could consider two different "complementary" pictures of the particle: one describing the spin in direction 1, the other the spin in direction 2. However, one should not try to combine the two pictures (spin in directions 1 and 2)

---

[2] In Fig. 2.2 left, we put one hole to the right instead of downwards, because the particles exiting through that hole go into the box on the right of the figure.

[3] This is discussed mathematically in more detail in Appendix 2.C. Bohr explained the "complementary, or reciprocal, mode of description" by emphasizing [71] "the relative meaning of every concept, or rather of every word, the meaning depending upon our arbitrary choice of view point, but also that we must, in general, be prepared to accept the fact that a complete elucidation of one and the same object may require diverse points of view which defy a unique description. Indeed, strictly speaking, the conscious analysis of any concept stands in a relation of exclusion to its immediate application." The reader may be forgiven for not understanding exactly what this means. We try here to give a plausible interpretation of that idea. See [181] for a discussion of different interpretations of Bohr's thinking. We will return to a discussion of Bohr's views, in relation to his debate with Einstein in Sect. 7.1.

together, to get a more complete description of the particle. One should rather choose to use one or the other picture but not both at the same time.[4]

As Bell stresses, the use of the word "complementary" by Bohr is not the usual one: one might say that an elephant, from the front, is "head, trunk, and two legs", while, from the back, "she is bottom, tail, and two legs" and yet something else from top or bottom. Bell adds: "These various views are complementary in the usual sense of the word. They supplement one another, they are consistent with one another, and they are all entailed by the unifying concept 'elephant'." But, for Bohr, according to Bell, "complementarity" means rather the reverse of that common usage: it means contradictariness, since the description of the spin in directions 1 and 2 exclude each other [49, p. 190].

This seems somewhat surprising, and the uncommon use of the word "complementarity" to describe the situation does not help, but in actual fact it may not be that surprising: one might think that these measuring devices, being big objects, necessarily perturb the microscopic system, the electron, being observed. By measuring in direction 1, we perturb the value of the spin in direction 2. If this view were tenable (we will see in Sect. 2.5 that it is not), there would be no big mystery in quantum mechanics, although the uncertainty principle is sometimes presented as *the* main new characteristic of the quantum world.

However, there is a far more perplexing situation to which we turn now.

2.2   The Mach–Zehnder Interferometer

Figure 2.3 describes an "interferometer".[5] The box inside carrying the label 2 measures the spin in direction 2 with the 2 ↑ and 2 ↓ particles exiting through different holes. After passing through the box measuring the spin in direction 2, there are two possible paths for the particle, one for 2 ↑, the other for 2 ↓. The particles are reflected by mirrors and their paths join at the black arrow (whose functioning is not described for the time being).

The reader may think, given what was said in Chap. 1, that in quantum mechanics there is no such thing as the path of a particle. This will be discussed in detail in Chap. 5, but let us now simply use the word "path" to mean that, if we try to detect the particle along any one of those paths, we will always find it along one and only one of them.

Let us start by sending particles that are 2 ↑ in the box with the label 2 (instead of 1 ↓ as in Fig. 2.3). Then 100 % of the particles will follow one path (the one indicated 2 ↑). If, after the black arrow, one measures the spin in direction 1, one obtains 50 % 1 ↑, 50 % 1 ↓ (unlike what is represented in Fig. 2.3). If one sends particles that are 2 ↓, one gets a similar result (with the path 2 ↓ being followed).

---

[4]Bohr did not use spin as an example, but rather the descriptions in terms of wave and particle, which we will discuss in Appendix 2.E.

[5]See Greenberger [239] for a more detailed discussion.

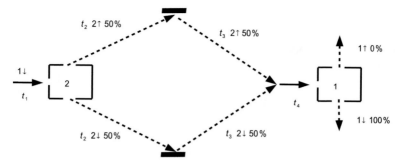

Fig. 2.3 Interference

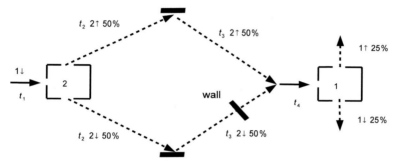

Fig. 2.4 Interference with a wall

Let us now send particles with the spin in direction 1 to be, say, down, as in Fig. 2.3. Then, 50 % follow one path (2 ↑), 50 % follow the other path (2 ↓). Since, in the previous setup, we had 50 % 1 ↑, 50 % 1 ↓, for the particles that follow either the path 2 ↑ or 2 ↓, one would expect to get 50 % 1 ↑, 50 % 1 ↓ here too, if one measures the spin in direction 1 after the black arrow. But one finds that, *after* the black arrow, 100 % are 1 ↓, so that, whichever path they follow, the particles "remember" that they were 1 ↓ to start with. Of course, the same would happen if we did the experiment with particles that are initially 1 ↑.

This result is surprising, since we just learned from trying to measure simultaneously the spin in the directions 1 and 2 that one aspect of quantum mechanics is that measurements tend to erase the memory of past states. The first surprise is that particles seem to have no memory, after a measurement, of their pre-measurement state, but now they seem to remember it completely.

And there are more surprises. Suppose we add a wall along the path 2 ↓, as in Fig. 2.4. If the wall is inserted across the path, it blocks the particles taking that path. What will we then observe after the black arrow?

1. 50 % fewer particles (which is to be expected since half of the particles follow the path 2 ↓ and are now blocked).
2. Without the wall, 100 % of those that take the path 2 ↑ are found to be 1 ↓ after the black arrow. The same is true for those that followed the path 2 ↓. If one blocks the path 2 ↓, one would think that it should not affect the particles that take the path 2 ↑. Thus, one should get 100 % 1 ↓ (of the remaining 50 % of particles that reach the black arrow).

And here is the big surprise: one gets 25 % 1 ↓ and 25 % 1 ↑ (that is, half of the remaining 50 % for each possibility). Therefore, one acts in a certain way on the particles that take the path 2 ↑ by blocking the path 2 ↓ *that they do not take!*

This leads to an apparent *dead end*. Let us go back to the experiment without the wall, sending particles that are 1 ↓. What does each particle do?

- Does it take path 2 ↑ ? No because if it did, one would have 25 % 1 ↑, 25 % 1 ↓ at the black arrow, as one sees when one puts a wall blocking the path 2 ↓.
- The path 2 ↓ ? No, for the same reason.
- Both paths? No, one always finds the particle along one of the paths if one tries to measure it.
- Neither of the paths? No, if both paths are blocked, nothing happens at the black arrow.

This phenomenon and other related phenomena are called *interference*, because whether one path is open or not seems to influence the behavior of the particles following the other path. This is the essence of the first quantum mystery!

It should be remembered that, in principle, the experiment is done by sending one particle at a time, so that no explanation can possibly be based on interactions between particles.

The way this experiment is usually described is by saying that the particle "follows both paths if they are both open" and only one path if one of them is blocked. But how does the particle know ahead of time, whether both paths are open or not?

Indeed, one might do a "delayed choice" experiment,[6] that is, introducing the wall *after* the passage of the particle through the first box measuring the spin in direction 2 (we can imagine both paths to be very long or put the wall just before the black arrow).

Alternatively, one could remove the black arrow while the particle is in flight and then, there would be no recombination of the paths and the particles would continue their trajectory and pass each other (see Fig. 2.5). Those following the path 2 ↑ continue downwards and those following the path 2 ↓ continue upwards. If we then measure the spin in direction 1, along any of these paths, we get 25 % 1 ↑, 25 % 1 ↓ in each case. Indeed, we have, along each path, particles that are only 2 ↑ or 2 ↓, and are measured in direction 1; the result is then as in Fig. 2.1 (right) and also for 2 ↓ instead of 2 ↑.

---

[6]See [507, 509] for the theoretical proposal of such experiments by Wheeler, and [280] for experimental realizations.

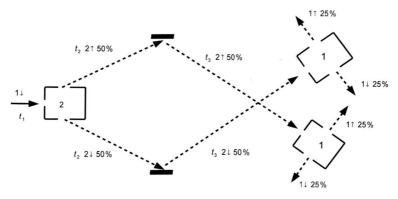

Fig. 2.5 No interference without the *black arrow*

This only deepens the mystery and is the basis of the claim by Wheeler, that "the past [meaning here whether the particle "has chosen" to follow both paths or only one] is not really the past until it has been registered" [107, p. 68]. Moreover, Wheeler invented an ingenious scheme where such "experiments" would not take place in the laboratory, but on a cosmic scale. If we accept his reasoning, this implies that we could decide *now*, by choosing which experiment to perform on the light coming from distant quasars, what happened billions of years ago[7] [507].

Another paradoxical consequence of the experiment described here is the Elitzur–Vaidman bomb-testing mechanism.[8] Suppose that we have a stock of bombs, some of which are active and some of which are duds. We want to find out which is which, but an active bomb will explode if it is hit by only one particle. On the other hand, by definition, a dud is totally insensitive to being hit by one or more particles, so that it does not affect those particles in any way. How could we tell, by classical means, which bombs are active without exploding them? There seems to be no way to do that.

But there is a trick, based on the Mach–Zehnder interferometer, that allows to identify at least a fraction of the active bombs as being active without exploding them. Let us replace the wall in Fig. 2.4 by a bomb. First, suppose that the bomb is a dud. Then, since it is insensitive to the particles, it is as if we had done nothing, i.e., as if we had not put a wall. The particle will behave as if there was no wall and therefore its spin at the black arrow will always be 1 ↓ if we measure the spin in direction 1.

On the other hand, if the bomb is active and detects the particle, it explodes and that's it—it is lost. That happens half of the time if the bomb is active. But suppose that the bomb is active and does not explode. This means that the particle took the path 2 ↑; if we then measure the spin at the black arrow in direction 1, we will get

---

[7]This will be clarified in Sect. 5.1.2. See also [38], where Bell discusses the delayed-choice experiment from the viewpoint of the de Broglie–Bohm theory.

[8]See [173] for the theory and [300] for experiments.

1 ↓ for half of those particles and 1 ↑ for the other half. If we get 1 ↓, we cannot conclude anything since that would also happen if the bomb were a dud. *But*, if we get 1 ↑, then we can be certain that the bomb was *not* a dud since that would *never* happen if the active bomb is replaced by a dud. Since each result 1 ↓, 1 ↑ happens half of the time (among the 50% that have not exploded), we can identify 25% of our initial stock of active bombs as being active without exploding them.

Altogether, half of the active bombs explode and are lost, but a quarter are "saved" (not exploded and known to be active). For the remaining quarter, we don't know. We can then repeat the operation (together with the duds, since we don't know which is which) and identify as active one quarter of that remaining quarter. Repeating the operation many times, we can get as close as we like to a total of one third of the initial stock of active bombs as being known to be active and not exploded,[9] since $1/3 = \sum_{n=1}^{\infty}(1/4)^n$.

2.3 The Quantum Formalism

We now describe a mathematical algorithm that allows us to predict these surprising results, without worrying yet about what it "means" physically.[10]

We associate with each particle a "state", which is simply a two-dimensional vector. In principle, the vector is complex, i.e., the vector space is $\mathbb{C}^2$ rather than $\mathbb{R}^2$, but this will not matter here. The association is as follows (there is of course some arbitrariness in the way this association is made, but let us put that aside):

$$|1 \uparrow\rangle = \begin{pmatrix} 1 \\ 0 \end{pmatrix}, \tag{2.3.1}$$

$$|1 \downarrow\rangle = \begin{pmatrix} 0 \\ 1 \end{pmatrix}, \tag{2.3.2}$$

$$|2 \uparrow\rangle = \frac{1}{\sqrt{2}}\begin{pmatrix} 1 \\ 1 \end{pmatrix}, \tag{2.3.3}$$

$$|2 \downarrow\rangle = \frac{1}{\sqrt{2}}\begin{pmatrix} 1 \\ -1 \end{pmatrix}. \tag{2.3.4}$$

---

[9]If one can modify the experiment so that a fraction p of particles follow the path 2 ↓ and a fraction $1 - p$ follow the path 2 ↑, then one can "save" a fraction $(1 - p)/2$ of the bombs in one operation and, repeating this many times, one can eventually identify a fraction $(1 - p)/(1 + p) = \sum_{n=1}^{\infty}[(1 - p)/2]^n$ of the active bombs, which is as close to 1 as one wants, for p small.

[10]For an elementary introduction to the quantum formalism, see also Susskind and Friedman [467].

We have the obvious relations

$$|2 \uparrow\rangle = \frac{1}{\sqrt{2}} (|1 \uparrow\rangle + |1 \downarrow\rangle) , \qquad (2.3.5)$$

$$|2 \downarrow\rangle = \frac{1}{\sqrt{2}} (|1 \uparrow\rangle - |1 \downarrow\rangle) , \qquad (2.3.6)$$

$$|1 \uparrow\rangle = \frac{1}{\sqrt{2}} (|2 \uparrow\rangle + |2 \downarrow\rangle) , \qquad (2.3.7)$$

$$|1 \downarrow\rangle = \frac{1}{\sqrt{2}} (|2 \uparrow\rangle - |2 \downarrow\rangle) . \qquad (2.3.8)$$

The states of the particles, i.e., the vectors, change according to the following rules:

1. When no measurements are made:

$$|\text{state}(t)\rangle = c_1(t) |1 \uparrow\rangle + c_2(t) |1 \downarrow\rangle = d_1(t) |2 \uparrow\rangle + d_2(t) |2 \downarrow\rangle , \qquad (2.3.9)$$

where $c_1(t)$, $c_2(t)$, $d_1(t)$, $d_2(t)$ are related by (2.3.7) and (2.3.8), and change *continuously* in time in such a way that, at all times, we have $|c_1(t)|^2 + |c_2(t)|^2 = 1$ and $|d_1(t)|^2 + |d_2(t)|^2 = 1$.

This evolution is *deterministic*, i.e., if, at some time, say 0, we give ourselves a state $|\text{state}(0)\rangle$, then this determines a unique state $|\text{state}(t)\rangle$ for all times.

To say more precisely what this evolution is, one would have to write down a differential equation for $c_i(t)$, $d_i(t)$, $i = 1, 2$, which in more general situations is called the Schrödinger equation.[11] But we will not need to go into that for the moment (the Schrödinger equation is discussed in Appendix 2.A).

This evolution is *linear*, i.e., if, at some time, say 0, we have

$$|\text{state}(0)\rangle = c_1|\text{state}_1(0)\rangle + c_2|\text{state}_2(0)\rangle , \qquad (2.3.10)$$

for two states $|\text{state}_1\rangle$ and $|\text{state}_2\rangle$ and numbers c_1, c_2, then, at all times, we have

$$|\text{state}(t)\rangle = c_1|\text{state}_1(t)\rangle + c_2|\text{state}_2(t)\rangle . \qquad (2.3.11)$$

2. But if a measurement is performed, the rule of evolution changes. Suppose the state is (2.3.9):

---

[11]Or, to be precise, the Dirac or the Pauli equations in order to deal with spin.

If one measures the spin in direction 1 at time t, one finds ↑ with probability $|c_1(t)|^2$ and ↓ with probability $|c_2(t)|^2$, bearing in mind that $|c_1(t)|^2 + |c_2(t)|^2 = 1$ so that the probabilities add up to 1.

If one measures the spin in direction 2, one finds ↑ with probability $|d_1(t)|^2$ and ↓ with probability $|d_2(t)|^2$, where $|d_1(t)|^2 + |d_2(t)|^2 = 1$.

After the measurement in direction 1, if one "sees" the result ↑, the state changes and becomes $|1 \uparrow\rangle$, and if one "sees" ↓, the state changes and becomes $|1 \downarrow\rangle$. The same thing holds for the spin in direction 2.

This second rule is called the "reduction", or the "collapse" of the state.

Before proceeding, let us note that, a priori, the two rules are mutually *incompatible*: one rule gives rise to a *continuous in time, deterministic and linear evolution*, the other to a *discontinuous in time, non-deterministic and nonlinear* one. It is discontinuous because, when the state is reduced, it "jumps" discontinuously to one state or the other. It is non-deterministic because, in the operation of reduction, one only assigns probabilities to each possible result. Finally, it is nonlinear because, if the state is $c_1|1 \uparrow\rangle + c_2|1 \downarrow\rangle$, and if one then measures the spin in direction 1 and the result is ↑, the state is reduced to $|1 \uparrow\rangle$, irrespective of the values of the coefficients c_1 and c_2, while if the operation were linear, the result should be a linear combination involving c_1 and c_2. The coefficients c_1, c_2 determine the probability of a given result, but do not affect the resulting state after the collapse.

Moreover, the theory does not tell us what a measurement is. In practice, one knows what it is, and therefore there are no practical problems here, but there is a serious problem of principle: what are the exact physical properties that define what a "measurement" is and why do we have to use one rule "between measurements", and another "during measurements"? It is of course because of this second rule that "measurements" play a central role in the quantum theory and this explains to some extent the strange statements quoted in Chap. 1.

Note that there are other quantities to which this algorithm can be applied. In classical physics, one introduces the momentum, the energy, the angular momentum, etc. and there are quantum analogue of these variables. In Appendix 2.B we give the definition of the quantum state, and explain its time evolution and the collapse rule when one "measures" those quantities.

What we have described here is the "spin" part of the quantum state. But there is also a position part, namely a function $\Psi(x, t)$ called the *wave function*, where $x \in \mathbb{R}^3$ is a variable like the spin above, but taking continuous rather than discrete values. $|\Psi(x, t)|^2$ is then the probability density for finding the particle at x at time t if one "measures" its position.

The wave function $\Psi(x, t)$ varies with time. The time evolution is governed by a differential equation, the Schrödinger equation (see Appendix 2.A), which is such that, if one has $\int |\Psi(x, 0)|^2 dx = 1$ at the initial time, then $\int |\Psi(x, t)|^2 dx = 1$ at all times.[12] This is analogous to the constraint $|c_1(t)|^2 + |c_2(t)|^2 = 1$ above. The evolution is continuous in time, deterministic and linear: if the initial wave function at time 0, $\Psi(x, 0) = c_1\Psi_1(x, 0) + c_2\Psi_2(x, 0)$, with c_1, c_2, complex numbers, then for all times,

$$\Psi(x, t) = c_1\Psi_1(x, t) + c_2\Psi_2(x, t) ,$$

with all three terms solving the Schrödinger equation.

What happens to $\Psi(x, t)$ if we measure the position of the particle? Suppose that we have a wave function $\Psi(x, t) = c_1\Psi_1(x, t) + c_2\Psi_2(x, t)$, where $\Psi_1(x, t)$ and $\Psi_2(x, t)$ have disjoint support,[13] meaning that $\Psi_1(x, t) = 0$ for all x such that $\Psi_2(x, t) \neq 0$ and vice-versa. Suppose also that $\int |\Psi_1(x, t)|^2 dx = \int |\Psi_2(x, t)|^2 dx = 1$, and $|c_1|^2 + |c_2|^2 = 1$ which implies, since the supports of Ψ_1 and Ψ_2 are disjoint, $\int |\Psi(x, t)|^2 dx = |c_1|^2 \int |\Psi_1(x, t)|^2 dx + |c_2|^2 \int |\Psi_2(x, t)|^2 dx = 1$. Then, we will find the particle in the region where $\Psi_1(x, t) \neq 0$ with probability $|c_1|^2$, and in the one where $\Psi_2(x, t) \neq 0$ with probability $|c_2|^2$. After the measurement, the wave function "collapses" to either $\Psi_1(x, t)$ or $\Psi_2(x, t)$, depending on the result. This change "after a measurement", like the one for the spin measurement, is also incompatible with the continuous in time, deterministic and linear evolution given by the Schrödinger equation.

The problem posed by this duality of rules is expressed ironically by Bell:

> Was the wavefunction of the world waiting to jump for thousands of millions of years until a single-celled living creature appeared? Or did it have to wait a little longer, for some better qualified system …with a PhD?

John S. Bell [46, p. 34]

However, we will put such questions aside for the time being and show how this formalism can predict the strange experimental results described in the previous section. Let us first introduce the following terminology:

1. We will call the function $\Psi(x, t)$ the *wave function* of the particle. We will extend this notion to a system comprising several particles at the end of Appendix 2.A and also in Chaps. 4 and 5.
2. We will call a product of the form $\Psi(x, t)|1 \uparrow\rangle$, or a linear combination of such products, e.g., $\Psi^\uparrow(x, t)|1 \uparrow\rangle + \Psi^\downarrow(x, t)|1 \downarrow\rangle$ or the corresponding expression with $|2 \uparrow\rangle$, $|2 \downarrow\rangle$, the *quantum state* [see (2.4.2) below for an example of such state]. It combines both the spatial and the spin parts discussed above. We will sometimes use the notation Ψ without the arguments (x, t) to denote the quantum state.

---

[12]Since $\Psi(x, t)$ is in general a complex number, $|\Psi(x, t)|^2 = \Psi(x, t)^*\Psi(x, t)$, where z^* is the complex conjugate of $z \in \mathbb{C}$.

[13]The support of a function is the closure of the set on which it is nonzero.

2.4 How Does It Work?

Let us first show how this formalism accounts for the fact that one cannot have particles with a well defined spin in directions 1 and 2 simultaneously. Indeed, if we measure the spin in direction 1 of a particle with its spin up in direction 2, as in Fig. 2.2 (left), we get either the state $|1 \uparrow\rangle$ or the state $|1 \downarrow\rangle$, because of the collapse rule, and each of these states is a superposition of $|2 \uparrow\rangle$ and $|2 \downarrow\rangle$ with equal coefficients [see (2.3.5) and (2.3.6)]. Hence, if we measure the spin in direction 2 for the particles that are now in the state $|1 \downarrow\rangle$ (see Fig. 2.2 right), we will get 50 % $|2 \downarrow\rangle$ and 50 % $|2 \uparrow\rangle$. The "memory" of the particle being $|2 \uparrow\rangle$ to start with is lost. So we see here that the collapse erases the memory.

Let us now go back to the data of Sect. 2.2 and consider Fig. 2.3. At time t_1, the state is [see (2.3.8)]

$$|1 \downarrow\rangle = \frac{1}{\sqrt{2}}\left(|2 \uparrow\rangle - |2 \downarrow\rangle\right) . \qquad (2.4.1)$$

At times t_2 and t_3, we have

$$\frac{1}{\sqrt{2}}\left(|2 \uparrow\rangle|\text{path } 2 \uparrow\rangle - |2 \downarrow\rangle|\text{path } 2 \downarrow\rangle\right) , \qquad (2.4.2)$$

where $|\text{path } 2 \uparrow\rangle$ and $|\text{path } 2 \downarrow\rangle$ are the spatial parts of the quantum state $\Psi(x, t)$, which propagate more or less along the paths indicated $2 \uparrow$ and $2 \downarrow$ (meaning that, if we detect where the particle is, we will always find it along one of those paths).

Let us assume that, at time t_4, the black arrow is a device that is able to recombine the paths of the two particles and send them in the direction $\rightarrow$. Then the state becomes

$$\frac{1}{\sqrt{2}}\left(|2 \uparrow\rangle - |2 \downarrow\rangle\right)|\text{path} \rightarrow\rangle = |1 \downarrow\rangle|\text{path} \rightarrow\rangle . \qquad (2.4.3)$$

Since the spin part of the state is now $|1 \downarrow\rangle$, we will get 100 % "down" results if we measure the spin in direction 1 after the black arrow.

If we now put a wall on the path $2 \downarrow$ (see Fig. 2.4), this means that we perform a measurement, since it allows us to know which path the particle follows[14]: if the particle is blocked by the wall, it means that it took the path $|\text{path } 2 \downarrow\rangle$, but if the particle is not blocked, its path must be $|\text{path } 2 \uparrow\rangle$. So if the particle is not blocked, the state "collapses" and becomes

$$|\text{state}\rangle \longrightarrow |2 \uparrow\rangle|\text{path } 2 \uparrow\rangle , \qquad (2.4.4)$$

which becomes at time t_4, at the "black arrow",

---

[14]The same is true if one puts an active bomb (as opposed to a dud, which is the same as not putting anything), along the path $2 \downarrow$. Depending on whether the bomb explodes or not, one knows which path the particle took.

$$\frac{1}{\sqrt{2}}\big(|1\uparrow\rangle + |1\downarrow\rangle\big)|\text{path}\rightarrow\rangle . \tag{2.4.5}$$

Thus, if we then measure the spin in direction 1, we will get 25 % 1 ↑ and 25 % 1 ↓.

Here, the strange role of "measurements" cannot be explained away by appealing to local disturbances caused by the measuring device (as we did in Sect. 2.1 when discussing the impossibility of simultaneous measurement of the spin in two different directions), since the wall has an effect on the particle even when it follows a path on which the wall is *not* inserted. Moreover, those two paths can in principle be arbitrarily far from each other, and the size of the separation has no effect on the final statistics (which always remain 25 % 1 ↑ and 25 % 1 ↓).

Finally, note that if we remove the black arrow (say, while the particles are moving past the first box, so that we are dealing with a "delayed choice" experiment), as in Fig. 2.5, there would be no recombination of the paths and the state would remain in the form

$$\frac{1}{\sqrt{2}}\big(|2\uparrow\rangle|\text{path }2\uparrow\rangle - |2\downarrow\rangle|\text{path }2\downarrow\rangle\big) .$$

The parts $|2\uparrow\rangle|\text{path }2\uparrow\rangle$ and $|2\downarrow\rangle|\text{path }2\downarrow\rangle$ then continue along their trajectories and pass each other: the part $|2\uparrow\rangle|\text{path }2\uparrow\rangle$ continues downwards and the part $|2\downarrow\rangle|\text{path }2\downarrow\rangle$ continues upwards. If we then measure the spin in direction 1, along either of these paths, we will get 25 % 1 ↑ and 25 % 1 ↓ in each case, since this is what is predicted by both the states

$$|2\uparrow\rangle = \frac{1}{\sqrt{2}}\big(|1\uparrow\rangle + |1\downarrow\rangle\big) \quad\text{and}\quad |2\downarrow\rangle = \frac{1}{\sqrt{2}}\big(|1\uparrow\rangle - |1\downarrow\rangle\big) .$$

So it looks as though we could affect the past behavior of the particle (whether it follows "both paths at once" or only one path) by inserting or not the black arrow, *after* they have already started their trip.

As we see, the mysterious behavior described in Sects. 2.1 and 2.2 can easily be predicted by the quantum formalism which, at least in this special case, is not mathematically complicated. However, a natural question arises: how are we supposed to understand all this talk about vectors or wave functions, and also, of course, this duality of rules?

2.5 What Is the Meaning of the Quantum State?

To me it seems like "quantum theory" is in a sense like a traditional herbal medicine used by "witch doctors". We don't REALLY understand what is happening, what the ultimate truth really is, but we have a "cook book" of procedures and rituals that can be used to obtain useful and practical calculations (independent of fundamental truth).

John Nash [347, p. 4]

There are several possible reactions to what has been described in the previous sections.[15]

A *first reaction* is to claim that one cannot understand the microscopic world and that one must content oneself with predicting the results of measurements, which are necessarily macroscopic, and are thus described in a "classical" (i.e., understandable) language. For example, one could say that, if one prepares the particle in state $|1 \uparrow\rangle$, say, it will have such and such probability of ending up in a given state if we do this or that experiment. The final state is described by the result of measurement and can be described in terms of where the particle goes (through which hole it leaves our measuring device of Fig. 2.1, for example). Since we know which hole the particle goes through by detecting it, and since the detection leads to a direct observation by us, humans, one can understand why human beings seem to be put at the center of a physical theory.

Those who do not like this return to anthropocentrism may of course claim that human beings are not essential here: all that matters is a macroscopic trace left on the measuring device, which could be seen by us, but which is there whether we look at it or not. This is certainly a possible understanding of the "orthodox" view. Landau and Lifshitz wrote in their standard textbook:

> [...] we emphasize that, in speaking of 'performing a measurement', we refer to the interaction of an electron with a classical 'apparatus', which in no way presupposes the presence of an external observer.
>
>   Lev Landau and Evgeny Lifshitz [302, p. 2], quoted in [46, p. 35]

A *second reaction* follows naturally from this idea, and consists in the hope that, by analyzing the measurement process in more detail, as a purely physical process (with no reference whatsoever to an outside "observer"), one may arrive at an understanding of what is going on.

A *third reaction* is to view the quantum state as representing, not an individual system, but an ensemble of systems and having thus a role similar to probabilities in classical physics,[16] with $|c_i(t)|^2$, $|d_i(t)|^2$, $i = 1, 2$, in (2.3.9) being the probabilities of the spin being either up or down in directions 1 and 2. According to this view, the quantum state provides us with incomplete information on individual microscopic systems. The latter do have properties such as a given spin value in all directions, or a position and a velocity, but the quantum state itself does not contain that information. We may not know how to prepare an individual system with those given properties, but they nevertheless exist. When we produce several particles in a given quantum state, we actually produce particles with different individual properties, whose statistics are encoded in the quantum state. This approach may be called the statistical interpretation of quantum mechanics (see, e.g., Ballentine [27], Blokhintsev [59], or Taylor [468] for a defense of this viewpoint).

A *fourth reaction* (sometimes motivated by the third) is to propose a more complete theory than quantum mechanics. One would not simply say, as in the third reaction,

---

[15]This section is in part an extension of [81].

[16]The latter are discussed in Sect. 3.4.3.

that particles do have properties not described by the quantum state, but one would try to say what these properties are and how they evolve in time. Einstein's reaction (which we will discuss in Chap. 7) was basically the third one, but he also hoped for a more complete theory.[17]

The first reaction is associated with the Copenhagen interpretation and is basically what was called the "official" position in Chap. 1. It was criticized there and will be discussed further in Chap. 3. The fourth reaction will be the topic of Chap. 5. The de Broglie–Bohm theory is a subtle form of the statistical interpretation, but based on a more complete theory than ordinary quantum mechanics.

We now discuss the second and the third reactions, one of which probably lies in the back of the minds of most physicists who don't see any problem with quantum mechanics. Of course, if one of those positions were tenable, they would be right not to worry, but, as we will try to show, neither of these "ways out" is compatible with either the quantum formalism (for the second reaction) or with experimental facts (for the third one).

These two reactions are different answers to the same question: does the measurement somehow create the result that is being observed or does it simply reveal some pre-existing property of the system? The word "measurement" suggests the latter meaning: if I measure the length of a table, I assume that the table has a certain length before I measure it. The same thing holds for more indirect measurements, like the distance between the Earth and the Sun. This view leads us to the statistical interpretation (since the quantum state does not assign a fixed value to the quantities being measured) and we will return to it in Sect. 2.5.2.

One natural objection to the idea that the quantum state is just like a classical probability comes from the bizarre interference effects described above (which have no equivalent in classical probabilities). In the Mach–Zehnder interferometer, if we assign a probability $1/2$ to the event "the particle follows the path 2 ↑" and $1/2$ to the event "the particle follows the path 2 ↓" when both paths are open, and a probability 1 to the event "the particle follows the path 2 ↑" when the path 2 ↓ is blocked, why do we get different results in those two situations if we measure the spin in direction 1 at the black arrow?[18] If probabilities reflect our ignorance (of the path being taken), simply knowing which path is taken should not have any physical effect. Does this not show that the quantum state is a physical quantity rather than a pure expression of our ignorance?

But if the quantum state is physical, should one not think that the measurement, viewed as a physical process, perturbs the quantum state in such a way as to "create"

---

[17]In [43], Bell considers six "possible worlds of quantum mechanics", i.e., six possible reactions with respect to the problems discussed here: the pragmatic attitude, the Bohr approach, introducing the mind into physics, the many-worlds interpretation, spontaneous collapse theories, and the de Broglie–Bohm theory. The last three theories pertain to what we call the fourth reaction and will be discussed in Chaps. 5 and 6. The pragmatic attitude and the Bohr approach both exemplify what we call the first reaction, while introducing the mind into physics could be considered as part of the fourth reaction, but we will not discuss it beyond a few words in Chap. 3.

[18]As se saw in Sect. 2.2, we get $100\,\%$ 1 ↓ if we measure the spin after the black arrow in direction 1 when both paths are open, and $25\,\%$ 1 ↑, $25\,\%$ 1 ↓, when path 2 ↓ is blocked.

the result? There is nothing a priori irrational or even strange about this idea: the measuring device is necessarily macroscopic (otherwise we would not be able to see the result) and the object being measured is microscopic. The huge difference in size between the measuring device and the system being "measured" leaves ample room for the macroscopic object to affect the microscopic one.

One could of course declare a priori that the measuring device, being macroscopic, "collapses" the quantum state during a measurement, and that view is also some- times associated with the Copenhagen interpretation. The two basic rules of the time evolution of a quantum state would then simply reflect the micro/macro distinction: a microstate evolves according to the Schrödinger equation, except when it interacts with a macroscopic object, in which case it may be reduced or collapsed.[19] How- ever, that view assumes that there is a sharp distinction between microscopic and macroscopic; but if the Schrödinger evolution applies to one particle, two particles, ten particles, and so on, where should we stop?

A way out of this problem would exist if one could treat the measuring device in a quantum mechanical way and obtain those reduced states for the microscopic systems at the end of a measurement.[20] We begin by considering this possibility.

2.5.1 The Measurement Process Within the Quantum Formalism

Let us see what happens if we analyze the measurement process within the quantum formalism, an analysis that goes back to von Neumann [496]. We want to see if, within the quantum formalism, one can avoid the dual nature of the evolution law for the quantum state.

We consider a very simplified measurement process. Let

$$\Psi_0 = \varphi_0(z) \left[c_1 \begin{pmatrix} 1 \\ 0 \end{pmatrix} + c_2 \begin{pmatrix} 0 \\ 1 \end{pmatrix} \right] ,$$

which describes the original state of a particle whose spin is going to be measured, viz.,

$$c_1 \begin{pmatrix} 1 \\ 0 \end{pmatrix} + c_2 \begin{pmatrix} 0 \\ 1 \end{pmatrix} ,$$

and the state $\varphi_0(z)$ of the measuring device. Here z is a macroscopic variable, indi- cating the position of the measuring device (for example, the position of its center of

---

[19]This would make the collapse rule somewhat analogous to the phenomenon of entropy increase in statistical mechanics. We will discuss this analogy further in Sect. 5.1.7.

[20]We will see in Sect. 5.1.6 that this is, in a sense, what happens in the de Broglie–Bohm theory; but it only works because, in that theory, one has a more complete description of the quantum system than the one in ordinary quantum mechanics.

Fig. 2.6 Evolution of the
pointer during a
measurement

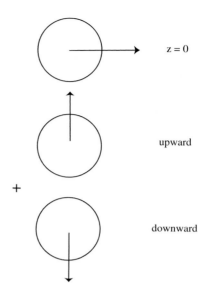

mass along the vertical axis), and $\varphi_0(z)$ is centered at $z = 0$, meaning that the pointer
is as in the first picture in Fig. 2.6. To simplify matters, we do not include here the
wave function of the particle whose spin is being measured, considering only the
"spin" part of its quantum state.

One might question the assignment of a quantum state to a macroscopic object.
But this is exactly what we mean by "working within the quantum formalism". In that
formalism, by assumption, every object is describable by such a state. Of course, the
pointer is composed of a great number of particles, not just the variable z introduced
here. However, if we measured the positions of all the particles composing the pointer
(which is of course impossible in practice, but we can at least imagine doing that),
we would also know the value of z, so that one may consider that variable as being
determined by all the other variables. Note that we say "if we measured" the positions.
We do not assume that particles *have* positions independently of whether we measure
them or not. The same holds of course for the value of z.

As we show in Appendix 2.D, the state resulting after the measurement is

$$c_1 \begin{pmatrix} 1 \\ 0 \end{pmatrix} \varphi^\uparrow(z) + c_2 \begin{pmatrix} 0 \\ 1 \end{pmatrix} \varphi^\downarrow(z) , \qquad (2.5.1.1)$$

where $\varphi^\uparrow(z)$ and $\varphi^\downarrow(z)$ correspond to the last two pictures in Fig. 2.6, i.e., the pointer
pointing upward or downward. Thus, the system is in a superposition of two macro-
scopically distinct states: one in which the pointer is pointing upward *and* one in
which it is pointing downward. The problem is that we never see the pointer in such
a superposed state: we see it *either* up *or* down, but not both. The ordinary quantum
formalism does not correctly predict the state of the measuring device at the end

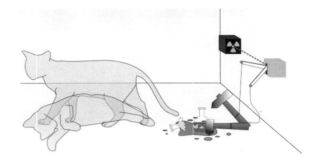

of the experiment, since it unambiguously predicts a superposed state, and this is
simply not what is observed.

Note that here we assume, somewhat naively, that we can identify a macroscopic
quantum state with a physical situation in the three-dimensional world, such as
the state $\varphi^\uparrow(z)$ and a pointer pointing up. This is often implicitly assumed in all
discussions about macroscopic superpositions, but we will see in Sect. 6.1 that this
assumption, far from being obvious, is in fact hard to justify.[21]

But since the situation is now macroscopic, one may just *look* at the result. If
the pointer points upward, we take the state to be $\begin{pmatrix} 1 \\ 0 \end{pmatrix} \varphi^\uparrow(z)$. If the pointer points
downward, the state becomes $\begin{pmatrix} 0 \\ 1 \end{pmatrix} \varphi^\downarrow(z)$. One thus reduces the quantum state, which
now describes a macroscopic object, just by looking at it.

One may also replace the pointer by a cat, as in Schrödinger's famous thought
experiment [441]: suppose a cat is in a sealed box and there is a purely classical
mechanism linking the pointer above to a hammer that will break a bottle containing
some deadly poison if the pointer is up, but not if it is down. If the poison is released,
it kills the cat (see Fig. 2.7). Then, following the same reasoning as above, including
now the state of the cat, we get after the measurement:

$$c_1 \begin{pmatrix} 1 \\ 0 \end{pmatrix} \varphi^\uparrow(z)|\text{cat dead}\rangle + c_2 \begin{pmatrix} 0 \\ 1 \end{pmatrix} \varphi^\downarrow(z)|\text{cat alive}\rangle \ .$$

The natural interpretation of the state of the cat is that it is "both alive and dead".
Of course, we never see a cat in such a state. We do not even know what that
could mean. But the cat example just dramatizes a problem that occurs already with
the pointer, namely the fact that ordinary quantum mechanics predicts macroscopic
superpositions that are simply not observed and that are even hard to conceive.

To better illustrate the problem, consider a coin. Once it has been thrown and has
fallen on one of its sides, it is either heads or tails. If we do not look at the result,
we may attribute probability one-half to each of these possibilities, but it would be

---

[21]See Maudlin [325] for a detailed discussion.

silly to say that these probabilities give a complete description of the state of the coin. The latter is heads or tails, not both, and our assignment of probabilities simply reflects our ignorance. Saying that the quantum mechanical description is complete is similar to saying that the probabilistic description of the coin is complete. While we may not know whether this is true or not for microscopic systems (about which we have no direct experience), it is manifestly absurd for a pointer or a cat.

Sometimes people seem to think that quantum mechanics has proven that the unfortunate cat is both alive and dead before anybody looks. But that was certainly not Schrödinger's idea when he introduced the cat example, which he called "quite ridiculous", as a *reductio ad absurdum* of the quantum formalism [441]. More precisely, Schrödinger wanted to show that the quantum formalism does not provide a complete description of at least *some* systems, because, at least in the case of the cat, we know that the latter is either alive or dead, and not both.

In order to turn Schrödinger's reasoning around and produce an argument in favor of the existence of macroscopic superpositions, one has to assume that the quantum formalism is absolutely true and applies to all objects, irrespective of their size. But there has never been any observable consequence of such an extension of the quantum formalism to macroscopic objects.[22]

One possible answer to this problem is to say that looking at the pointer or the cat changes its state and thus collapses its quantum state. But what does "looking" mean? There are many different ways to look at an object. With the help of binoculars, or with a telescope, one could look from far away. One could peek through a small hole made in the box where the cat is, etc. None of this changes anything regarding the result of course: the pointer is always up or down, the cat is alive and dead. Since all the different physical ways of looking do not make any difference, isn't it reasonable to think that looking does not have any physical effect on the system itself and that by "looking" we simply *learn* something about the state of the system, without changing it? In other words, this situation would be analogous to throwing a coin and first hiding the result; then when we later look at the coin, we see whether it is heads or tails, but of course the coin was heads or tails before we looked. This analogy is the common-sensical solution to the cat problem, and it is the one that Schrödinger had in mind.

Of course, one may also hold the view that, as long as "looking" is described in physical terms, with eyes, brains, etc., it only produces more macroscopic superpositions: in the end, the whole universe has a quantum state like (2.5.1.1), with $\varphi^{\uparrow}(z)$ and $\varphi^{\downarrow}(z)$ corresponding to the pointer being up or down and the observer (at least as long as it is considered to be a physical object) seeing it up or down, and everything linked to the observer being in such a superposed state. Then one arrives at a sort of infinite regress (everything physical being in a superposed state) and one has to appeal to a nonphysical consciousness to collapse the quantum state. But even if we accept the existence of a conscious mind entirely independent of the laws of physics,

---

[22]Of course, there are macroscopic phenomena like superconductivity or superfluidity whose explanations appeal directly to quantum mechanics, but these are different from the examples of the cat or the pointer.

which consciousness should it be? Mine when I observe, someone else's when I tell them the results, a universal mind? This leads us to "theories" that are very poorly defined, assuming that they make any sense at all.[23]

Another line of thought is to say that the quantum state never collapses and that the two terms in (2.5.1.1) simply correspond to two different universes. This is called the "many-worlds interpretation" and will be discussed in Sect. 6.1. But in any case, it can be considered as part of the fourth reaction, since it modifies ordinary quantum mechanics, in which the collapse is part of the theory.

It is sometimes suggested that decoherence solves the cat problem: decoherence means that it is, in practice, impossible to produce interference effects between the states of the live and the dead cat or between the states of the upward and downward pointers, as was done for a single particle in the interferometer described in Sect. 2.2, between the state of the particle following one path and the state following the other path.[24] This is due to the large number particles composing a pointer or a cat.[25]

Therefore, we don't see, for cats or pointers, the strange effects that were seen in that experiment. This is something on which all sides agree: indeed, if it wasn't so, the macroscopic world would look very different from what it does, because if the cat is alive, its future behavior could interfere with the state of the dead cat, as the states following the two different paths in the Mach–Zehnder interferometer do, and that could lead to strange behavior for the live cat.

But decoherence does not change the fact that one needs to look at the result, and then "collapse" the state according to what *we see* in order for the formalism to work. Therefore, it does not remove the centrality of observations in the quantum theory. In [46], John Bell analyzes the books by Dirac [137], Landau and Lifshitz [302], Gottfried [236], and an article by van Kampen [494] and shows that in each case there is a subtle but unacknowledged transition between *and* and *or*: quantum mechanics predicts unambiguously that the cat is alive *and* dead, while we always see it either alive *or* dead. It is only by switching from *and* to *or* that one can "eliminate" that problem.

What decoherence does show is that there are no empirical consequences of quantum mechanics for macroscopic objects, at least in situations such as those described here. So for the pointer, the most natural interpretation of the state (2.5.1.1) is that of a classical probability: the pointer is up or down and the state simply reflects our ignorance: hence we say that it is up with probability one-half and down with

---

[23]This should not be confused with the familiar "mind–body" problem: how can the material body produce mental states, and in particular, conscious ones (see [345] for a good explanation of the problem)? Even those, like Colin McGinn [329], who regard the link between the body or the brain and the qualitative aspects of consciousness (e.g., pain) as being an unsolvable mystery, given the limitations of the human mind, admit that they can be caused by, or at least correlated with, physical events in the brain. But here, when we consider the possibility that the mind collapses the quantum state, we are envisaging a direct action of the mind on matter, and this entails a radical form of dualism, since the mind would then act totally independently of the brain!.

[24]See the double-slit experiment described in Appendix 2.E for another example of interference between states of one particle.

[25]This notion of decoherence will be explained in more detail in Sect. 5.1.6 and Appendix 5.E.

the same probability. This is just like a coin which is heads or tails before we look at it and where we simply learn which of the two is the case when we do actually look at it.

But this means that, for macroscopic objects, the quantum state is not a complete description of the physical situation. There is no way within the quantum formalism to make a definite macroscopic state emerge as result of a measurement, because of the linearity of the Schrödinger equation. So that the idea, implicit in the statement of Landau and Lifshitz quoted at the beginning of this section and in many others analyzed by Bell in [46], that the interaction with a macroscopic device creates the results of measurements is simply untenable if we stick to the existing quantum formalism.

On the other hand, if we give a classical probabilistic interpretation to the state (2.5.1.1), why not do the same for microscopic objects described by quantum states like those in Sects. 2.3 and 2.4? All this leads us naturally to the third reaction, namely the ensemble or statistical interpretation of quantum mechanics.

2.5.2 The "Naive" Statistical Interpretation

The statistical approach assumes that systems of all sizes have properties, such as the spin being up or down, before they are measured and that a measurement simply reveals something pre-existing (as the word "measurement" suggests). Of course, that does not mean that we have a theory about those properties or that we can predict or control them. It simply means that we can think of them as existing prior to our measurements and as being revealed by them.

If one adopts that view, probabilities in quantum mechanics are not very different from classical probabilities: they just reflect our ignorance, and so does the quantum state. When we measure a certain physical quantity, we simply learn something about the system, and we thus modify our state (or our probabilities) accordingly. This view means that quantum mechanics is incomplete; the very definition of incomplete is that each individual system is characterized by variables other than the quantum state and that the latter has only a statistical meaning. To come back to the analogy with a coin, if we learn that it is heads or tails (a property that the coin had before looking at it), we change our probabilities for heads and tails from $(1/2, 1/2)$ (before looking) to $(1, 0)$ or $(0, 1)$, depending on the result. The collapse of the quantum state would be similar to that adjustment of probabilities.

This is the basic idea behind what are called "hidden variables". They would be variables that characterize an individual system and whose statistical distribution would be determined by the quantum state. For example, if we prepare particles in a state such as (2.3.9), the assumption of hidden variables means that a fraction $|c_1(t)|^2$ of such particles have spin up in direction 1 and a fraction $|c_2(t)|^2$ of such particles have spin down in direction 1, and similarly for direction 2, with $|c_i(t)|^2$ replaced by $|d_i(t)|^2$.

Of course, the time evolution of the quantum state, which allows for the interference effects, is very different from the one encountered in the application of probability theory to classical physics. But the time evolution of the state and its status are two separate issues and there is nothing a priori inconsistent in thinking that the spin is up or down before we measure it or that the particle goes through one slit or the other in the double-slit experiment (see Appendix 2.E). We simply have to take into account the fact that the behavior of the particle after passing through one slit is affected by whether the other slit is open or not, and likewise, in the interferometer experiment, whether the other path is blocked or not.

The uneasiness in treating the quantum state as a classical probability also follows from the fact that one cannot simultaneously measure the spin in two different directions, or the position and the velocity. But if we think that the measurement disturbs the system being measured, then there is nothing *a priori* implausible in the statistical interpretation. Note, however, that this "perturbation" view of measurements cannot be incorporated within ordinary quantum mechanics, as we just saw above, simply by analyzing measurements within quantum mechanics. But since assuming the existence of variables other than the quantum state means going beyond quantum mechanics, it is perfectly logical to assume that measurements affect those variables in ways that are not covered by the ordinary quantum theory. Then, of course, affecting those variables would modify the state and we would have to change our probabilities accordingly (and this would then be the reason behind the collapse rule).

Consider, for example, what Heisenberg wrote in his famous paper on the uncertainty relations:

> At the instant of time when the position is determined — therefore, at the moment when the photon is scattered by the electron — the electron undergoes a discontinuous change in momentum. This change is the greater the smaller the wavelength of the light employed — that is, the more exact the determination of the position. At the instant at which the position of the electron is known, its momentum therefore can be known only up to magnitudes which correspond to that discontinuous change; thus, the more precisely the position is determined, the less precisely the momentum is known, and conversely.

> Werner Heisenberg [510, p. 64]. Originally published in [256]

This quote clearly expresses a "perturbation" view of measurements: if I want to measure the position of the electron, then I must perturb it in such a way that its momentum is affected. Therefore, I cannot measure both its position and the momentum it had when I did that measurement. But this statement certainly suggests that the electron *has* both a position and a momentum.

However, there is a serious problem for the statistical interpretation, namely that it is inconsistent. To see how this comes about, let us define the idea of "hidden variables" more precisely. There are many physical quantities besides spin that can in principle be measured: for example, the angular momentum, the energy, and the momentum.[26] The statistical interpretation means that, in each individual system,

---

[26]These quantities are called observables and are represented mathematically by matrices or operators acting on the quantum states. The eigenvalues of these operators are the possible results of

each of these quantities will have a well defined value, which may be unknown or even unknowable, and uncontrollable, but which nevertheless *exists*.

Let us denote by A a physical quantity and by $v(A)$ the value that this quantity has for a particular system, which of course varies from system to system, but in such a way that the quantum state gives the statistical distribution of those values: going back to the examples of Sect. 2.3, a state like (2.3.9) would mean that, if we prepare a large number of particles with that same state, then a fraction $|c_1|^2$ of them will have the value v(spin in direction 1) $= \uparrow$ and a fraction $|c_2|^2$ of them will have the value v(spin in direction 1) $= \downarrow$, and similarly for direction 2 (the situation with more general quantities is discussed in Appendix 2.B).

To make the statistical interpretation interesting, we have to assume that $v(A)$ exists for more than one A. For example, it would be quite arbitrary to assume that the spin values exist, but only in one direction, since our definition of directions is completely conventional. Now, if we assume that $v(A)$ exists for a reasonable class of quantities A, and that those values agree with quantum mechanical predictions, we can derive a contradiction:

Theorem 2.5.1 No Hidden Variables Theorem[27]

(1) There does not exist a function $v : \mathcal{O} \to \mathbb{R}$ where $\mathcal{O}$ is a collection of quantities related to "spin", such that $\forall A \in \mathcal{O}$, $v(A)$ agrees with the predictions of quantum mechanics.

(2) There does not exist a function $v : \mathcal{O} \to \mathbb{R}$ where $\mathcal{O}$ is the set of functions of the four quantities representing the positions and the momenta of two particles moving on a line, such that $\forall A \in \mathcal{O}$, $v(A)$ agrees with the predictions of quantum mechanics.

It is important to stress that the requirement of "agreeing with the predictions of quantum mechanics" means only that certain constraints, inherent to the quantum algorithm for predicting results of measurements and discussed in Appendix 2.F, have to be satisfied. This requirement is thus totally independent of any quantum state or of the need to reproduce any particular quantum statistics.

A possible, but misleading, reaction to this theorem is to say that there is nothing new here, since it is well known that there is no quantum state that assigns a given

---

(Footnote 26 continued)

the measurement of these observables, but we do not need these notions here. They are explained in Appendix 2.B.

[27]We refer to Appendices 2.B, 2.C and 2.F for a more precise formulation of this theorem in terms of "observables" and "operators", and for a definition of the sets $\mathcal{O}$. The original version of this theorem is due to Bell [36] and to Kochen and Specker [291] for the first part (the proofs were based on a theorem of Gleason [215]), and to Clifton [98] for the second part. The version given here is simpler than the original ones and is due to Mermin (see [335] and reference therein) and Perez [392, 393] for the first part and to Myrvold for the second [344]. The proofs are given in Appendix 2.F. We will discuss another no hidden variables theorem in connection with nonlocality in Chap. 4. We will discuss other variants of this theorem in Sect. 5.3.4 and in Sect. 6.3 and Appendix 6.C. We will also discuss the famous but misleading no hidden variables theorem due to von Neumann in Sect. 7.4.

value to all the spin variables simultaneously or to both position and momentum, as we saw, at least for the spin (for the position and momentum, this is discussed in Appendix 2.C). But that misses the point: the theorem considers the possibility that there be other variables characterizing an individual system than its quantum state (in other words, that the quantum description is incomplete), variables whose values would be revealed by proper measurements. The theorem shows that, at least if the class of those variables is large enough, merely assuming the existence of those variables is impossible. Note that we are not assuming that there exists a theory about those "hidden" variables, telling us how they evolve in time for example, but merely that these variables exist and that their values agree with the quantum mechanical predictions.

Of course, this result *does not mean* that a theory introducing hidden variables cannot exist (we will discuss such a theory in Chap. 5), but it does mean that one cannot introduce such variables for all the "observables" or at least for a sufficiently large class of them at once. If we want to express precisely what the statistical interpretation of quantum mechanics means, we have to assume that individual systems possess values corresponding to a physical property A, which we denote $v(A)$, before being measured, and that a measurement simply reveals that value $v(A)$. But that leads to contradictions and therefore the "naive" statistical view is untenable.

2.6   Conclusions

As we already said (and it would be interesting to make a sociological study of this issue), it is probable that the creation of a definite quantum state by the interaction with a macroscopic apparatus, or the statistical interpretation, lies in the back of the mind of many physicists who are not bothered by the problems raised by quantum mechanics. It is unlikely that most physicists literally believe that the cat suddenly becomes alive or dead, simply because we look at it, especially if "looking" refers to the action of a mind independent of all physical laws.

But now we face a serious conundrum. There are two positions that would naturally justify the "no worry" attitude with respect to the meaning of the quantum state: *either* a proper quantum treatment of the measurement process would lead to a collapsed state and the need for two different laws of evolution would be eliminated, *or* the quantum state does not represent a single system but an ensemble of systems, each having its own individual properties that a measurement would simply reveal. But neither of these positions are defensible, either because the linearity of the Schrödinger equation leads necessarily to macroscopic superpositions, or because of the no hidden variables theorems.

We are left with the first and the fourth reactions. We will deal with the last one, i.e., look for a more complete theory than ordinary quantum mechanics, in Chap. 5. But before doing that, we have to discuss a host of philosophical arguments trying to present the first reaction as not simply making the best of a bad deal, but as a necessity, independently of quantum mechanics, or even as a positive development. In the next chapter, we will examine those arguments.

Appendices

2.A The Wave Function and the Schrödinger Equation

In this appendix, we will describe some of the mathematical properties of Schrödinger's equation, without discussing in detail its physical meaning, something already done in the main text of this chapter and in Chaps. 4 and 5.

2.A.1 Linear Differential Equations

Let us start with the simplest differential equation[28]:

$$\frac{dz(t)}{dt} = az(t) , \qquad (2.A.1.1)$$

where $t \in \mathbb{R}$, $z : \mathbb{R} \to \mathbb{R}$, and $a \in \mathbb{R}$. By definition, a solution of this equation is a function satisfying it for all t. It is easy to see that all solutions are of the form

$$z(t) = Ae^{at} , \qquad (2.A.1.2)$$

for some constant A.

We obtain a unique solution if we fix some initial condition, that is, if we fix the value of $z(t)$ at a given time t. To simplify the notation, let $t = 0$ and let us look for a solution such that $z(0) = z_0$. Then we obtain a unique solution:

$$z(t) = z_0 e^{at} . \qquad (2.A.1.3)$$

In this simple example, we see that (2.A.1.1) has a class of solutions (2.A.1.2) and a unique solution (2.A.1.3) once an initial condition is fixed. This is true for more general equations of the type

$$\frac{dz(t)}{dt} = f(z(t)) , \qquad (2.A.1.4)$$

---

[28]For an introduction to differential equations, see, e.g., [18, 267].

for fairly general functions $f : \mathbb{R} \to \mathbb{R}$, at least for short intervals of time[29] (but we will not use those more general equations here).

Equation (2.A.1.1) is said to be linear because, if $z_1(t)$ and $z_2(t)$ are solutions of (2.A.1.1), then the function $z(t) = c_1 z_1(t) + c_2 z_2(t)$, with $c_1, c_2 \in \mathbb{R}$, is also a solution.

We now generalize this simple example. First, we could replace $z(t)$ by a complex-valued function: $z : \mathbb{R} \to \mathbb{C}$, with $a \in \mathbb{C}$ in (2.A.1.1). Nothing changes except that A and z_0 in (2.A.1.2) and (2.A.1.3) are also complex.

Next, we replace $z(t)$ by an n-component complex vector:

$$\mathbf{z} : \mathbb{R} \to \mathbb{C}^n , \quad \mathbf{z}(t) = \begin{pmatrix} z_1(t) \\ \vdots \\ z_n(t) \end{pmatrix} , \quad z_i(t) \in \mathbb{C} , \quad i = 1, \ldots, n .$$

Equation (2.A.1.1) is replaced by

$$\frac{d\mathbf{z}(t)}{dt} = \mathcal{A}\mathbf{z}(t) , \tag{2.A.1.5}$$

where $\mathcal{A}$ is an $n \times n$ complex matrix. The general solution is of the form

$$\mathbf{z}(t) = e^{\mathcal{A}t}\mathbf{A} , \tag{2.A.1.6}$$

where $\mathbf{A} \in \mathbb{C}^n$,

$$e^{\mathcal{A}t} = \sum_{n=0}^{\infty} \frac{\mathcal{A}^n t^n}{n!} , \tag{2.A.1.7}$$

and $\mathcal{A}^n$ denotes the n th product of $\mathcal{A}$ with itself. Equation (2.A.1.5) is again linear.

If we fix an initial condition $\mathbf{z}(0) = \mathbf{z}_0 \in \mathbb{C}^n$, we get a unique solution:

$$\mathbf{z}(t) = e^{\mathcal{A}t}\mathbf{z}_0 . \tag{2.A.1.8}$$

When $\mathcal{A}$ possesses a basis of eigenvectors, i.e.,

$$\mathcal{A}\mathbf{e}_i = \lambda_i \mathbf{e}_i , \tag{2.A.1.9}$$

where $(\mathbf{e}_i)_{i=1}^n$ is an orthonormal basis of $\mathbb{C}^n$, the solution (2.A.1.8) can be written more explicitly. Indeed, (2.A.1.9) and (2.A.1.7) imply that

$$e^{\mathcal{A}t}\mathbf{e}_i = e^{\lambda_i t}\mathbf{e}_i , \tag{2.A.1.10}$$

and if we expand $\mathbf{z}_0$ in the basis $(\mathbf{e}_i)_{i=1}^n$, i.e.,

---

[29] See, e.g., [267, Chap. 7] for more details.

$$\mathbf{z}_0 = \sum_{i=1}^{n} c_i \mathbf{e}_i \, , \qquad (2.A.1.11)$$

where

$$c_i = \langle \mathbf{e}_i | \mathbf{z}_0 \rangle \, , \qquad (2.A.1.12)$$

and $\langle \cdot | \cdot \rangle$ is the scalar product in $\mathbb{C}^n$ ($\langle z_1 | z_2 \rangle \equiv \sum_{n=1}^{N} z_{1n}^* z_{2n}$), we find by linearity that (2.A.1.8) can be written as

$$\mathbf{z}(t) = \sum_{i=1}^{n} c_i e^{\lambda_i t} \mathbf{e}_i \, . \qquad (2.A.1.13)$$

So the "recipe" for solving (2.A.1.5) is to solve the eigenvalue/eigenvector problem for $\mathcal{A}$ (assuming that $\mathcal{A}$ has a basis of eigenvectors), compute the coefficients using (2.A.1.12), and insert the result in (2.A.1.13).

2.A.2 The Schrödinger Equation

Let us start with the equation for one particle in three-dimensional space:

$$i\hbar \frac{d}{dt} \Psi(x, t) = H \Psi(x, t) \, , \qquad (2.A.2.1)$$

where $t \in \mathbb{R}$, $x \in \mathbb{R}^3$, and $\hbar = h/2\pi$, with h the Planck constant. The unknown here is Ψ, which is a complex-valued function of x and t.

One can think of Ψ as playing the role of $\mathbf{z}$ in (2.A.1.5), with the index $i = 1, \ldots, n$ being replaced by a continuous variable x. The factor $i = \sqrt{-1}$, while essential for the physics of (2.A.2.1), does not make much difference at this stage with respect to (2.A.1.5), since Ψ, like $\mathbf{z}$, is complex anyway. H plays the role of $\mathcal{A}$ in (2.A.1.5) and is a linear operator: it transforms a given function $\Psi(x, t)$ into a new function $(H\Psi)(x, t)$ and does it in a linear way:

$$H(\alpha \psi_1 + \beta \psi_2) = \alpha H \psi_1 + \beta H \psi_2 \, , \qquad (2.A.2.2)$$

which implies that a linear combination of solutions of (2.A.2.1) is again a solution of (2.A.2.1).

The detailed form of H or the justification of (2.A.2.1) will not matter very much and they can be found in any textbook on quantum mechanics,[30] but for one particle of mass m moving in $\mathbb{R}^3$, the operator H has the form

---

[30]For a discussion close to our point of view, see [152, Chap. 7].

$$H = \frac{\hbar^2}{2m} \left(-\frac{d^2}{dx_1^2} - \frac{d^2}{dx_2^2} - \frac{d^2}{dx_3^2} \right) + V(x) , \qquad (2.A.2.3)$$

where $x = (x_1, x_2, x_3)$ and $V(x)$ is simply the classical potential [so that the force $F(x)$ in classical mechanics equals $F(x) = -\nabla V(x)$, ∇ denoting the gradient]. The first term, viz.,

$$\frac{\hbar^2}{2m} \left(-\frac{d^2}{dx_1^2} - \frac{d^2}{dx_2^2} - \frac{d^2}{dx_3^2} \right) ,$$

is the kinetic energy term. To simplify notation, we will often consider the situation in one spatial dimension, where H is given by:

$$H = -\frac{\hbar^2}{2m} \frac{d^2}{dx^2} + V(x) , \qquad (2.A.2.4)$$

with $x \in \mathbb{R}$. Classically, the Hamiltonian is (again, in one dimension)

$$H = \frac{p^2}{2m} + V(x) ,$$

and corresponds to the energy of an isolated system. The quantum version (2.A.2.4) is obtained by replacing the momentum variable p (equal to the mass times the velocity) in the classical Hamiltonian by the operator[31] $P = -i\hbar d/dx$ and the variable x by the operator Q of multiplication by x [and hence $V(x)$ by the operator of multiplication by $V(x)$]. We will not justify this replacement now, but we will explain why the statistical distribution of results of measurements of momenta is related to the operator P in Appendices 2.A.3 and 2.B. For the variable x, we have already said that $|\Psi(x, t)|^2$ is the probability density of the results of position measurements.

In the rest of these appendices, we will choose units in which $\hbar = 1$. Then, given an initial condition $\Psi(x, 0) = \Psi_0(x)$, the solution of (2.A.2.1) is (remember that $1/i = -i$):

$$\Psi(x, t) = (e^{-iHt}\Psi_0)(x) , \qquad (2.A.2.5)$$

where the operator e^{-iHt} can be defined through a power series as in (2.A.1.7) when the series converge, and in more subtle ways otherwise.[32] We will give concrete examples of what this solution looks like below, and also in Appendix 2.D and Chap. 5.

An important property of (2.A.2.5) is

$$\int_{\mathbb{R}^3} |\Psi(x, t)|^2 dx = \int_{\mathbb{R}^3} |\Psi_0(x)|^2 dx , \qquad (2.A.2.6)$$

---

[31]Operators are linear functions that map "ordinary" functions into other functions. The space of functions on which they act is infinite dimensional. We will not give a rigorous or detailed treatment of these operators; see, e.g., [152, Chaps. 13–15] or [412, Chaps. 7 and 8] for such a treatment.

[32]See, for example, [412, 413] and [152, Chap.14].

for all t, which allows us to consider $|\Psi(x,t)|^2$ as the probability density of finding the particle at x if one measures its position at time t, provided one normalizes $\int_{\mathbb{R}^3} |\Psi_0(x)|^2 dx = 1$.

What about H having a basis of eigenvectors? For that to make sense, we have to define a space of functions Ψ and explain what a basis in that space means, but a simple example is provided by Fourier series.[33] Let $f(x)$, $x \in \mathbb{R}$, be a complex-valued integrable periodic function of period 2π :

$$f(x + 2\pi) = f(x) , \quad \forall x \in \mathbb{R} . \tag{2.A.2.7}$$

Then $f(x)$ can be written as

$$f(x) = \sum_{n=-\infty}^{+\infty} c_n \frac{e^{inx}}{\sqrt{2\pi}} , \tag{2.A.2.8}$$

with

$$c_n = \frac{1}{\sqrt{2\pi}} \int_0^{2\pi} f(x) e^{-inx} dx , \tag{2.A.2.9}$$

at least when the series converge, which happens, in different senses, given certain properties of $f(x)$. If $f(x)$ is square integrable over $[0, 2\pi]$, i.e., $\int_0^{2\pi} |f(x)|^2 dx < \infty$, then

$$\lim_{N \to \infty} \int_0^{2\pi} \left| f(x) - \sum_{n=-N}^{n=N} c_n \frac{e^{inx}}{\sqrt{2\pi}} \right|^2 dx = 0 \tag{2.A.2.10}$$

and

$$\sum_{n=-\infty}^{n=\infty} |c_n|^2 < \infty , \tag{2.A.2.11}$$

which means, by definition, that the family of functions $(e^{inx}/\sqrt{2\pi})_{n=-\infty}^{+\infty}$ is a basis of the space[34] of square integrable functions over $[0, 2\pi]$.

These relations are similar to those in spaces of N dimensions (with $N < \infty$), the main difference being that in (2.A.2.8), (2.A.2.10), and (2.A.2.11) one has to take a limit $N \to \infty$ and not simply write algebraic identities.

Now if H has a basis of eigenvectors, viz.,

$$H|e_n(x)\rangle = \lambda_n |e_n(x)\rangle , \tag{2.A.2.12}$$

---

[33] See, for example, Dym and McKean [155] for the properties of Fourier series and integrals used here.

[34] This space, denoted $L^2([0, 2\pi], dx)$, is a Hilbert space and the family $(e^{inx}/\sqrt{2\pi})_{n=-\infty}^{+\infty}$ is a Hilbert basis, but we will not need any detailed property of such spaces in this book. The basis here is orthonormal, which will be implicit when we use the word basis.

with $n \in \mathbb{N}$ (in general, the family of eigenvectors will be infinite but countable, as in the example of the Fourier series, so that it can be indexed by $\mathbb{N}$), then one can apply the same recipe that led to (2.A.1.13). We thus write

$$\Psi(x, 0) = \Psi_0(x) = \sum_{n=0}^{\infty} c_n |e_n(x)\rangle , \qquad (2.A.2.13)$$

and the solution of (2.A.2.1) is

$$\Psi(x, t) = \sum_{n=0}^{\infty} c_n \exp(-i\lambda_n t)|e_n(x)\rangle . \qquad (2.A.2.14)$$

Since $|e^{-i\lambda_n t}| = 1$, one can show that the solution converges for all times, provided that we have $\int_{\mathbb{R}^3} |\Psi_0(x)|^2 dx < \infty$ (which implies, as for the Fourier series, $\sum_{n=0}^{\infty} |c_n|^2 < \infty$).

To illustrate what precedes with a simple example, consider a free particle, i.e., with $V(x) = 0$ in (2.A.2.4), on a circle of radius 1. This means that we take $\Psi(x, t)$ to be periodic of period 2π in $x \in \mathbb{R}$ [see (2.A.2.7)], for all t. The operator H being given by $H = -(1/2m)d^2/dx^2$, the eigenvalue/eigenvector problem is easy to solve. We have the following periodic eigenvectors:

$$H \frac{e^{inx}}{\sqrt{2\pi}} = -\frac{1}{2m} \frac{d^2}{dx^2} \frac{e^{inx}}{\sqrt{2\pi}} = \frac{1}{2m} n^2 \frac{e^{inx}}{\sqrt{2\pi}} , \qquad (2.A.2.15)$$

and, applying what we just said about Fourier series (2.A.2.7) and using (2.A.2.13) and (2.A.2.14), we get

$$\Psi(x, t) = \sum_{n=-\infty}^{+\infty} c_n \exp\left(-\frac{in^2 t}{2m}\right) \frac{e^{inx}}{\sqrt{2\pi}} , \qquad (2.A.2.16)$$

where the coefficients c_n come from (2.A.2.8) for $f(x) = \Psi(x, 0)$.

Sometimes the operator H does not have a basis of eigenvectors but the Schrödinger equation nevertheless has a more or less explicit solution. One example that we will refer to is given by a free particle [$V(x) = 0$ in (2.A.2.3)] in the whole d-dimensional space $\mathbb{R}^d$, instead of the circle (but we will set $d = 1$ for simplicity). We want to solve the Schrödinger equation (2.A.2.1), with $H = -(1/2m)d^2/dx^2$, so we want to solve:

$$i \frac{d}{dt} \Psi(x, t) = -\frac{1}{2m} \frac{d^2}{dx^2} \Psi(x, t) . \qquad (2.A.2.17)$$

It is convenient to introduce the *Fourier transform* of $\Psi(x, t)$:

$$\hat{\Psi}(p, t) = \frac{1}{(2\pi)^{1/2}} \int_{\mathbb{R}} \Psi(x, t) e^{-ipx} dx . \qquad (2.A.2.18)$$

This is an invertible operation (for suitable functions $\Psi(x, t)$, for example those satisfying $\int_{\mathbb{R}} |\Psi(x, t)|^2 dx < \infty$):

$$\Psi(x, t) = \frac{1}{(2\pi)^{1/2}} \int_{\mathbb{R}} \hat{\Psi}(p, t) e^{ipx} dp \ . \tag{2.A.2.19}$$

This last formula defines the *inverse Fourier transform*. Inserting (2.A.2.19) into (2.A.2.17), we see that $\hat{\Psi}(p, t)$ satisfies the equation

$$i\frac{d}{dt}\hat{\Psi}(p, t) = \frac{p^2}{2m}\hat{\Psi}(p, t) \ , \tag{2.A.2.20}$$

whose solution is $\hat{\Psi}(p, t) = \exp(-itp^2/2m)\hat{\Psi}(p, 0)$. So the solution of (2.A.2.17) is

$$\Psi(x, t) = \frac{1}{(2\pi)^{1/2}} \int_{\mathbb{R}} \exp\left(-\frac{itp^2}{2m}\right) \hat{\Psi}(p, 0) e^{ipx} dp \ , \tag{2.A.2.21}$$

where $\hat{\Psi}(p, 0)$ is given in terms of the initial wave function by

$$\hat{\Psi}(p, 0) = \frac{1}{(2\pi)^{1/2}} \int_{\mathbb{R}} \Psi(x, 0) e^{-ipx} dx \ .$$

To see what happens in a concrete example, let us start with a Gaussian wave function, in $d = 1$:

$$\Psi_0(x) = \Psi(x, 0) = \frac{1}{\pi^{1/4}} \exp\left(-\frac{x^2}{2}\right) \ ,$$

which is normalized so that $\int_{\mathbb{R}} |\Psi_0(x)|^2 dx = 1$. Then one easily computes that

$$\hat{\Psi}(p, 0) = \frac{1}{\pi^{1/4}} \exp\left(-\frac{p^2}{2}\right) \ .$$

Inserting this in (2.A.2.21), one gets

$$\Psi(x, t) = \frac{1}{(2\pi)^{1/2}} \frac{1}{\pi^{1/4}} \int_{\mathbb{R}} \exp\left(-\frac{itp^2}{2m}\right) \exp\left(-\frac{p^2}{2}\right) e^{ipx} dp \ , \tag{2.A.2.22}$$

and the integral can again be computed to yield

$$\Psi(x, t) = \frac{1}{(1 + it/m)^{1/2}} \frac{1}{\pi^{1/4}} \exp\left[-\frac{x^2}{2(1 + it/m)}\right] \ . \tag{2.A.2.23}$$

The important property of $\Psi(x, t)$ is its spreading:

$$|\Psi(x, t)|^2 = \frac{1}{\sqrt{\pi\left[1 + (t/m)^2\right]}} \exp\left[-\frac{x^2}{1 + (t/m)^2}\right] . \tag{2.A.2.24}$$

Note that we have $\int_{\mathbb{R}} |\Psi(x, t)|^2 = 1$, in conformity with (2.A.2.6).

This means that the variance[35] of the Gaussian $|\Psi(x, t)|^2$ which was equal to $1/2$ at $t = 0$, becomes equal to $[1 + (t/m)^2]/2$ as time goes by. So the Gaussian becomes more and more "flat", which means that, if $|\Psi(x, t)|^2$ represents the probability density of finding the particle in some region of space, then that probability becomes less and less localized as time increases, and in a sense more and more "uncertain".

We will end this appendix with a remark which, although well known, is at the root of the most revolutionary aspect of quantum mechanics, as we will see in Chap. 4.

Suppose we have a system of N particles, each of them in $\mathbb{R}^3$. Then the wave function[36] is a function $\Psi(x_1, \ldots, x_{3N}, t)$ of $\mathbb{R}^{3N} \times \mathbb{R}$ with values in $\mathbb{C}$. It still satisfies the Schrödinger equation (2.A.2.1), but H now has the form

$$H = -\frac{1}{2} \sum_{i=1}^{N} \frac{1}{m_i} \Delta_i + V(x_1, \ldots, x_N) , \tag{2.A.2.25}$$

where

$$\Delta_i = \frac{d^2}{dx_{i_1}^2} + \frac{d^2}{dx_{i_2}^2} + \frac{d^2}{dx_{i_3}^2} , \tag{2.A.2.26}$$

m_i is the mass of the i th particle, and V is again the classical potential. What is "revolutionary" or at least has revolutionary consequences, is that Ψ is defined on what is called the *configuration space* of the system, i.e., the set of all possible positions of all the N particles, where N is arbitrary and could in principle include all the particles in the universe.

So there is a sense (although not very precise at this stage) in which all the particles of the universe are linked with one another. What this implies will be clarified in Chaps. 4 and 5.

2.A.3 The Probability Distribution for Results of Momentum Measurements

We want to show here that the results of measurements of the momentum p (which classically is just the mass times the velocity of the particle) are distributed with a

---

[35] See the definition of variance in Appendix 2.C, (2.C.1.1)–(2.C.1.3).

[36] Ignoring here the issue of symmetry or antisymmetry, for bosons and fermions.

probability density given by $|\hat{\Psi}(p)|^2$, where $\hat{\Psi}$ is the Fourier transform of Ψ, defined by (2.A.2.18) (without the time variable).

More precisely,

$$\int_A |\hat{\Psi}(p)|^2 dp \qquad (2.A.3.1)$$

is the probability that the value obtained by a measurement of momentum will belong to $A \subset \mathbb{R}$.

In order to prove (2.A.3.1), we will measure p by measuring $x(t)$ at time t, using $p = mx(t)/t$, since p is the mass times the velocity. Since we want the result to be independent of t, we will consider the asymptotic position, which means letting $t \to \infty$. We will set $m = 1$ here and consider one dimension for simplicity.

We already know that the probability density of finding $x(t) = x$, when one measures the position at time t, is given by $|\Psi(x, t)|^2$. Then, the probability of the momentum being observed to belong to a subset $A \subset \mathbb{R}$ is $\int_{At} |\Psi(x, t)|^2 dx$. Now, by a change of variable $x = pt$, we get

$$\int_{At} |\Psi(x, t)|^2 dx = t \int_A |\Psi(pt, t)|^2 dp . \qquad (2.A.3.2)$$

Suppose that we have an initial wave function $\Psi_0(x) = \Psi(x, 0)$ supported in a bounded region $B \subset \mathbb{R}$. We will prove that, $\forall A \subset \mathbb{R}$,

$$\lim_{t \to \infty} t \int_A |\Psi(pt, t)|^2 dp = \int_A |\widehat{\Psi}(p, 0)|^2 dp . \qquad (2.A.3.3)$$

Combining with (2.A.3.2), this means that, if we measure the asymptotic position x as $t \to \infty$, we will obtain the quantum mechanical predictions (2.A.3.1).

To prove[37] (2.A.3.3), we consider the free evolution, which should hold for t large, and use (2.A.2.21). Since the inverse Fourier transform of a product of functions is the convolution of their inverse Fourier transforms, divided by $\sqrt{2\pi}$, we get

$$\Psi(x, t) = \left(\frac{1}{2\pi it}\right)^{1/2} \int_{\mathbb{R}} \exp\left[\frac{i(x - y)^2}{2t}\right] \Psi(y, 0) dy , \qquad (2.A.3.4)$$

using the fact that $\sqrt{1/it} \exp(ix^2/2t)$ is the inverse Fourier transform of $\exp(-itp^2/2)$. Set $x = pt$ in (2.A.3.4) and write it as

$$\Psi(pt, t) = \left(\frac{1}{2\pi it}\right)^{1/2} \exp\left(\frac{ip^2t}{2}\right) \int_{\mathbb{R}} \exp\left(-ipy + i\frac{y^2}{2t}\right) \Psi(y, 0) dy . \qquad (2.A.3.5)$$

---

[37] We follow here [268, Sect. 8.6] and proceed informally; for a more rigorous treatment, see [152, pp. 306–310].

Since $\Psi(y, 0)$ vanishes outside a bounded region B, we have, $\forall y \in B$, $\lim_{t \to \infty}$ $\exp(iy^2/2t) = 1$, which implies

$$\lim_{t \to \infty} \left(\frac{1}{2\pi}\right)^{1/2} \int_B \exp\left(-ipy + \frac{iy^2}{2t}\right)\Psi(y, 0)dy = \left(\frac{1}{2\pi}\right)^{1/2}\int_{\mathbb{R}} \exp\left(-ipy\right)\Psi(y, 0)dy$$

$$(2.A.3.6)$$

$$= \widehat{\Psi}(p, 0),$$

where, in the last equality, we use the fact that $\Psi(y, 0)$ is supported in B. Obviously,

$$\left|\left(\frac{1}{it}\right)^{1/2} \exp\left(\frac{ip^2t}{2}\right)\right|^2 = t^{-1}.$$

$$(2.A.3.7)$$

Inserting (2.A.3.5)–(2.A.3.7) in the left-hand side of (2.A.3.3) proves (2.A.3.3).

2.B Quantum States, "Observables" and the "Collapse" Rule

We have already encountered in Sect. 2.3 the special role of measurements within the quantum formalism. As we saw, we can have two different bases in $\mathbb{C}^2$, $(|1\uparrow\rangle, |1\downarrow\rangle)$ or $(|2\uparrow\rangle, |2\downarrow\rangle)$, and using (2.3.5)–(2.3.8), we can write any given state in terms of those different bases. A measurement of the spin in direction 1 or 2 is associated with a given basis, and after a measurement, the state collapses onto one vector of the basis, depending on the result. Let us explain now the general quantum formalism.

In quantum mechanics, the space of states is a complex vector space, of finite dimension, $\mathbb{C}^N$, or of infinite dimension (we will discuss that situation below). The finite dimensional case generalizes the states associated with spin of Sect. 2.3. The state is endowed with a scalar product $\langle z_1 | z_2 \rangle \equiv \sum_{n=1}^N z_{1n}^* z_{2n}$, where, for $z \in \mathbb{C}$, z^* denotes its complex conjugate.

The state is also endowed with a norm associated with that scalar product: $\|z\|^2 = \langle z | z \rangle$. The quantum state $|$state $(t)\rangle$ is a vector in that space and evolves in time, when no measurements are made, according to a deterministic equation: a given state at time 0, $|$state $(0)\rangle$, determines a unique state at time t, $|$state $(t)\rangle$, for all times. This evolution is continuous in time and linear, see (2.3.10), (2.3.11). The norm of that vector $\|$state $(t)\rangle\|$ is constant in time.

In classical physics, one introduces various physical quantities such as angular momentum, energy, etc. (all of which are functions of the positions and the velocities). In quantum mechanics, one associates with each such physical quantity a basis of vectors $(|e_n\rangle)$ of the state space and a set of numbers (λ_n), where n runs over $\{1, \ldots, N\}$. The choice of these numbers λ_n is conventional. When there is a measurement of the quantity associated with those vectors and numbers at a certain time t, one writes the state as a linear combination of the basis vectors:

$$|\text{state (t)}\rangle = \sum_{n=1}^{N} c_n(t)|e_n\rangle \,, \qquad\qquad (2.B.1)$$

where $c_n(t) = \langle e_n|\text{state (t)}\rangle$.

The recipe for computing probabilities of results of measurements, which generalizes what we discussed in Sect. 2.3, is that a measurement at time t yields a value λ_k with probability $|c_k(t)|^2$. Since $\|\text{state (t)}\rangle\|^2 = \sum_{n=1}^{N} |c_n(t)|^2$ is constant in time,[38] if we normalize $\|\text{state (0)}\rangle\| = 1$, we have $\sum_{n=1}^{N} |c_n(t)|^2 = 1$ for all times, so that the sum of the probabilities of all the results equals 1.

This assignment of probabilities to results of measurements is called *Born's rule*. Moreover, after the measurement, the quantum state collapses to $|e_k\rangle$. As we explained in Sect. 2.3, that collapse is neither continuous in time, nor deterministic nor linear, contrary to the time evolution when no measurements are made.

To simplify matters, we assume here that each eigenvalue is non-degenerate, i.e., it corresponds to a unique eigenvector. In general, if there are several eigenvectors with the same eigenvalue λ_k, the collapsed state is the projection of the original state on the subspace spanned by those eigenvectors, and the probability of occurrence of λ_k is the norm of that projected vector.

A correspondence can be made with the example of the spin measurement by associating $\lambda = +1$ with the up result and $\lambda = -1$ with the down result, but other conventions could be chosen.

The more advanced reader may find the above presentation somewhat unusual. Indeed, the standard approach is to associate a matrix with any physical quantity when N is finite, these having a basis of eigenvectors, viz.,

$$A|e_n\rangle = \lambda_n|e_n\rangle \,, \qquad\qquad (2.B.2)$$

where the λ_n are real.[39] But this is just a way to repeat what we said above: what matters is the basis of vectors ($|e_n\rangle$), while the choice of the numbers λ_n as the real eigenvalues of a self-adjoint matrix A is a matter of convenience.[40]

In the spin example, for direction 1, we could introduce the matrix[41]

---

[38]That last formula comes from:

$$\|\text{state (t)}\rangle\|^2 = \langle\text{state (t)}|\text{state (t)}\rangle = \sum_{n,m=1}^{N} c_n^*(t)c_m(t)\langle e_n|e_m\rangle = \sum_{n=1}^{N} |c_n(t)|^2,$$

since, by orthonormality of the basis vectors, $\langle e_n|e_m\rangle = 0$ if $n \neq m$ and equals 1 if $n = m$.

[39]This is automatic if we assume that A is self-adjoint. For matrices, this means that its matrix elements satisfy $A_{ij} = A_{ji}^*$.

[40]There is a more general notion associated with measurements, namely, the positive operator-valued measure (POVM), discussed further in [147] and [152, Chap. 12].

[41]These are the usual Pauli matrices: $\sigma_1 = \sigma_z$, $\sigma_2 = \sigma_x$, while $\sigma_y = \begin{pmatrix} 0 & -i \\ i & 0 \end{pmatrix}$.

$$\sigma_1 = \begin{pmatrix} 1 & 0 \\ 0 & -1 \end{pmatrix} \qquad\qquad (2.B.3)$$

and for direction 2 the matrix

$$\sigma_2 = \begin{pmatrix} 0 & 1 \\ 1 & 0 \end{pmatrix} . \qquad\qquad (2.B.4)$$

It is easy to check, using the definitions (2.3.1)–(2.3.4), that

$$\sigma_1|1 \uparrow\rangle = |1 \uparrow\rangle , \quad \sigma_1|1 \downarrow\rangle = -|1 \downarrow\rangle , \quad \sigma_2|2 \uparrow\rangle = |2 \uparrow\rangle , \quad \sigma_2|2 \downarrow\rangle = -|2 \downarrow\rangle ,$$
$$(2.B.5)$$

so that our basis vectors are indeed eigenvectors of the corresponding matrices with eigenvalues $+1$ and -1. But all that we really need conceptually are the eigenvectors and the associated numbers, even though the language of operators is very useful in practice.

Now we must also consider the spaces of wave functions, that are infinite dimensional.[42] One introduces (let's say for a physical system consisting of one particle in one dimension)[43] the space of complex-valued functions $\Psi : \mathbb{R} \to \mathbb{C}$ that are square-integrable: $\int_{\mathbb{R}} |\Psi(x)|^2 dx < \infty$.

One can define a scalar product on that space: $\langle\Psi|\Phi\rangle = \int_{\mathbb{R}} \Psi^*(x)\Phi(x)dx$, and, therefore, one can also define the notion of orthonormal sets of vectors and a norm associated to the scalar product: $\|\Psi\|^2 = \langle\Psi|\Psi\rangle = \int_{\mathbb{R}} |\Psi(x)|^2 dx$.

The wave function is a vector in that space that depends on time, $\Psi(x, t)$. When no measurements are made, that vector evolves according to a deterministic equation, like Schrödinger's equation, the evolution is continuous in time and linear. Moreover, $\|\Psi(t)\|^2 = \int_{\mathbb{R}} |\Psi(x, t)|^2 dx$ is constant in time, as in (2.A.2.6).

Again, one associates to physical quantities linear operators (see (2.A.2.2)) that act on functions, like matrices act on vectors.[44] If a quantity is associated to an operator A satisfying (2.B.2), with n running now over $\mathbb{N}$, we have the same rule as above when one measures A, except that the sum (2.B.1) has to be replaced by a limit, as in (2.A.2.10). We have again $\int_{\mathbb{R}} |\Psi(x, t)|^2 dx = \sum_{n\in\mathbb{N}} |c_n(t)|^2$, and if we normalize $\|\Psi(0)\|^2 = \int_{\mathbb{R}} |\Psi(x, 0)|^2 dx = 1$, we have $\sum_{n\in\mathbb{N}} |c_n(t)|^2 = 1$ for all times.

For example, suppose that we measure the quantity associated with H, defined in (2.A.2.3), (2.A.2.4) (this quantity corresponds classically to the energy). Suppose also that H has a basis of eigenvectors, see (2.A.2.12), and that the state is of the form

---

[42]This fact is intuitively understandable since a set of functions defined on $\mathbb{R}$ cannot be characterized by a finite set of parameters, which would be the case if the space was finite dimensional (the parameters would be the coefficients of the expansion of a function in a basis of the space).

[43]The extension to more dimensions or more particles is straightforward: for M particles in a physical space of k dimensions, the wave functions are functions $\Psi : \mathbb{R}^N \to \mathbb{C}$, where $N = kM$, and the integrals are over $\mathbb{R}^N$.

[44]We proceed formally here; see, e.g., [152, Chap. 15] or [412] for more details on the definition of operators.

(2.A.2.14). Then we get the result λ_k with probability $|c_k|^2$, and after the measurement the wave function becomes $|e_k(x)\rangle$.[45]

We will also need, but only in Appendix 2.F, operators that do not have a basis of eigenvectors. We introduced these operators Q and P in Appendix 2.A. The operator Q is called the position operator, and acts as

$$Q\Psi(x) = x\Psi(x) \, , \tag{2.B.6}$$

and its eigenvectors are formally Dirac delta functions $\delta(q - x)$.[46] We have

$$Q\delta(q - x) = q\delta(q - x) \, , \tag{2.B.7}$$

with eigenvalue q. If we write $\Psi(x, t) = \int \delta(q - x)\Psi(q, t)dq$, we can see this as a sort of continuous version of (2.B.1), and the interpretation of $|c_k|^2$ as the probability of finding the eigenvalue λ_k upon measurement of A, translates here into considering $|\Psi(q, t)|^2$ as the probability density of finding the particle at q, upon measurement of its position.

The momentum operator P is defined by:

$$P\Psi(x) = -i\frac{d}{dx}\Psi(x) \, , \tag{2.B.8}$$

and we have the eigenvectors[47]

$$\frac{1}{(2\pi)^{1/2}} \exp(ipx) \, ,$$

with eigenvalue p. Indeed, one checks that

$$P\frac{1}{(2\pi)^{1/2}} \exp(ipx) = p\frac{1}{(2\pi)^{1/2}} \exp(ipx) \, . \tag{2.B.9}$$

It we consider the inverse Fourier transform formula (2.A.2.19), we can see it as the continuous version of (2.B.1), with eigenvectors $[1/(2\pi)^{1/2}] \exp(ipx)$, and the interpretation of $|c_k|^2$ as the probability of finding the eigenvalue λ_k upon measurement of A, translates here into considering $|\hat{\Psi}(p, t)|^2$ as the probability density of

---

[45]In the concrete example (2.A.2.15), (2.A.2.16), we get the result $k^2/2m$ with probability $|c_k|^2$ and, after the measurement, the wave function becomes $e^{ikx}/\sqrt{2\pi}$ (the factor of $1/\sqrt{2\pi}$ coming from the requirement that $\int_0^{2\pi} |\Psi(x, t)|^2 dx = 1$ at all times).

[46]These are not real functions but can be thought as limits of functions whose integrals are always equal to one and that tend to 0 for all $x \neq q$, for example the sequence $f_n(x) = \sqrt{\frac{n}{2\pi}} \exp(-\frac{n(x-q)^2}{2})$, as $n \to \infty$. In that limit, the function becomes more and more concentrated on $x = q$, and tends to 0 elsewhere. This explains Eq. (2.B.7) below.

[47]The factor $1/(2\pi)^{1/2}$ plays the role of a normalization factor.

finding the value p upon measurement of its momentum, see (2.A.3.1), derived in Appendix 2.A.3.

We see that, for both Q and P, the set of possible results of measurements is the set $\mathbb{R}$ of real numbers. This set plays the same role here as the one played by the eigenvalues for matrices.[48]

The "collapse rule" in the case of measurements of Q and P works as follows: since a measurement whose result can be any real number is never infinitely precise, but is rather an interval of real numbers, the collapsed wave function will be the original wave function restricted to that interval and normalized so that $\int_{\mathbb{R}} |\Psi(x, t)|^2 dx = 1$ holds after the collapse.

All this may sound terribly abstract and "unphysical", but the goal of this presentation is precisely to emphasize how much the quantum algorithm is an unambiguous method for accurately predicting results of measurements, and nothing else. In particular, it should not be associated with any mental picture of what is "really" going on. The main issue of course is whether one should consider this algorithm as satisfactory or as being, in some sense, the "end of physics", or whether one should try to go beyond it.

2.C "Uncertainty" Relations and "Complementarity"

An easy remark about the uncertainty relations is that there is a great deal of uncertainty about what exactly they mean: indeed, are they uncertainty relations or indeterminacy relations, and what are the differences between these two terms?

The first derivation of these relations by Heisenberg [256], which was more a heuristic argument than a real derivation,[49] was entirely compatible with a disturbance view of measurement, as expressed, for example, in the statement by Heisenberg [256] quoted in Sect. 2.5.2. This way of speaking assumes that electrons have a position and a velocity, even when they are not measured. It only shows that there are limits to how much we can know about one of these quantities without disturbing the other.

However, more radical conclusions are sometimes drawn, namely, that those uncertainty relations are really indeterminacy relations, i.e., that the positions and the velocities are indeterminate or do not exist before we measure them, or even that it does not make sense to speak of quantities that we cannot measure simultaneously. Here, we will leave aside these issues, which ultimately depend on our views about the meaning of the quantum state, discussed in Sect. 2.5, and simply give some precise versions of those relations.

---

[48]The set $\mathbb{R}$ is called the spectrum of the operators Q and P and is also called "continuous". See, e.g., [152, Chap. 15] or [412] for more details on the spectra of operators.

[49]The first real derivation is due to Kennard [287], see, e.g., [266] for the history of the uncertainty principle.

2.C.1 A Statistical Relation

Consider a random variable x that can take values $a_1, \ldots, a_n$ with respective probabilities p_i, $i = 1, \ldots, n$. The variance of x, $\text{Var}(x)$, is a way to measure how much the distribution of x is spread around its mean. For $f : \{a_1, \ldots, a_n\} \to \mathbb{R}$, we define the mean or the average of $f(x)$ by

$$\langle f(x) \rangle = \sum_{i=1}^{n} f(a_i) p_i \; . \tag{2.C.1.1}$$

Then $\text{Var}(x)$ is defined as

$$\text{Var}(x) = \langle x^2 \rangle - \langle x \rangle^2 = \langle (x - \langle x \rangle)^2 \rangle \; , \tag{2.C.1.2}$$

where the second equality is checked by expanding the binomial. The quantity $|x - \langle x \rangle|$ expresses the deviation of the variable x from its mean, so (2.C.1.2) gives a measure of the size of that deviation.

If x is a continuous random variable on $\mathbb{R}$ (we work in one dimension for simplicity), with probability density $p(x)$, then the definition (2.C.1.2) is still valid, with (2.C.1.1) replaced by

$$\langle f(x) \rangle = \int_{\mathbb{R}} f(x) p(x) dx \; . \tag{2.C.1.3}$$

A precise statement of the uncertainty relations is as follows. Given a wave function $\Psi(x)$, we know that the probability distribution density of results of measurements of the position x is $|\Psi(x)|^2$, meaning that $\int_A |\Psi(x)|^2 dx$ is the probability that, when the position of the particle is measured, the result belongs to $A \subset \mathbb{R}$. We also showed in Appendix 2.A.3 that the results of measurements of the momentum p (which classically is just the mass times the velocity of the particle) are distributed with a probability density given by $|\hat{\Psi}(p)|^2$, where $\hat{\Psi}$ is the Fourier transform of Ψ, defined by (2.A.2.18) (without the time variable), see (2.A.3.1).

We note that, since $\int_{\mathbb{R}} |\Psi(x)|^2 dx = 1$, then by Plancherel's theorem $\int_{\mathbb{R}} |\hat{\Psi}(p)|^2 dp = 1$.

Given this, we have a variance $\text{Var}(x)$ for the distribution of x and a variance $\text{Var}(p)$ for the distribution of p. Their product satisfies a lower bound:

$$\text{Var}(x) \text{Var}(p) \geq \frac{1}{4} \; , \tag{2.C.1.4}$$

bearing in mind that we choose units where $\hbar = 1$. The bound (2.C.1.4) is a rather simple mathematical relation between a function and its Fourier transform and its proof can be found in many textbooks on Fourier transforms (see, e.g., [155]), as well as those on quantum mechanics.

One can give a concrete example of Heisenberg's inequality (2.C.1.4) by considering Gaussian wave functions. For $d = 1$, let $\Psi(x) = (a/\pi)^{1/4} \exp(-ax^2/2)$, which is normalized so that $\int_{\mathbb{R}} |\Psi(x)|^2 dx = 1$. Then, using (2.A.2.18), it is easy to show that

$$\hat{\Psi}(p) = \frac{1}{(\pi a)^{1/4}} \exp\left(-\frac{p^2}{2a}\right) .$$

If one computes the respective variances, one obtains: $\mathrm{Var}(x) = 1/2a$, $\mathrm{Var}(p) = a/2$, whose product is $1/4$, namely the lower bound in (2.C.1.4).

This illustrates the impossibility of "measuring both the position and the momentum" simultaneously with arbitrary precision. Indeed, assume that, after a position measurement, the "collapsed" wave function is a "narrow" one (assumed to be Gaussian for simplicity), $\Psi(x) = (a/\pi)^{1/4} \exp(-ax^2/2)$, with a large, which means that the position measurement is precise, since $\mathrm{Var}(x) = 1/2a$ is small. Then, the variance of the distribution of future measurements of momenta, $\mathrm{Var}(p) = a/2$, will necessarily be large.

Since (2.C.1.4) is a lower bound on variances of results of measurement, it implies nothing whatsoever about the intrinsic properties of quantum particles. One could perfectly think, in accordance with the statistical interpretation, that each individual particle has a well-defined position and momentum; but, when we prepare a large number of particles having the same quantum state, then the positions and momenta of those particles vary and have certain statistical distributions whose variances satisfy (2.C.1.4).

This statistical view is untenable, but not because of the uncertainty relations. The problem for that view comes, as we saw in Sect. 2.5.2, from the no hidden variables theorems.

However there is another, more qualitative, version of "uncertainty" in quantum mechanics.

2.C.2 A Qualitative Argument and Its Relation to "Complementarity"

Let us consider finite-dimensional systems for simplicity. As we saw in Appendix 2.B, a physical quantity (such as the spin) is associated with a self-adjoint matrix. Consider two such matrices A and B. Let us define their commutator:

$$[A, B] = AB - BA , \qquad\qquad (2.C.2.1)$$

where AB is the matrix product. Suppose $[A, B] = 0$. If $|e\rangle$ is an eigenvector of A, i.e.,

$$A|e\rangle = \lambda|e\rangle ,$$

then it is easy to see that $B|e\rangle$ is also an eigenvector of A, with the same eigenvalue:

$$AB|e\rangle = BA|e\rangle = \lambda B|e\rangle . \tag{2.C.2.2}$$

This holds also if we exchange A and B. Using this remark, one shows that, if $[A, B] = 0$, then A and B have a common basis of eigenvectors (with different eigenvalues).[50]

Conversely, if A and B have a common basis of eigenvectors, then $[A, B] = 0$. Since the only physically meaningful quantity are the basis vectors (and the associated numbers) corresponding to a physical quantity, if $[A, B] = 0$, A and B just associate different numbers to the same basis.

Measuring A will reduce the quantum state to one of the eigenvectors of A. But if we then measure B, we will reduce the state to one of the eigenvectors of B, which is also an eigenvector of A if A and B commute. Hence, if we remeasure A after having measured B, the result will be with certainty the same eigenvalue of A as before and the state will not change, unlike when one tries to measure the spin in two different directions (see Fig. 2.2). It is in this sense that, if $[A, B] = 0$, one can measure A and B simultaneously (and also the products AB or BA).

But if $[A, B] \neq 0$, there will be some eigenvector of A that is not an eigenvector of B. Suppose that one measures A when the state is an eigenvector of B with eigenvalue b. If, after the measurement of A, the state is an eigenvector of A that is not an eigenvector of B, then the result of a later measurement of B will not give back the original value b, since the state produced by the measurement of A is no longer an eigenvector of B.

This is what happened with the spin in directions 1 and 2, as was observed phenomenologically in Sect. 2.1 and described by the quantum formalism in Sect. 2.3. If we start with an eigenstate of the spin in direction 1 and then measure it in direction 2, we "lose" the memory of what value the spin had in direction 1, since the result of the spin measurement in direction 2 is an eigenvector of the matrix σ_2 (2.B.4) and hence a superposition of states in direction 1 [see (2.3.5) and (2.3.6)].

This is one possible meaning of the word "complementarity" which was so fundamental to Niels Bohr. The measurement of A or B gives us a "classical" description of reality where "classical" does not refer to classical physics but means "expressible in ordinary language" or "representable" or "macroscopic". But since the two quantities cannot be measured simultaneously (i.e., without the measurement of one quantity disturbing the measurement of the other), one cannot "combine" the picture coming from the measurement of A and the one coming from the measurement of B into a coherent picture.

One can check that the operators Q and P, introduced in Appendices 2.A.2 and 2.B, do not commute:

$$(PQ\Psi)(x) = -i\hbar\frac{d}{dx}\big[x\Psi(x)\big] = -i\hbar\left[\Psi(x) + x\frac{d}{dx}\Psi(x)\right] \neq -i\hbar x\frac{d}{dx}\Psi(x)$$
$$= (QP\Psi)(x).$$

---

[50]Typically, some of these eigenvalues will be degenerate for A or B (or both).

This can then be interpreted in terms of "complementarity" between a "picture" based on positions and one based on momenta. But what this means depends on how we understand (2.C.1.4), and therefore how we understand the quantum formalism. The non-commutation of Q and P does not have an obvious meaning.

Let us remark finally that there is also a generalization of (2.C.1.4) that expresses quantitively this incompatibility between A and B. Given a quantum state Ψ, one obtains a probability distribution for the results of the measurements of A and of B (described in Appendix 2.B: an eigenvalue λ_k occurs with probability $|c_k|^2$). Thus we can define the variances $\text{Var}_\Psi(A)$, $\text{Var}_\Psi(B)$, associated with those probability distributions. The generalization of (2.C.1.4) is[51]

$$\text{Var}_\Psi(A)\,\text{Var}_\Psi(B) \geq \frac{1}{4}\left|\langle\Psi|[A, B]\Psi\rangle\right|^2 , \qquad (2.\text{C}.2.3)$$

which is similar to (2.C.1.4).[52]

2.D The Quantum Mechanical Description of Measurements

Let us consider a very simple measurement of the spin.[53] We start with a quantum state for the combined system composed of the particle and the measuring device:

$$\Psi_0 = \varphi_0(z)\left[c_1\begin{pmatrix}1\\0\end{pmatrix} + c_2\begin{pmatrix}0\\1\end{pmatrix}\right] , \qquad (2.\text{D}.1)$$

where z denotes a macroscopic variable, namely the position of the center of mass of the measuring device, and $\varphi_0(z)$ is centered at $z = 0$, meaning that the pointer is as in the first picture of Fig. 2.6. We leave aside here the spatial part of the quantum state of the particle, since we are only interested in what happens to the measuring device.

Let the Hamiltonian be

$$H = -i\sigma\frac{\partial}{\partial z} , \quad \text{where} \quad \sigma = \begin{pmatrix}1 & 0\\0 & -1\end{pmatrix} ,$$

which corresponds to the introduction of an inhomogeneous magnetic field. One neglects here the kinetic energy term (corresponding to the free evolution) $-(1/2m)\partial^2 \Psi(z, t)/\partial z^2$. With these simplifications, the Schrödinger equation is

---

[51]For a proof, see, e.g., [236, Sect. 24] or [447, Chap. 9].

[52]Note that, for states that are eigenstates of A or of B, both sides of (2.C.2.3) vanish [in contrast to (2.C.1.4)], but the impossibility of a simultaneous measurement of A and B holds nevertheless.

[53]We follow here Bell [49, p. 130]. See also Bohm and Hiley [70, Chap. 6].

$$i \frac{\partial}{\partial t} \Psi = -i\sigma \frac{\partial}{\partial z} \Psi \, ,$$

and one can easily check that its solution is

$$c_1 \begin{pmatrix} 1 \\ 0 \end{pmatrix} \varphi_0(z - t) + c_2 \begin{pmatrix} 0 \\ 1 \end{pmatrix} \varphi_0(z + t) \, . \qquad (2.D.2)$$

Since $\varphi_0(z)$ is centered at $z = 0$, $\varphi_0(z \pm t)$ is centered at $z = \mp t$, corresponding to the last two pictures in Fig. 2.6 (for a suitable t), which is the result mentioned in Sect. 2.5.1, where we wrote $\varphi^\uparrow(z)$ for $\varphi_0(z - t)$ and $\varphi^\downarrow(z)$ for $\varphi_0(z + t)$.

We can discuss the Mach–Zehnder interferometer in the presence of a wall in a similar way. Once the wall is inserted as in Fig. 2.4 and the state of the particle is (2.4.2), we get, for the combined system particle plus wall, the state

$$\frac{1}{\sqrt{2}} \Big[|2 \uparrow\rangle \, |\text{path2} \uparrow\rangle \varphi_0(z) - |2 \downarrow\rangle \, |\text{path2} \downarrow\rangle \varphi_1(z) \Big] \, , \qquad (2.D.3)$$

where $\varphi_0(z)$ denotes the wave function of the wall not having absorbed the particle and $\varphi_1(z)$ that of the wall having absorbed the particle. If we replace the wall by an active bomb, as in the Elitzur–Vaidman bomb testing mechanism, $\varphi_0(z)$ will be the wave function of the unexploded bomb and $\varphi_1(z)$ that of the bomb having exploded. In both cases, we have a macroscopic object (the wall or the bomb) that plays the same role as the pointer in (2.D.2).

Consider now the more general situation described in Appendix 2.B, where the operator A is associated with a given physical quantity having a basis of eigenvectors:

$$A|e_n\rangle = \lambda_n |e_n\rangle \, , \qquad (2.D.4)$$

and the state of the system to be measured is

$$|\text{state}\rangle = \sum_n c_n |e_n\rangle \, , \qquad (2.D.5)$$

where n runs over a finite set or over $\mathbb{N}$. Consider a quantum state for the combined system plus measuring device:

$$\Psi_0 = \varphi_0(z) \sum_n c_n |e_n\rangle \, , \qquad (2.D.6)$$

where z and $\varphi_0(z)$ are as above, i.e., $\varphi_0(z)$ is localized around 0.

Introducing a coupling between the system and the measuring device of the form $H = -i A \partial/\partial z$ (A being the matrix σ in the example of the spin measurement above), one gets, neglecting again the kinetic energy term, the Schrödinger equation

$$i \frac{\partial}{\partial t} \Psi = -i A \frac{\partial}{\partial z} \Psi \, , \qquad (2.D.7)$$

whose solution is

$$\sum_n c_n \varphi_0(z - \lambda_n t)|e_n\rangle \, , \qquad (2.D.8)$$

which generalizes (2.D.2), with $\varphi_0(z - \lambda_n t)$ having macroscopically disjoint supports for different λ_n when t is not too small, since $\varphi_0(z)$ is localized around 0. One obtains a situation similar to the pointer in the last two pictures in Fig. 2.6, but now with more possible positions [one for each n in the sum (2.D.8)].

2.E The Double-Slit Experiment

A standard way to introduce interference effects in quantum mechanics, such as the ones we saw in the Mach–Zehnder interferometer, is via the double-slit experiment [184]: particles are sent (one by one) through slits in a wall and the pictures below show how the particles are distributed when they are detected on another wall somewhere behind the slits. If only one slit is open, one gets the curves (a) and (b) of Fig. 2.8, representing the densities of particles being detected behind the slits (which is not surprising), while if both slits are open, one gets the interference effects shown in the last picture (c). One might expect that, with both slits open, the distribution of the particles would be the sum of those detected when only one slit is open. But instead we get the wavy line of Fig. 2.8c, with *fewer* particles at some places than there would be with only one slit open. So opening or closing one slit seems to influ-

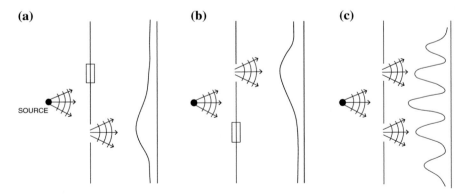

Fig. 2.8 The double-slit experiment. In all three figures, there is a source of particles going towards a screen in which one or two slits are open. There is second screen behind the first one on which particles are detected. In **a** the curve represents the density of particles detected on the second screen when one slit is open, and in **b** likewise, when the other slit is open. **c** The result when both slits are open. This is clearly not the sum of the first two results

ence the particles going through the other slit. And this remains true, qualitatively, even if the open slit is very far from the closed one.

This experiment illustrates once again the role of measurements in quantum mechanics: it is often described by saying that, if we close one slit, then we *know* which slit the particle went through, hence its behavior will be affected by our measurement. The same phenomenon (suppression of interference) would occur if we put a small light behind one of the slits that would allow us to detect the slit the particle went through.

This double-slit experiment is similar to the experiment with the Mach–Zehnder interferometer described in Sect. 2.2, but in the latter we dealt with sharper figures (100 % vs. 50 %) rather than the interference patterns. The calculus with spin (i.e., vectors in two dimensions) is also easier than it would be for the double-slit experiment, where one would have to solve the Schrödinger equation with initial conditions located around each of the slits in order to deduce the interference pattern of Fig. 2.8c. Figure 5.1 in Chap. 5 shows a numerical solution yielding the interference pattern (within the de Broglie–Bohm theory).

This experiment is often considered as the essence of the quantum mechanical mystery. On the basis of this experiment, one often denies that it makes sense to speak of particles going through one slit or the other. One also sometimes says that, if both slits are open, quantum objects behave as waves, and if only one slit is open, they behave as particles, which is another instance of Bohr's "complementarity": one can have a "wave picture" or a "particle picture", but not both simultaneously.

After describing the double-slit phenomenon, Feynman wrote:

> Nobody knows any machinery. Nobody can give you a deeper explanation of this phenomenon than I have given; that is, a description of it.
>
> Richard Feynman, [185, p. 145]

And in a well known classical textbook on quantum mechanics, Landau and Lifshitz said:

> It is clear that [the results of the double-slit experiment] can in no way be reconciled with the idea that electrons move in paths. [...] In quantum mechanics there is no such concept as the path of a particle.
>
> Lev Landau and Evgeny Lifshitz [302, p. 2]

We will discuss these statements in Chap. 5 in the light of the de Broglie–Bohm theory.

2.F Proof of the No Hidden Variables Theorem

We will now state more precisely and prove the theorem given at the the end of Sect. 2.5. The theorem is divided into two parts, and so are the proofs, which are similar, but using different background notions. We first state each part of the theorem precisely and then give its proof.

Precise Statement of Part 1

Let $\mathcal{O}$ be the set of self-adjoint matrices on a complex vector space of dimension four. Then, there does not exist a function v :

$$v : \mathcal{O} \to \mathbb{R} \tag{2.F.1}$$

such that:

(1)

$$\forall A \in \mathcal{O} , \quad v(A) \in \{\text{eigenvalues of A}\} , \tag{2.F.2}$$

(2)

$$\forall A, B \in \mathcal{O} , \text{ with } [A, B] = AB - BA = 0 , \quad v(AB) = v(A)v(B) . \tag{2.F.3}$$

Remarks

We use here the formulation of quantum mechanical "measurements" in terms of matrices and eigenvalues, see Appendix 2.B. The first condition is natural if a measurement is supposed to reveal a pre-existing value corresponding to the quantity A. However, it should be stressed that we do not use the first condition very much in the proof. In fact, we *only* use it for $A = -\mathbf{1}$, $\mathbf{1}$ being the unit matrix, in the form $v(-\mathbf{1}) = -1$.

The second condition is necessary if the values $v(A)$ are supposed to be in agreement with the quantum predictions, since, when A and B commute (i.e., when $AB - BA = 0$), it is in principle possible to measure A, B, and AB simultaneously, and the product of the results of the first two measurements must be equal to the result of the last one, i.e., they must satisfy (2.F.3) (see Appendix 2.C.2).[54] This condition, unlike the first one, will be used repeatedly in the proof. Indeed, by choosing suitable pairs of commuting matrices, and applying (2.F.3) to each pair, we will derive a contradiction.

There are similar no hidden variables theorems in any space of dimension at least 3, see Bell [36], Kochen and Specker [291], and Mermin [335], but the proof given here works only in a four dimensional space (or in any space whose dimension is a multiple of four, by considering matrices that are direct sums of copies of the matrices used here).

It should be emphasized that, even though the set $\mathcal{O}$ contains matrices that do not commute with each other, we use relation (2.F.3) *only* for commuting matrices, so that the only assumptions of the theorem are the quantum mechanical predictions for the results of possible measurements.

---

[54]As Mermin suggests [335, pp. 811], if the eigenvalues of the matrices were all 0 or 1 (unlike the situation here, but one could easily adapt the argument), then measuring the "observable" $A + 2B + 4AB$ alone would give the values of all three quantities, A, B, and AB, and they would have to satisfy $v(AB) = v(A)v(B)$.

Sometimes people think that this theorem rules out only "non-contextual" hidden variables: what this means is that, if we consider three matrices, A, B and C, where A commutes with B and C, but B and C do not commute, then we are assuming that the result of measuring A does not depend on whether we choose to measure B *or* C simultaneously with A.[55] To be precise, we could write $v(AB) = v(A)v(B)$ or $v(AC) = v(A)v(C)$, since A commutes with both B and C, and we assume here that one has the same value $v(A)$ in both equations.

Hidden variables would be called contextual if they depended on that choice (so, here, the hidden variables are non-contextual). But this is not a way to "save" the possibility of hidden variables, at least those considered here: if measuring A is supposed to reveal an intrinsic property of the particle pre-existing to the measurement (and this is what is meant here by hidden variables), then it cannot possibly depend on whether I choose to measure B or C simultaneously with A, since I could measure A and nothing else. If someone has an age, a height and a weight (those being intrinsic properties of that person), then how could the result of measuring one of those properties depend on whether I measure or not another property together with that one, or on which property I would choose to measure?

The second condition (2.F.3) is necessary to derive a proof of the Theorem, but it does not affect its meaning.[56]

Proof

We use the standard Pauli matrices σ_x [equal to σ_2 in (2.B.4)], σ_y, and σ_z [equal to σ_1 in (2.B.4)]:

$$\sigma_x = \begin{pmatrix} 0 & 1 \\ 1 & 0 \end{pmatrix}, \qquad \sigma_y = \begin{pmatrix} 0 & -i \\ i & 0 \end{pmatrix}, \qquad \sigma_z = \begin{pmatrix} 1 & 0 \\ 0 & -1 \end{pmatrix}.$$

We consider a couple of each of those matrices, σ_x^i, σ_y^i, $i = 1, 2$, where tensor products are implicit: $\sigma_x^1 \equiv \sigma_x \otimes \mathbf{1}, \sigma_x^2 \equiv \mathbf{1} \otimes \sigma_x^2$, etc., with $\mathbf{1}$ the unit matrix. These operators act on $\mathbf{C}^4$. The following identities are well known and easy to check:

(i)

$$(\sigma_x^i)^2 = (\sigma_y^i)^2 = (\sigma_z^i)^2 = \mathbf{1}, \qquad (2.F.4)$$

for $i = 1, 2$.

(ii) Different Pauli matrices anticommute:

$$\sigma_\alpha^i \sigma_\beta^i = -\sigma_\beta^i \sigma_\alpha^i, \qquad (2.F.5)$$

---

[55]This is discussed by Bell [49, pp. 8–9] and Mermin [335, pp. 811–812].

[56]To avoid creating some later confusion in the reader's mind, we should already mention here that the de Broglie-Bohm theory, discussed in Chap. 5, is, in some sense, a "contextual" hidden variables theory. This is explained in Sects. 5.1.4, 5.1.5 and 5.3.4. But in that theory, one does *not* introduce the hidden variables ruled out by the no hidden variables theorems (otherwise the theory would be inconsistent!).

for $i = 1, 2$, and $\alpha, \beta = x, y, z, \alpha \neq \beta$. And they have the following commutation relations:

$$[\sigma_\alpha^i, \sigma_\beta^i] = 2i\sigma_\gamma^i \,, \tag{2.F.6}$$

for $i = 1, 2$, and α, β, γ a cyclic permutation of x, y, z.
(iii) Finally,

$$[\sigma_\alpha^1, \sigma_\beta^2] = \sigma_\alpha^1\sigma_\beta^2 - \sigma_\beta^2\sigma_\alpha^1 = \mathbf{0} \,, \tag{2.F.7}$$

where $\alpha, \beta = x, y, z$ and $\mathbf{0}$ is the matrix with all entries equal to zero.

Consider now the identity

$$\sigma_x^1\sigma_y^2\sigma_y^1\sigma_x^2\sigma_x^1\sigma_x^2\sigma_y^1\sigma_y^2 = -\mathbf{1} \,, \tag{2.F.8}$$

which follows, using first (ii) and (iii) above to move σ_x^1 in the product from the first place (starting from the left) to the fourth place, a move that involves one anticommutation (2.F.5) and two commutations (2.F.7), viz.,

$$\sigma_x^1\sigma_y^2\sigma_y^1\sigma_x^2\sigma_x^1\sigma_x^2\sigma_y^1\sigma_y^2 = -\sigma_y^2\sigma_y^1\sigma_x^2\sigma_x^1\sigma_x^1\sigma_x^2\sigma_y^1\sigma_y^2 \,, \tag{2.F.9}$$

and then using (i) repeatedly, to see that the right-hand side of (2.F.9) equals $-\mathbf{1}$.
We now define the operators

$$A = \sigma_x^1\sigma_y^2 \,, \quad B = \sigma_y^1\sigma_x^2 \,, \quad C = \sigma_x^1\sigma_x^2 \,, \quad D = \sigma_y^1\sigma_y^2 \,, \quad X = AB \,, \quad Y = CD \,.$$

Using (ii) and (iii), we observe:
(α) $[A, B] = 0$
(β) $[C, D] = 0$
(γ) $[X, Y] = 0$

The identity (2.F.9) can be rewritten as

$$XY = -\mathbf{1} \,. \tag{2.F.10}$$

But, using (2.F.3), (α), (β), (γ), and (2.F.7), we get:
(a) $v(XY) = v(X)v(Y) = v(AB)v(CD)$
(b) $v(AB) = v(A)v(B)$
(c) $v(CD) = v(C)v(D)$
(d) $v(A) = v(\sigma_x^1)v(\sigma_y^2)$
(e) $v(B) = v(\sigma_y^1)v(\sigma_x^2)$
(f) $v(C) = v(\sigma_x^1)v(\sigma_x^2)$
(g) $v(D) = v(\sigma_y^1)v(\sigma_y^2)$

Since the only eigenvalue of the matrix $-\mathbf{1}$ is -1, by combining (2.F.10) with (2.F.2) in the theorem and (a)–(g), we get

$$v(XY) = -1 = v(\sigma_x^1)v(\sigma_y^2)v(\sigma_y^1)v(\sigma_x^2)v(\sigma_x^1)v(\sigma_x^2)v(\sigma_y^1)v(\sigma_y^2) \,, \qquad (2.F.11)$$

where the right-hand side equals $v(\sigma_x^1)^2 v(\sigma_y^2)^2 v(\sigma_y^1)^2 v(\sigma_x^2)^2$, since all the factors in the product appear twice. But this last expression, being the square of a real number, is positive, and so cannot equal -1. ∎

Part (2) of the Theorem

The proof of part (2) of the theorem is very similar to the proof of part (1) and is taken from a paper by Wayne Myrvold [344], which is a simplified version of a result due to Robert Clifton [98]. We need to introduce here operators Q_1, Q_2 that act as multiplication on functions[57]:

$$Q_j \Psi(x_1, x_2) = x_j \Psi(x_1, x_2) \,, \quad j = 1, 2 \,, \qquad (2.F.12)$$

and operators P_1, P_2 that act by differentiation:

$$P_j \Psi(x_1, x_2) = -i \frac{\partial}{\partial x_j} \Psi(x_1, x_2) \,, \quad j = 1, 2 \,. \qquad (2.F.13)$$

We already mentioned these operators, for one variable, in our discussion of Schrödinger's equation in Appendices 2.A.2 and 2.B.

We will also need the operators $U_j(b) = \exp(-ibQ_j)$, $V_j(c) = \exp(-icP_j)$, with Q_j, P_j defined by (2.F.12), (2.F.13), and $b, c \in \mathbb{R}$. They act as

$$U_j(b)\Psi(x_1, x_2) = \exp(-ibx_j)\Psi(x_1, x_2) \,, \quad j = 1, 2 \,, \qquad (2.F.14)$$

which follows trivially from (2.F.12), and

$$V_1(c)\Psi(x_1, x_2) = \Psi(x_1 - c, x_2) \,, \qquad (2.F.15)$$

and similarly for $V_2(c)$. Equation (2.F.15) follows from (2.F.13) by expanding both sides in a Taylor series, for functions Ψ such that the series converge, and by extending the unitary operator $V_2(b)$ to more general functions Ψ (see, e.g., [412, Chap. 8] for an explanation of that extension).

Precise Statement of Part 2
Let $\mathcal{O}$ be the set of functions of the operators Q_1, Q_2, P_1, or P_2. Then, there does not exist a function

$$v : \mathcal{O} \to \mathbb{R} \qquad (2.F.16)$$

---

[57]This proof is the only place in this book where we use operators that act on an infinite-dimensional space of functions and that are not simply reduced to matrices. See, e.g., [152, Chaps. 13–15] or [412, Chaps. 7 and 8] for a rigorous treatment of operators.

such that

(1)

$$\forall A \in \mathcal{O}, \quad v(A) \in \{\text{eigenvalues of A}\}, \qquad (2.F.17)$$

(2)

$$\forall A, B \in \mathcal{O}, \quad \text{with } [A, B] = AB - BA = 0, \quad v(AB) = v(A)v(B), \qquad (2.F.18)$$

where AB is the operator product.

Remark

Since, for $A \in \mathcal{O}$, the set of possible results of measurements of A is $\mathbb{R}$ and the function $v : \mathcal{O} \to \mathbb{R}$, we do not need to specify a condition like (2.F.2) in part 1 of the theorem for all $A \in \mathcal{O}$ (that is why the condition only refers to eigenvalues). And, as in the proof of the first part of the theorem, the first condition (2.F.17) is only used for $A = -\mathbf{1}, \mathbf{1}$ being the unit operator, in the form $v(-\mathbf{1}) = -1$.

Proof

We choose the following functions of the operators Q_i, P_i :

$$A_1 = \cos(a Q_1), \quad A_2 = \cos(a Q_2), \quad B_1 = \cos \frac{\pi P_1}{a}, \quad B_2 = \cos \frac{\pi P_2}{a},$$

where a is an arbitrary constant, and the functions are defined by (2.F.14), (2.F.15), and the Euler relations:

$$\begin{aligned}
\cos(a Q_j) &= \frac{\exp(i a Q_j) + \exp(-i a Q_j)}{2}, \\
\cos \frac{\pi P_j}{a} &= \frac{\exp(i \pi P_j/a) + \exp(-i \pi P_j/a)}{2},
\end{aligned} \qquad (2.F.19)$$

for $j = 1, 2$. By applying (2.F.18) several times to pairs of commuting operators, we will derive a contradiction.

We have the relations

$$[A_1, A_2] = [B_1, B_2] = [A_1, B_2] = [A_2, B_1] = 0, \qquad (2.F.20)$$

since these operators act on different variables, and

$$A_1 B_1 = -B_1 A_1, \quad A_2 B_2 = -B_2 A_2. \qquad (2.F.21)$$

To prove (2.F.21), note that, from (2.F.14) and (2.F.15), one gets

$$U_j(b) V_j(c) = \exp(-ibc) V_j(c) U_j(b), \qquad (2.F.22)$$

for $j = 1, 2$, which, for $bc = \pm\pi$, means

$$U_j(b)V_j(c) = -V_j(c)U_j(b) \ . \tag{2.F.23}$$

Now use (2.F.19) to expand the product $\cos(a Q_j) \cos(\pi P_j/a)$ into a sum of four terms; each term will have the form of the left-hand side of (2.F.22) with $b = \pm a$, $c = \pm\pi/a$, whence $bc = \pm\pi$. Then applying (2.F.23) to each term proves (2.F.21).
The relations (2.F.20) and (2.F.21) imply

$$A_1 A_2 B_1 B_2 = B_1 B_2 A_1 A_2 \ , \qquad A_1 B_2 A_2 B_1 = A_2 B_1 A_1 B_2 \ . \tag{2.F.24}$$

Let $v(Q_1) = q_1$, $v(Q_2) = q_2$, $v(P_1) = p_1$, and $v(P_2) = p_2$. Since the functions A_1, A_2, B_1, and B_2 can be defined by their Taylor series and we have $v(Q_1^n) = v(Q_1)^n = q_1^n$ by (2.F.18) (Q_1 commutes with itself), and similarly for Q_2, P_1, P_2, it follows that

$$v(A_1) = \cos(aq_1) \ , \quad v(A_2) = \cos(aq_2) \ , \quad v(B_1) = \cos\frac{\pi p_1}{a} \ , \quad v(B_2) = \cos\frac{\pi p_2}{a} \ . \tag{2.F.25}$$

Since A_1 and A_2 commute, we get from (2.F.18),

$$v(A_1 A_2) = v(A_1)v(A_2) \ ,$$

and similarly,

$$v(B_1 B_2) = v(B_1)v(B_2) \ , \quad v(A_1 B_2) = v(A_1)v(B_2) \ , \quad v(A_2 B_1) = v(A_2)v(B_2) \ . \tag{2.F.26}$$

Consider now the operators $X = A_1 A_2 B_1 B_2$ and $Y = A_1 B_2 A_2 B_1$. Using $B_2 A_2 = -A_2 B_2$, from (2.F.21), we get

$$X = -Y \ . \tag{2.F.27}$$

On the other hand, since by (2.F.24) $A_1 A_2$ commutes with $B_1 B_2$, we have from (2.F.18),

$$v(X) = v(A_1 A_2 B_1 B_2) = v(A_1 A_2)v(B_1 B_2) = v(A_1)v(A_2)v(B_1)v(B_2) \ , \tag{2.F.28}$$

where, in the last equality, we use (2.F.26). Similarly, since by (2.F.24) $A_1 B_2$ commutes with $A_2 B_1$,

$$v(Y) = v(A_1 B_1)v(A_2 B_2) = v(A_1)v(B_2)v(A_2)v(B_1) \ . \tag{2.F.29}$$

Comparing (2.F.28) and (2.F.29), we see that

$$v(X) = v(Y) \ ,$$

while (2.F.27) implies $v(X) = v(-Y) = v(-\mathbf{1}Y) = v(-\mathbf{1})v(Y) = -v(Y)$. This means that $v(X) = v(Y) = 0$ and hence that one of the four quantities $v(A_1)$, $v(A_2)$, $v(B_1)$, or $v(B_2)$ vanishes, and this obviously cannot hold for all values of a in (2.F.25) and given values of q_1, q_2, p_1, p_2.   ∎

Chapter 3
"Philosophical" Intermezzo

This chapter is something of a digression in this book. Indeed, it does not deal with physics, and one of the main ideas defended here is that the conceptual problems of quantum mechanics are internal to the physical theory itself and do not have a "philosophical" solution. However, there is an enormous literature arguing that the "lesson" of quantum mechanics, or its "main innovation" in the history of science is that we must abandon realism or determinism or both. It may therefore be useful to discuss these ideas, to define them precisely, and to see what arguments one can give for and against them, both independently of quantum mechanics and by taking it into account.

3.1 Realism and Idealism

> I still hold that any proposition other than a tautology, if it is true, is true in virtue of a relation to *fact*, and that facts in general are independent of experience. I see nothing impossible in a universe devoid of experience. On the contrary, I think that experience is a very restricted and cosmically trivial aspect of a very tiny portion of the universe.
>
> Bertrand Russell [429, pp. 49–50]

First of all, the words *realism* and *idealism* are meant here as attitudes with respect to knowledge, or epistemology, not in their moral or political sense. We will not discuss metaphysics very much either, for example, the question as to whether the world is fundamentally made of matter or contains also some non-material entities.

Realism is both the attitude of common sense and of most scientists (in fact, of almost all scientists, except sometimes those involved in discussions about quantum mechanics).[1] What is meant here by realism is the following combination of ideas:

- There exists a world independent of human consciousness and this world is structured; it has its own properties. Even idealist philosophers know (or should at least admit) that, if they were born, it is because they had parents and that the process of childbirth requires the existence of highly structured organisms.
- A proposition is true or false depending on whether it reflects or not the properties of that world. This means in particular that the truth or falsehood of a proposition is independent of the person who expresses it, or of the group to which he/she "belongs".
- We can know true propositions about the world, for example, through our sensory experiences. Everyday life is full of such experiences that tell us how the world is (to some extent, of course).
- However, our senses can deceive us; therefore, we can never be absolutely certain that our knowledge is true and even the most common experiences could be illusory. I see myself typing on a computer, but I could, in principle, be dreaming.
- Our knowledge is *human*. It is the result of a specific interaction between "us" and the world; that interaction depends on our biology, but also on our history and our culture. Other species have other types of interactions, with the same world. Other cultures or people living in other historical periods will have a different knowledge from ours of the same world. The fact that our knowledge is human implies also that our knowledge can have limits, just as our means of perception or our physical abilities have limits: we see only a small part of the electromagnetic spectrum and nobody is ever going to run a kilometer in less than one second.

However, even if realism, with these caveats, may seem obvious to many, realism has been challenged in several ways throughout history. There is, first of all, an idealist challenge to realism; then there are problems related to scientific realism, i.e., realism about scientific theories, in particular the issue of "underdetermination of theories by data", of "incommensurability of paradigms", and of the status of "unobservable entities" (we will explain those expressions below). Let us discuss these topics one by one.

Idealism can stem from a critical reflection on realism; it may start with the following (obvious) observations:

- In order to observe the world, we need our senses.
- In order to talk about it, we need our languages.
- In order to name things, we need concepts.
- In order to have theories about the world, we need conceptual schemes.

---

[1] For defences of realism more or less related to the content of this section, see, e.g., Boghossian [60], Devitt [126], Ghins [207], Haack [251], Maudlin [324], Psillos [408], Sankey [431], and Stove [461, 462]. In [354], Norris defends realism in quantum mechanics from a philosophical perspective.

The next step is to ask whether what we call the "structure of the world" isn't in reality an effect produced by our senses, languages, or concepts? Thus, when we talk about "the world", aren't we in reality talking only about our senses, languages, or concepts?[2] Are our discourses about the "outside world" or are they rather reflecting the "inside world" (what is in our minds)?

The following quotes,[3] three from philosophers, one from a mathematician, all very famous, illustrate that last idea:

> The mind [...] is deluded to think it can and does conceive of bodies existing unthought of, or without the mind, though at the same time they are apprehended by, or exist in, itself.

> George Berkeley [53, p. 270]

> Are the perceptions of the senses produced by external objects that resemble them? This is a question of fact. Where shall we look for an answer to it? To experience, surely, as we do with all other questions of that kind. But here experience is and must be entirely silent. The mind never has anything present to it except the perceptions, and can't possibly experience their connection with objects.

> David Hume [278, p. 80]

> If we treat outer objects as things in themselves, it is quite impossible to understand how we could arrive at a knowledge of their reality outside of us, since we have to rely merely on the representation which is in us. For we cannot be sentient [of what is] outside ourselves, but only [of what is] in us, and the whole of our self-consciousness therefore yields nothing save merely our own determinations.

> Immanuel Kant [286, p. 351]

> All that is not thought is pure nothingness; since we can think only thought and all the words we use to speak of things can express only thoughts, to say that there is something other than thought, is therefore an affirmation which can have no meaning.

> Henri Poincaré [403, p. 355] (original [402])

The gist of those arguments is always the same: we have direct access only to our perceptions, not to perceptions *of things* but to perceptions *tout court*. Nothing guarantees that those perceptions correspond to or are produced by outer objects.

In his polemical but trenchant critique of idealism *The Plato Cult and Other Philosophical Follies*, the Australian philosopher David Stove characterizes these arguments, which he calls the *gem* of idealism, as being always a non sequitur of the following form:

---

[2]In [396], the assistant of Bohr, Aage Petersen stresses the role of language in Bohr's approach to the problems of quantum mechanics. For Bohr, we are "suspended" in language. For example, Petersen quotes Bohr as saying: "We are suspended in language in such a way that we cannot say what is up and what is down." [396, p. 188].

[3]The ones from Berkeley, Kant, and Poincaré are quoted by David Stove [462]. We refer to his work for a further discussion of these statements. We do not want here to discuss the whole philosophy of these authors, but only to examine those statements critically.

You cannot have trees-without-the-mind in mind, without having them in mind. Therefore, you cannot have trees-without-the-mind in mind.

David Stove [462, p. 139]

By "trees-without-the-mind", Stove means what we ordinarily call trees, namely things that exist "out there", independently of our minds.

A gem starts from a true but tautological premise, "You cannot have trees-without-the-mind in mind, without having them in mind", and "arrives" at an interesting, in fact radical, but false and unwarranted conclusion: "Therefore, you cannot have trees-without-the-mind in mind." In other words, to see a tree outside of my mind, I need my eyes and my brain, so the tree must be represented in my brain; but that does not imply that there is no tree nor that I cannot think about it as existing outside of my mind.

Consider the quote from Poincaré that we just gave: he starts from a tautology, "We can think only thought and all the words we use to speak of things can express only thoughts", which is similar to "You cannot have trees-without-the-mind in mind, without having them in mind", and "arrives" at a radical conclusion: "To say that there is something other than thought, is therefore an affirmation which can have no meaning", which is similar to: "Therefore, you cannot have trees-without-the-mind in mind." But where is the logic? Why can't I think about things that are not thought?[4]

A colleague of David Stove, Alan Olding, has put the matter even more bluntly [363]: "We have eyes, therefore we cannot see."

Here is another example of a gem, coming from a famous contemporary physicist, Anton Zeilinger, whom we already quoted in Chap. 1:

[…] the distinction between reality and our knowledge of reality, between reality and information, cannot be made.

And why is that? Because:

There is no way to refer to reality without using the information we have about it.

Anton Zeilinger [526]

In a critique of that reasoning, almost identical to Stove's critique of idealism, one reads:

In other words, what we can say about reality, or better what we can know about reality, must correspond to our information about reality. In other words, what we know about reality must conform to what we know about reality. Does Zeilinger really believe that a tautology such as this can have interesting consequences?

Martin Daumer, Detlef Dürr, Sheldon Goldstein, Tim Maudlin, Roderich Tumulka, Nino Zanghì [106]

---

[4]The reader is invited to re-read the other quotes above and to verify that they all have the form of a gem.

The issue of idealism is of course an old one. Here is how another famous mathematician reacted to it:

> Thus when my brain excites in my soul the sensation of a tree or of a house I pronounce without hesitation that a tree or a house really exists out of me of which I know the place, the size and other properties. Accordingly we find neither man nor beast who calls this truth in question. If a peasant should take it into his head to conceive such a doubt, and should say, for example, he does not believe that his bailiff exists, though he stands in his presence, he would be taken for a madman and with good reason; but when a philosopher advances such sentiments, he expects we should admire his knowledge and sagacity, which infinitely surpass the apprehensions of the vulgar.

> Leonard Euler [178, pp. 428–429]

The French philosopher Denis Diderot was equally dismissive:

> Those philosophers, madam, are termed idealists who, conscious only of their own existence and of a succession of internal sensations, do not admit anything else; an extravagant system which should to my thinking have been the offspring of blindness itself; and yet, to the disgrace of the human mind and philosophy, it is the most difficult to combat though the most absurd.

> Denis Diderot [133, p. 104]

In a more humoristic version, the French philosopher (by training) and journalist Jean-François Revel, remembers an exercise he had to do while studying philosophy, namely answering the following questions:

> Given that a rock is a creation of my understanding, how is it possible that I might be killed by a falling rock, since in that case I would be smashed by one of my own notions? Can one commit suicide with the help of one's own concepts? Or be surprised and assassinated by it?

> Jean-François Revel [415, p. 513]

And Revel adds [415, p. 513]: "A problem to be urgently solved, as everyone can see."

This may sound ridiculous, but the logic of idealism leads to such "problems". There are however several caveats to be made about idealism. One is that a radical version of it, namely solipsism, the idea that everything is an illusion, that there is nothing outside my mind (not even my body) and that there is a sort of movie going on in my mind corresponding to my "experiences", cannot be refuted. But of course, it is not clear that solipsism can be formulated coherently—why assume that my mind existed a year ago or 5 min ago, rather than thinking that my memories (about my past "experiences") are also an illusion? Maybe only solipsism of the instant (the only thing that exists is my mind right now) is really consistent. But who cares? Nobody really believes this sort of thing. As Bell says:

> Solipsism cannot be refuted. But if such a theory were taken seriously it would hardly be possible to take anything else seriously. So much for the social implications. It is always interesting to find that solipsists and positivists,[5] when they have children, have life insurance.

John Bell [49, p. 136]

One can invent many variants of solipsism. For example, "I" could be simply a "brain in a vat", that is, "I" could be reduced to my brain, which would be manipulated from the outside by some ingenious scientist who has "wired up" my brain and sends me just the right electrical signals to ensure that I have all the sensations that I actually have. But, in reality, it is all an illusion.[6] The movie *The Matrix* has popularized this idea.

Another irrefutable idea is radical skepticism: maybe there is something outside of my mind, but I cannot obtain any reliable information about it. Taken literally, radical skepticism would imply that we have no reason to look left or right when crossing a street (one could not even invoke a "habit", since knowledge of the habit would already be knowledge of something and moreover, there is nothing that would justify the habit, if we accept radical skepticism). Hume adequately characterized this skepticism:

> Sceptical principles may flourish and triumph in the philosophy lecture-room, where it is indeed hard if not impossible to refute them. But as soon as they come out of the shadows, are confronted by the real things that our beliefs and emotions are addressed to, and thereby come into conflict with the more powerful principles of our nature, sceptical principles vanish like smoke and leave the most determined sceptic in the same believing condition as other mortals.

David Hume [278, p. 83]

Idealism often presents itself as a sort of third way between realism (labelled naive) and solipsism or radical skepticism. The (recurring) idea is that it is maybe more careful or more rigorous or more "scientific" to speak only of our perceptions and not of what is "out there", since we have only "direct" access to our perceptions. But there is no way to escape the following dilemma: either we think that some of our perceptions are perceptions of something existing outside of our minds and we agree with realism or we don't and we are solipsists or radical skeptics. Idealism is not a well defined doctrine, but rather a rhetoric of the "middle ground", in a situation where there is no middle ground.

This rhetoric is also at the basis of the idea that the goal of science is to "save the phenomena", that is, to limit ourselves to accounting for the latter or for our "observations" or our "experiments". But the easiest way to save the phenomena is to have a minimum number of them; close your eyes and block your ears. Then any

---

[5]Here Bell refers to some versions of logical positivism (or of the Vienna circle) that were trying to reduce the outside world to a series of impressions of the senses. Positivism will be discussed further in Sect. 7.7 and in Chap. 8. (Note by J.B.).

[6]This argument can be expressed in many different ways. Descartes already suggested that we might be manipulated by an evil demon [123]. For a modern discussion, see Putnam [409, Chap. 1].

theory will "save the phenomena". Of course, that is not what scientists do. There is no way to explain why one makes new experiments all the time if one does not accept the idea that the goal of science is to find out how the world really is and that, in order to do so, one tries to invent ways to test those theories, and to test hem as severely as possible. Experiments are not given to us, like everyday experience might be, they are constructed to check that our theories about the world are true.

However, idealism is related to a serious question: how does the cognitive interaction between ourselves and the world work? How do we form representations? Where do concepts come from? And, ultimately, how do conscious sensations arise? All these questions are difficult.[7] But parts of them can be investigated scientifically, through cognitive science or neurophysiology.

It is important to stress that any study of our means of acquiring knowledge, whether that study refers to biology, psychology, sociology, or history, presupposes the truth of realism. We have to assume, for example, that there are biological or psychological facts that are independent of us and that explain how we perceive things.

Idealists often suggest that we have to analyze our means of interactions with the world in order to better understand it—for example, that the problems of quantum mechanics are caused by our "language" or our concepts, and that we have to better analyze them in order to solve those problems. Of course, it is sometimes useful to reflect on one's concepts or one's language, but this is only part of the general investigation of how the world is. In particular, it is difficult to define a priori what are the mental categories that are "necessary" for science to work, because, as science evolves, we tend to revise those supposedly necessary categories.[8]

In any case, it is an empirical question to know how our minds work, what categories they use, and what limits those categories may impose on our ability to know the world. Therefore, when one investigates such questions, one is in the same boat as the physicist: one has to assume that one can get reliable information about the world (say, in biology), and there is no reason to think that such an enquiry can be carried out a priori, nor that it should be a presupposition (or a "condition of possibility") of other empirical studies.

The mistake that idealism commits is to jump from the fact that our ways to acquire knowledge are to some extent mysterious, to doubting that we are able to acquire knowledge. Stove draws a parallel with "gastronomical idealism":

> We can eat oysters only insofar as they are brought under the physiological and chemical conditions which are the presuppositions of the possibility of being eaten. Therefore, we cannot eat oysters as they are in themselves.

David Stove [462, p. 161]

---

[7] And the last one, called the mind–body problem, may lie beyond the reach of our understanding.

[8] Euclidean geometry being probably the most famous example of this process, since it was thought to be a necessary truth before the invention of non-Euclidean geometries in the 19th century; and the latter even became relevant to physics because of their role in Einstein's general theory of relativity.

Obviously nobody is a gastronomical idealist, even if one had no idea whatsoever about how digestion works.

Our knowledge of mid-size objects in the "outside" world is a brute fact that science takes for granted. A mistake sometimes made is to think that physics predicts our experiences or that it deals somehow with our knowledge of the world. But physics is entirely silent about our cognitive processes. As Einstein explained in a letter to Maurice Solovine [171], physics starts from "axioms", as he called it, or what one might call laws of Nature, that are the results of "free creations of the human mind" as he says elsewhere [166]. This means that they cannot be inferred from data either inductively or deductively. Then, one deduces mathematically empirical "observable" consequences of those laws.

For example, Newton's laws of motion predict the motion of projectiles, satellites and planets. Other parts of physics may predict where needles end up on a screen or traces of ink on a computer printout. But the next and last step, where humans acquire knowledge of these observable consequences is not explained by physics; it is presupposed by it. As Einstein emphasized, if this last step were not reliable (even in the absence of a theory explaining why it is reliable), then nothing would be.[9] Physicists have no alternative but to accept that we can reliably know, by direct observation, certain facts about the world, at least the macroscopic data that are the results of experiments.

It is of some importance to realize this, because it shows that all the loose talk about physics not dealing with the world but with our knowledge of it cannot be taken literally. Maybe physics deals only with meter readings, positions of needles on a screen etc., i.e., "results of experiments", but these are nevertheless objective facts about the world that have nothing to do with our knowledge of them (we will return to that question in Sect. 3.3).

Of course, the arguments given here deal only with what one might call common sense or everyday life realism: there really are tables and chairs and people and computers, etc.[10] But it says nothing about the realism of scientific theories, about which other objections can be raised.

3.2 Scientific Realism

Let us admit that there are objective facts about the world, such as those discussed above. But what about *scientific theories*?[11] Should they be understood realistically and if so how? The "naive" view is that we can have empirical data that support our theories and, if there are enough data and if they are sufficiently impressive, then

---

[9]See the discussion by Maudlin in [325] of Einstein's 1952 letter to Maurice Solovine [171].

[10]It is sometimes argued that tables and the like do not really exist because they are just an assembly of atoms. That will be discussed at the end of Sect. 3.2.

[11]This section is in part already in [82, 454]; see also [453, Chap. 4].

the theory is probably approximately true. Moreover, as time progresses, we obtain theories that are closer and closer to the truth.

There are a priori and a posteriori challenges to those ideas. The a posteriori ones are based on the study of the history of science. The a priori ones rely on the idea that theories are underdetermined by data. Let us start with the latter.

3.2.1 Underdetermination

In its most common formulation, the underdetermination thesis says that, for any finite, or even infinite, set of data, there are infinitely many mutually incompatible theories that are "compatible" with those data.[12] Here is how Quine stated his thesis:

> Any statement can be held true come what may, if we make drastic enough adjustments elsewhere in the system. Even a statement very close to the periphery [i.e., close to direct experience] can be held true in the face of recalcitrant experience by pleading hallucination or by amending certain statements of the kind called logical laws.

> Willard Van Orman Quine [411, p. 43]

This thesis, if not properly understood, can easily lead to radical conclusions. The scientist who believes that a disease is caused by a virus presumably does so on the basis of some "evidence" or some "data". Saying that a "disease is caused by a virus" presumably counts as a "theory" (e.g., it involves, implicitly, lots of counterfactual statements[13]). But if one is able to convince the scientist that there are infinitely many distinct theories that are compatible with those "data", he or she may well wonder on what basis one can rationally choose between those theories.

In order to clarify the situation, it is important to understand how the underdetermination thesis is established; then, its meaning and its limitations become much clearer. Here are some examples of how underdetermination works. One may claim that:

- The past did not exist: the universe was created 5 min ago with all the documents and all our memories referring to the past in the state they were 5 min ago. Alternatively, it could have been created 100 or 1000 years ago.
- The stars do not exist: instead, there are spots on a distant sky that emit exactly the same signals as those we receive.

---

[12]This is often called the Duhem–Quine thesis. In what follows, we will use Quine's version, which is more radical than Duhem's. The latter emphasized the theory dependence of observations or, as Einstein said to Heisenberg [261, p. 63]: "It is the theory which decides what we can observe." For Duhem's views, see [139]. We will return to the discussion between Einstein and Heisenberg in Sect. 3.2.3.

[13]This means statements that depend on premises that are not realized. For example, one could say: "if that population had been infected by the virus, many people would have been sick." This can be considered as a true statement, even if the population in question has not been infected.

- All criminals ever put in jail were innocent: take any given criminal, explain away all testimonies by deliberate desire to harm the accused, declare that all evidence was planted by the police and all confessions were obtained by force.

Of course, all these "theses" may have to be elaborated, but the basic idea is clear: given any set of facts, just make up a story, no matter how implausible, to "account" for them without running into contradictions.[14] Quine would just say that these claims amount to making "drastic enough adjustments elsewhere in the system".

It is important to realize that this is all there is to the underdetermination thesis. This thesis is not very different from the observation that radical skepticism or even solipsism cannot be refuted: all our knowledge about the world is based on some sort of inference from observations, and no such inference can be justified by deductive logic alone. However, it is clear that, in practice, nobody ever takes seriously "theories" such as those mentioned above. Let us call them "crazy theories", or, as the physicist David Mermin would say, "Duhem–Quine monstrosities" [337]. Note that those theories require no work. They can be formulated entirely a priori.

On the other hand, the difficult problem is, given a set of data, to find even one non-crazy theory that accounts for them. Consider for example a police enquiry about some crime: it is easy enough to make up a story that "accounts for the facts" in an ad hoc fashion (sometimes, lawyers do just that), but it is hard to discover who actually committed the crime and to obtain evidence showing that "beyond reasonable doubt".[15] Reflecting on this elementary example clarifies the meaning of the underdetermination thesis. It may be that there is a unique "theory" (i.e., a unique story about who committed the crime) that is plausible and compatible with the facts; in that case, one will say that the criminal has been discovered, even though one can always freely make up theories that will reach different conclusions.

The same is true for scientific theories: in many instances, there is a unique theory that predicts the relevant data. If it predicts new data, as is often the case, then there are even more reasons to believe in that theory. This is sometimes called the "no miracle argument" (in favor of realism): it would be a miracle if theories that have many empirical successes (as the best scientific theories do) were not approximately true.

This common sense idea has also been challenged by philosophers—indeed it does not literally prove anything (and if we admit that radical skepticism is irrefutable, it could not prove anything). But if I have a leak in my bathroom, a plumber comes, and the leak disappears, it is natural to assume that he has fixed the leak. Of course, it is always conceivable that the plumber did nothing and that the leak disappeared

---

[14]In his famous paper quoted here, Quine allows himself even to "plead hallucination" or to change the rules of logic, in order to show that any statement can be held true, "come what may" [411]. Of course, if one is willing to change the rules of logic then, "proving" underdetermination of theories by data becomes even easier. On the other hand, Quine said that he did not want to encourage radical relativism, although this is what a "naive" reading of what he wrote leads to.

[15]This expression shows that people who use it are aware of the fact that, by making enough *ad hoc* assumptions, one could always reject the evidence no matter how strong it is.

by accident. But conceivable does not mean reasonable and objections to the "no miracle argument" are of the same level as what we just said.

A thesis close to the underdetermination thesis concerns the "theory dependence of observations" and the related fact that one cannot check scientific statements one by one. If I want, for example, to verify Newton's theory of gravitation, I may need a telescope to see where the planets are, but the functioning of the telescope depends on other principles than those of the theory of gravitation. Therefore, if one found a discrepancy between some theory and observations, it could be due to different factors, including the possibility that the means used in that observation (telescope or microscope, for example) do not work as advertised.

Once again, Quine formulated a radical version of the idea that observations are theory dependent:

> [O]ur statements about the external world face the tribunal of sense experience not individ-
> ually but only as a corporate body. […] The unit of empirical significance is the whole of
> science.

> Willard Van Orman Quine [411]

The philosopher Thomas Nagel offers a rather amusing reply:

> Suppose I have the theory that a diet of hot fudge sundaes will enable me to lose a pound a
> day. If I eat only hot fudge sundaes and weigh myself every morning, my interpretation of the
> numbers on the scale is certainly dependent on a theory of mechanics that explains how the
> scale will respond when objects of different weights are placed on it. But it is not dependent
> on my dietary theories. If I concluded from the fact that the numbers keep getting higher that
> my intake of ice cream must be altering the laws of mechanics in my bathroom, it would
> be philosophical idiocy to defend the inference by appealing to Quine's dictum that all our
> statements about the external world face experience as a corporate body, rather than one by
> one. Certain revisions in response to the evidence are reasonable; others are pathological.

> Thomas Nagel [346]

Again, either the idea "theory dependence of observations" is yet another formulation of radical skepticism and we can treat it with "neglect and inattention" (Hume) or it is a reminder of the fact, well-known to scientists, that observations are never pure and can be contaminated in various ways. After all, why do we do double blind experiments in medicine, or use controls in experiments, duplicate them, etc., if not to try to avoid such problems?

3.2.2 Incommensurability of "Paradigms"

Another attack on scientific realism comes from the study of the history of science, mostly the one inspired by Thomas Kuhn and his extremely influential book *The Structure of Scientific Revolutions* [298]. The basic idea of Kuhn is that science changes through revolutions, each one introducing its own paradigm, which defines a certain framework through which we think about the world and which is incommensurable with the previous ones. Now, what incommensurable means has been the subject of intense discussions. In [320] Maudlin distinguishes between a "moderate" and an "immoderate" Kuhn. The moderate Kuhn simply observes that, when a "revolution" occurs, the evidence that forces scientists to change may not be compelling, but he does not deny that we now have good evidence to justify the changes that took place.

It is the immoderate or radical Kuhn that we will discuss here, the one for whom "incommensurable" means that the different frameworks through which we look at the world cannot be directly compared because they define in different ways what counts as evidence or observations.[16] Since observations are (in the sense discussed in Sect. 3.2.1) "theory dependent", there is no way to appeal to pure or naked observations to discriminate between different theories. Here is an example of Kuhn in the radical mode, speaking of the introduction of atoms by John Dalton in the early 19th century:

> Chemists could not, therefore, simply accept Dalton's theory on the evidence, for much of that was still negative. Instead, even after accepting the theory, they had still to beat nature into line, a process which, in the event, took almost another generation. When it was done, even the percentage composition of well- known compounds was different. The data themselves had changed. That is the last of the senses in which we may want to say that after a revolution scientists work in a different world.
>
> There is, I think, no theory-independent way to reconstruct phrases like "really there"; the notion of a match between the ontology of a theory and its "real" counterpart in Nature now strikes me as illusive in principle. Besides, as a historian, I am impressed with the implausibility of the view.

Thomas Kuhn [298]

This seems to suggest that atoms (even atoms!) are not "really there", but are the results of a certain way of looking at the world (or of creating a "different world").[17] But this is hard to take seriously: even assuming that atoms were introduced at the time of Dalton on the basis of insufficient evidence, we *now* possess so many different and independent sources of evidence for the existence of atoms that it would be highly irrational to deny their existence. The same is true for many other scientific discoveries, like Newton's laws, Maxwell's equations, the theory of evolution, etc.

One must therefore distinguish two different questions: one is purely historical and is to see whether the evidence was rationally compelling at the time when a new "paradigm" (if one wants to use this word) like the atomic theory or heliocentrism was introduced; the other is to see whether the evidence that we now have in support of those ideas is compelling, and these are totally different issues. It might be that a theory was introduced initially for the wrong reasons (or for not totally good ones), yet that it is objectively correct.

---

[16]The adjectives "moderate" and "immoderate" may refer to different ways to *read* Kuhn, the immoderate one having become popular among some sociologists of science. What the real Kuhn thought is a separate question.

[17]See Krivine and Grossman [297] for a critique of relativism based on the history of the atomic theory.

An easy objection to the radical version of Kuhn's ideas, the one that conflicts with realism, is that, if physicists cannot establish objective theories about the world, why should one believe that historians can?[18] After all, history is also an empirical investigation, and if physicists just move from one paradigm to the next, without ever knowing what is "really there", why believe that historians don't suffer the same fate? Tim Maudlin has a nice way to refute the "radical" Kuhn[19]:

> Perhaps evidence is never "decisive" in the sense of showing how the world actually is. If so, then our earlier claim that the disputes between Ptolemy and Copernicus, Galen and Harvey, etc., were rightly decided could be brought into question. The Copernicans, it might be argued, having accepted the new paradigm came to live in a Copernican world, but Ptolemaic astronomers would never have experienced the supposedly decisive evidence. Similarly, Aristotle could not have experienced the world as refuting his cosmology until he had already accepted a new paradigm, but at that point the choice would already have been made, and so could not be induced by the evidence available given the new paradigm. The obvious objection to this strong reading of the third form of incommensurability[20] is that it is silly. If presented with a moon rock, Aristotle would experience it as a rock, and as an object with a tendency to fall. He could not fail to conclude that the material of which the moon is made is not fundamentally different from terrestrial material with respect to its natural motion. Similarly, ever better telescopes revealed more clearly the phases of Venus, irrespective of one's preferred cosmology, and even Ptolemy would have remarked the apparent rotation of a Foucault pendulum. The sense in which one's paradigm may influence one's experience of the world cannot be so strong as to guarantee that one's experience will always accord with one's theories, else the need to revise theories would never arise.
>
> [...]
>
> Perhaps the appearance of the success and progress of science is only an illusion, the ultimate propaganda produced by the winner. Of course the successful paradigm will regard itself as being right, will proclaim to the world that it tells the truth, will manufacture "evidence" of its correctness — but perhaps this is merely the will of the stronger, not the voice of nature. But no acceptable argument could possibly lead us to such a conclusion. After all, we have much stronger, direct, irrefutable evidence that the Earth rotates, that the blood circulates, that matter is made of atoms than we have grounds to believe any epistemology or account of scientific practice. Like Euripides in *The Frogs*, the social constructivist can pile all his works in one pan of the balance, and climb in with his family and friends as well, while Foucault or Harvey or Cavendish outweighs him on the other side with a scrap of paper from a laboratory journal.

Tim Maudlin [320]

---

[18] We raised the same objection above, when discussing idealism. It applies to *all* skeptical arguments based on empirical statements, as opposed to a priori reasonings.

[19] The article from which this quote is taken was only published in a French translation [320]. We thank Tim Maudlin for providing us with the original English version and for giving us the permission to quote it.

[20] In [320], Maudlin distinguishes three different senses of the word "incommensurability", which is the fundamental Kuhnian concept applied to the relations between "paradigms": "differences over which problems a theory must solve and over the standards of solution; the unnoticed shifts in meaning that occur when a new theory takes over the vocabulary of an old one; and changes in one's fundamental perception of the world that come from accepting a new paradigm." It is the "changes in one's fundamental perception of the world that come from accepting a new paradigm" that characterizes the radical Kuhn and that is discussed here. (Note by J.B.).

The problem with Kuhn's views is that their radical version has been taken to new extremes by a certain number of sociologists of science as well as philosophers of science (see [461] for a good critique of that trend), leading to the idea that all our scientific knowledge is "socially constructed" or even that reality itself is "constructed". Kuhn is not responsible for the radicalization of his views, but it cannot be denied that his ambiguities made his views both "interesting", because of their implications for the (im)possibility of discovering objective truths (the moderate Kuhn would have been of interest only to historians) and exploitable by others exactly in the way that they have been exploited.

Social constructivism applied to all our forms of knowledge can be considered as a sort of socialization of idealism, since it is our social environment (classes, historical periods, etc.), rather than some abstract human mind, that determines our way of "looking" at the world and therefore our "knowledge" of it. Classical idealists and social constructivists have many disagreements, but, in both cases, the link is severed between our representations of the world and the world itself, and our "knowledge" reflects something inside us (the mind or societies) rather than outside.

3.2.3   The Status of "Unobservable Entities"

There is a further problem for realism, namely the status of "unobservable entities", such as the gravitational force, for example. When Newton introduced his theory, he assumed that there were forces attracting bodies towards each other, even if the latter were far apart, while ignoring the ultimate cause that produced these forces.

Ever since that discovery, people have wondered whether these forces are "real" or at least in what sense one should consider them real. After all, nobody sees these forces, one only sees their consequences through the motion of bodies. And they are not comparable to contact forces (like one body colliding with another). In fact, Newton was accused by his critics of having reintroduced medieval "occult qualities" into science. He himself did not "feign hypotheses" on the ultimate nature of those forces, and simply observed that they allowed him to deduce the visible motion of bodies:

> I have not as yet been able to discover the reason for these properties of gravity from phenomena, and I do not feign hypotheses. For whatever is not deduced from the phenomena must be called a hypothesis; and hypotheses, whether metaphysical or physical, or based on occult qualities, or mechanical, have no place in experimental philosophy. In this philosophy particular propositions are inferred from the phenomena, and afterwards rendered general by induction.

> Isaac Newton [351, p. 943]

After Newton, physicists have introduced other unobservable quantities like electromagnetic fields "propagating in vacuum" or curved spacetime in general relativity.

The argument in favor of the existence of those unobservable entities is rather straightforward: we cannot formulate our theories without postulating those entities[21] and the evidence for the truth of the theory counts therefore as evidence for the existence of those unobservable entities. After all, we have never directly seen dinosaurs, distant galaxies, or the inside of the Sun. Yet, we infer the existence and properties of those objects from more direct observations, so why not do the same with forces and other unobservable entities?

Of course, this argument works only for entities that are postulated by sufficiently successful scientific theories; otherwise, we might have to believe in the phlogiston, a substance that was thought to be released during combustion, before modern chemistry was developed, and in many similar substances. The fact that one used to believe in the existence of things like the phlogiston, that were later abandoned, is the basis of the argument by the philosopher Larry Laudan [305] in favor of inductive pessimism: since we were wrong in the past about those objects, how can we be so sure about the reality of objects, like atoms, that are postulated by our current scientific theories? The answer is that it is a question of degree: of course, we might be wrong even about the existence of atoms (we know that already from the Duhem–Quine thesis or from the irrefutability of radical skepticism), but the evidence for the existence of atoms is vastly greater than any evidence that we ever had in favor of the existence of the phlogiston or other discarded substances.[22]

However, the real problem is not whether those unobservable quantities exist but rather, what do we mean when we use those words? They are *unanschaulich* (not representable) to use a German word often used in the early discussions about quantum mechanics. We can visualize dinosaurs but how to visualize forces acting at a distance? As noted by the philosopher of science Bas van Fraassen, realists tend to use arguments involving mid-size objects, while instrumentalists (meaning people who want to reduce scientific statements to be only about measurements or results of experiments) tend to argue their case by focusing on fundamental entities like forces or fields [493, p. 268]. But this is connected with the problem of meaning: if we say "X exists", we must know what "X" means. And since we can form pictures of mid-size objects like dinosaurs, the meaning of the words referring to them is pretty clear intuitively, whereas this is not necessarily the case for the unobservable entities introduced in physics.

It is important to note that, although many debates on quantum mechanics were centered on the non-representability of Nature according to that theory, the problem did occur long before its discovery, since it goes back to Newton. The supposed novelty in the case of quantum mechanics is that, if particles have no position and no trajectory and, in general, if they don't have any definite properties (like energy or speed of rotation), then nothing is representable, while in classical physics, the

---

[21] One might argue that one could replace forces, for example, by potentials or by Hamiltonians, but that would just introduce other unobservable entities. We will not enter into this rather technical discussion.

[22] See [207, 408, 431] for more critical discussions of Laudan's pessimistic induction.

motion of bodies was representable, even if what causes those motions, the forces, were not.[23]

One common temptation, in order to solve this problem, is to give a purely operationalist or instrumentalist meaning to the words referring to unobservable entities: let us forget completely about those "metaphysical" entities, and let us formulate our physical theories solely in terms of "observable" quantities.

The first problem with that idea is that the notion of "observable" is not clear at all, as we already mentioned in Chap. 2. Surely, there are observations made with our unaided senses, but should one limit oneself to those? Can one use eyeglasses, magnifying glasses, telescopes, or (electronic) microscopes?

The second, deeper problem, is that the meaning of the words used by scientists goes far beyond what is "observable". To take a simple example, should paleontologists be allowed to speak about dinosaurs? Presumably, yes. But in what sense are they "observable"? After all, everything we know about them is inferred from fossil data, which are the only quantities ever directly "observed". Of course, all those inferences are based on some kind of evidence, but the point is that the evidence is evidence for something other than itself, e.g., bones of dinosaurs are evidence for the existence of dinosaurs, but the latter are not made only of their bones, and the meaning of the word "dinosaur" is not easily expressible in a language that would refer only to their bones.

Besides, we cannot say that if an instrument measures an electric current, the meaning of the word "electric current" is given by the fact that the instrument readings are what they are, because those readings could be due to other causes, for example to a malfunctioning of the instrument. This shows that what we mean by "electric current", even if it is not entirely clear, is not reducible to a meter reading.[24]

It is because we have a theory of electric currents that we can predict that certain observations will be made, with adequate instruments. Those observations do of course count as evidence in favor of the theory, but they do not replace it. If we did

---

[23]We will see in Chap. 5 that the de Broglie–Bohm theory has, from this point of view, a status comparable to that of classical mechanics: particles have "representable" trajectories, but they are guided by a much less representable wave function.

[24]This point is also made by Feynman in his Lectures on Physics. Referring to a lady who is given a ticket for speeding, he wrote:

Many physicists think that measurement is the only definition of anything. Obviously, then, we should use the instrument that measures the speed – the speedometer – and say, "Look, lady, your speedometer reads 60." So she says, "My speedometer is broken and didn't read at all." Does that mean the car is standing still? We believe that there is something to measure before we build the speedometer. Only then can we say, for example, "The speedometer isn't working right," or "the speedometer is broken." That would be a meaningless sentence if the velocity had no meaning independent of the speedometer. So we have in our minds, obviously, an idea that is independent of the speedometer, and the speedometer is meant only to measure this idea.

Richard Feynman [183, Sect. 8.2]

not have the theory first, with its unobservable entities, there would be no meaning assigned to the observations in the first place.

This point was stressed by Einstein when he met Heisenberg for the first time in 1926. Einstein started by asking Heisenberg if he seriously believed "that none but observable magnitudes must go in a physical theory". Heisenberg replied that this is what Einstein had done with relativity: "that it is impermissible to speak of absolute time, simply because absolute time cannot be observed". Although Einstein admitted that he "did use this kind of reasoning", he said that "it is nonsense all the same".[25] He then explained "more diplomatically":

> [...] it is quite wrong to try founding a theory on observable magnitudes alone. In reality the very opposite happens. It is the theory which decides what we can observe. You must appreciate that observation is a very complicated process. The phenomenon under observation produces certain events in our measuring apparatus. As a result, further processes take place in the apparatus, which eventually and by complicated paths produce sense impressions and help us to fix the effects in our consciousness. Along this whole path [...] we must be able to tell how Nature functions [...] before we can claim to have observed anything at all. Only theory, that is, knowledge of natural laws, enables us to deduce the underlying phenomena from our sense impressions. When we claim that we can observe something new, [...] we nevertheless assume that the existing laws — covering the whole path from the phenomenon to our consciousness — function in such a way that we can rely upon them and hence speak of 'observations'.
>
> When it comes to observation, you behave as if everything can be left as it was, that is, as if you could use the old descriptive language.[26] In that case, however, you will also have to say: in a cloud chamber we can observe the path of an electron. At the same time, you claim that there are no electron paths inside the atom. This is obvious nonsense [...].

Albert Einstein speaking to Werner Heisenberg [261, pp. 63, 65]

Einstein came back to this theme much later, in his 1949 *Reply to criticisms*. After mentioning that "the quantum theorist" would object to postulating "something not observable as 'real'", Einstein writes:

> What I dislike in this kind of argumentation is the basic positivistic attitude, which from my point of view is untenable, and which seems to me to come to the same thing as Berkeley's principle, *esse est percipi*.[27] "Being" is always something which is mentally constructed by

---

[25] Heisenberg says [261, p. 64]: "I was completely taken aback by Einstein's attitude, though I found his arguments convincing." However, it is not clear that Heisenberg really incorporated Einstein's attitude in his views on quantum mechanics.

[26] This "old descriptive language" is at the basis of the idea, in quantum mechanics, that a "measurement" measures some real property of the quantum system. As we emphasized in Sect. 2.5, this is untenable; we will come back to this question in Sect. 5.1, where we will explain what measurements really are, in the framework of the de Broglie–Bohm theory. (Note by J.B.).

[27] "To be is to be perceived", which is a standard motto of idealism, as we saw in Sect. 3.1. The science writer John Horgan quotes John Wheeler as saying [273]: "[...] quantum phenomena are neither waves nor particles but are intrinsically undefined until the moment they are measured. In a sense the British philosopher Bishop Berkeley was right when he asserted two centuries ago that 'to be is to be perceived'." This is a surprising position to hold for a physicist, but maybe not that surprising, given Wheeler's views on quantum mechanics, and in particular on the delayed choice experiment, discussed in Chaps. 1 and 2. (Note by J.B.).

us, that is, something which we freely posit (in the logical sense). The justification of such constructs does not lie in their derivation from what is given by the senses. Such a type of derivation (in the sense of logical deducibility) is nowhere to be had, not even in the domain of pre-scientific thinking. The justification of the constructs, which represent "reality" for us, lies alone in their quality of making intelligible what is sensorily given (the vague character of this expression is here forced upon me by my striving for brevity).

Albert Einstein [170, p. 669]

For Einstein, observations were always "dependent on theories", in some sense; which does not mean that they are arbitrarily "constructed" by us (socially or otherwise), but that we cannot use the concept "observation" as a fundamental, unproblematic concept on which to base our scientific theories.

Maybe a hint of the solution to our problems is provided by the following comment by Einstein:

Science without epistemology is — insofar as it is thinkable at all — primitive and muddled. However, no sooner has the epistemologist, who is seeking a clear system, fought his way through such a system, than he is inclined to interpret the thought-content of science in the sense of his system and to reject whatever does not fit into his system. The scientist, however, cannot afford to carry his striving for epistemological systematic that far. [...] He therefore must appear to the systematic epistemologist as an unscrupulous opportunist.

Albert Einstein [170, p. 684]

Opportunism in epistemology would mean that we try to give meaning to the unobservable entities by combining rather than opposing the various tools that we have to grasp them: they are partly defined mathematically, partly defined by our indirect observations, and partly by the imperfect pictures that we can make of them. None of these tools is perfect and even when they are combined, we do not reach an understanding of unobservable entities similar to what we can do with mid-size objects, but there is nothing better that we can do, given our means of perception and the constitution of our minds. However, there is no reason, as instrumentalists tend to do, to restrict ourselves to using only one tool, i.e., direct macroscopic observations.

Returning to the unobservable entities, it is useful to keep in mind the following picture, which is basic to most thinking in modern physics,[28] and which allows us to make rather precise the idea of one theory being an approximation of another. In this view, reality is composed of a hierarchy of "scales", ranging from quarks to galaxies, and going through atoms, fluids, gases, and so on. The theory on one scale emerges from the one on a finer scale, by ignoring some of the (irrelevant) details of the latter. All our present theories are approximations to more basic theories: continuum and fluid mechanics are approximations to classical particle mechanics; classical physics is an approximation to non-relativistic quantum mechanics, which is itself an approximation to quantum field theories. Whether this process stops somewhere

---

[28]One might call this the "renormalization group view of the world", after the work in statistical mechanics and quantum field theory carried out during the 1970s (but too technical to be explained in detail here).

at some fundamental "final theory", or whether there are theories "all the way down", no one knows.[29] In each of those theories, the existence of some basic "unobservable" entities (forces, fields, etc.) is postulated. But all of these are supposed to be higher-level effects caused by the basic entities of a more fundamental theory (for example, classical forces do not enter in quantum field theories, and should be thought of as emerging from them as some sort of approximations). Since none of those theories is (yet) final, there is no reason to consider them as literally true or to worry too much whether the entities they postulate "really exist". In summary, all physical theories have to refer to some basic "unobservable" ontology, but there is some unavoidable vagueness in what that fundamental ontology consists of.

On the other hand, since we do not know what the ultimate entities are, it makes little sense to think that objects on a higher scale, like tables and chairs, are not real, because they are just, say, assemblies of atoms. If one reasons like this, atoms are not real either because they are made of nuclei and electrons and one can continue like that, from one scale to the next, without knowing whether this will ever stop. It is more reasonable to consider that objects on all scales exist, provided that we have good evidence for their existence.

Moreover, and this is also an important point, since we know that, for example, classical mechanics is only an approximation, we can expect it to live forever (with that status). There are no examples in the history of science of a theory that has been abandoned after it has been so well studied that its status has become one of approximate truth. For example, most of ballistics and celestial mechanics use the "approximately true" Newtonian mechanics. And the 19th century equations which approximately govern the motion of fluids have largely survived the quantum revolution.

As pointed out by Weinberg in his very interesting critique of Kuhn:

> If you have bought one of those T-shirts with Maxwell's equations[30] on the front, you may have to worry about its going out of style, but not about its becoming false. We will go on teaching Maxwellian electrodynamics as long as there are scientists.
>
> Steven Weinberg [506]

3.3 Realism and Quantum Mechanics

Sometimes idealism seems to be supported by arguments based on quantum mechanics. We quoted already in Sect. 1.3 some statements illustrating this line of thought, for example:

> We now know that the moon is demonstrably not there when nobody looks.
>
> David Mermin [331]

---

[29]See Weinberg [504] and Bohm [68, Chap. 5] for in-depth discussions of this issue, reaching different conclusions.

[30]These equations govern the motion of electromagnetic waves. (Note by J.B.).

But if someone asserts that quantum mechanics should lead us to doubt the existence of ordinary macroscopic objects, like the moon, why believe quantum mechanics in the first place? As every scientific theory, quantum mechanics is justified by "experiments" or "observations", and the latter presuppose that objects (at least laboratory instruments) exist independently of human consciousness and of whether we look at them or not. Of course, one could argue that a scientific theory shows that some objects or properties do not exist in themselves, but are somehow produced by a specific interaction with our senses (optical illusions would be an example). But that sort of statement becomes self-refuting when it applies to the entire "outside world".

David Stove has a particularly ironical, but entirely suitable reaction to the statement from Mermin quoted above:

> [...] if it is true, then it would be irrational to believe these physicist's best theories. Fundamental physical theories never say anything about a particular macroscopic object, such as the moon; but if they did say something about the moon, then they would say the same thing about all macroscopic physical objects, hence about all land mammals, hence about the particular land mammal, Professor N. D. Mermin, who wrote the sentence I have just quoted. [...] [Mammals] depend for their existence on a great many things; but somebody's looking at them is not among those things and everyone knows this.
>
> David Stove [462, pp. 99–100]

Considering what precedes, one may well ask: What do people *mean* when they say that quantum mechanics forces us to abandon realism? Which kind of realism? The one about ordinary objects? Should we believe that Schrödinger's cat is both alive and dead before we look? If a measurement is made in a laboratory and the result is recorded by a computer, does the record only appear when someone looks at it? It is hard to think that anybody could take such proposals seriously, despite all the statements suggesting that one does.

But then, what can the expression "abandoning realism" possibly mean? Maybe it refers to scientific realism: quantum mechanics would not be a theory about the microscopic world, but only about the effects that the latter has on the macroscopic world (i.e., on measurements), and no reality can be attached to the quantum description of the microscopic world. That is of course a possibility and it is a coherent position, but it is hardly a satisfactory one. There is no way to make a distinction, in principle, between the macroscopic world and the microscopic one. More importantly, macroscopic objects are constituted by microscopic ones—atoms for example. If the latter have no objective properties whatsoever (no position, no velocity, no angular momentum, etc.), how come that macroscopic objects do have such properties? It is sometimes thought that a classical description emerges from a quantum one, when some limit is taken—for example, if the mass or the number of particles of the quantum system increases. But nothing emerges out of nothing, just because some limit is taken in an equation. If there is nothing real on the microscopic scale, then there is nothing real on the macroscopic one either.[31]

---

[31] See the example of Einstein's boxes in Sect. 4.2 for a further discussion of this problem.

One could of course reply that what happens on the micro scale is unknowable and that only the macroscopic manifestations of what happens on that scale can be known. But some argument must be given for that view and the fact that ordinary quantum mechanics does not give us a description of what happens on the micro scale does not prove that such a description is impossible.

Sometimes, when arguments are given, they rely on the no hidden variables theorems,[32] and the ones discussed in Sect. 2.5 do prove that one cannot attribute a value independently of "measurements" to certain classes of "observables". That means of course that one should not be realist with respect to those values. But this has nothing to do with realism understood as a general or philosophical concept. After all, we know that colors are not "real", in the sense that they are the result of certain interactions between light of a given wavelength and our brains and that there is nothing really red or green in Nature. But there is nevertheless something real, namely the light rays and their interaction with matter.

What a realist wants is a theory describing what happens at the micro scale that accounts for the predictions made at the macro scale. As we will see in Chap. 5, such a theory actually exists. But even if it didn't, the mere fact, implied by the no hidden variables theorems, that one cannot attribute a reality to the values of all "observables" implies absolutely nothing whatsoever as far as a realist viewpoint is concerned.

However, the sort of antirealism discussed here is probably not what is in the mind of most physicists, who are of course realists with respect to ordinary objects, but also often have an "implicitly" realistic picture of the micro world, either because they think that the quantum state spontaneously collapses or because they have, in the back of their minds, a statistical interpretation of quantum mechanics (both of these views have been discussed in Sect. 2.5 and were shown there to be untenable).

It should be stressed that, among those who have held orthodox "Copenhagen" views, there has sometimes been an explicit rejection of the sort of idealism that we find in Mermin. In Chap. 2, we quoted Landau and Lifshitz, who were very explicit:

> [...] we emphasize that, in speaking of 'performing a measurement', we refer to the interaction of an electron with a classical 'apparatus', which in no way presupposes the presence of an external observer.

> Lev Landau and Evgeny Lifshitz [302, p. 2], quoted in [46, p. 35]

Besides, Bohr himself, according to Rosenfeld, rejected the accusation of "positivism" that was leveled against him by some Soviet Marxists.[33] Bohr is quoted by Rosenfeld as saying:

---

[32] Some no hidden variables theorems do not prove what they actually claim. This holds, in particular, for the most famous of them, the one due to von Neumann [496], which will be discussed in Sect. 7.4.

[33] In Sect. 7.7, we will discuss further the relationship between Marxist philosophy and various views of quantum mechanics. For the relationship between Bohr and positivism, see also Beller [50] and Howard [275].

"Why are those Russians dissatisfied?" And I [Rosenfeld] tried to explain, "They accuse you of being a positivist," and so on. Then he said, "Is that it? But those things are so trivial; they are not of interest to physicists. Physicists are beyond that point and that is not the thing that we are interested in. We are struggling with real problems, not with those trivial statements about our living in an external world."

Thomas S. Kuhn and J.L. Heilbron, interview with Léon Rosenfeld [299]

However, given their realist statements, it is not clear what Landau, Lifshitz, and Bohr thought about the objective properties of the microscopic world.

3.4   Determinism

Like realism, determinism is one of those words that leads to endless discussions, partly because it is not precisely defined. Hence, it is often unclear what the discussion is about.[34] We will first propose two definitions. According to one definition, determinism is trivially false. According to the other, it is very probably true. However, both definitions are rather uninteresting, and it is not clear how to formulate the issue of determinism so that the question becomes interesting. We will then discuss the status of probabilities in the framework of deterministic laws and how the law of large numbers connects probabilities and physics.[35]

3.4.1   Definitions

In his famous essay on probability, Laplace expressed the idea of universal determinism in a particularly clear way:

Given for one instant an intelligence which could comprehend all the forces by which nature is animated and the respective situation of the beings who compose it — an intelligence sufficiently vast to submit these data to analysis — it would embrace in the same formula the movements of the greatest bodies of the universe and those of the lightest atom; for it, nothing would be uncertain and the future, as the past, would be present to its eyes.[36]

Pierre Simon Laplace, [303, p. 4]

Laplace immediately added that *we* shall "always remain infinitely removed" from this imaginary "intelligence" and its ideal knowledge of the "respective situation of the beings who compose" the natural world; that is, in modern language, ideal

---

[34]See Earman [156] for a detailed discussion of determinism, in particular in physics. His views do not exactly coincide with those defended here, at least about quantum mechanics.

[35]The content of this section comes partly from [83].

[36]This intelligence is often referred to as the "Laplacian demon". (Note of J.B.).

knowledge of the precise initial conditions of all the particles.[37] He distinguished clearly between what Nature does and the knowledge we have of it. Moreover, he stated this principle at the beginning of an essay on *probability*. But, as we will discuss below, probability for Laplace is nothing but a method that allows us to reason in situations of partial ignorance. The meaning of Laplace's quote is completely misrepresented if one imagines that *he* hoped that one could arrive someday at a perfect knowledge and a universal predictability, for the aim of his essay was precisely to explain how to proceed in the absence of such a perfect knowledge.

However, determinism is often confused with predictability. So, according to that view, a process is deterministic if we, humans, can predict it, or, maybe, if we, humans, will be able to predict it in the future. For example, in an often quoted lecture[38] to the Royal Society, on the three hundredth anniversary of Newton's *Principia*, the distinguished British mathematician Sir James Lighthill gave a perfect example of how to confuse predictability and determinism:

> We are all deeply conscious today that the enthusiasm of our forebears for the marvelous achievements of Newtonian mechanics led them to make generalizations in this area of *predictability* which, indeed, we may have generally tended to believe before 1960, but which we now recognize were false. We collectively wish to apologize for having misled the general educated public by spreading ideas about *determinism* of systems satisfying Newton's laws of motion that, after 1960, were to be proved incorrect [...].

> James Lighthill, [310] (italics added by J.B.)

Of course, nobody who has ever defended universal determinism (in particular Laplace) meant it to be equated with predictability. Everybody agrees that not everything in the world is predictable, and it is somewhat surprising to see how many people present that truism as if it was a recent discovery.

To illustrate the problem posed by the conflation of the two terms, consider, for example, a perfectly regular, deterministic *and* in principle predictable mechanism, like a clock, but put it on the top of a mountain, or in a locked drawer, so that its state (its initial conditions) become inaccessible to us. This renders the system trivially unpredictable, yet it seems difficult to claim that it becomes non-deterministic.

So one has to admit that *some* physical phenomena could obey deterministic laws and yet are not predictable, possibly for "accidental" reasons. But, once this is admitted, how does one show that *any* unpredictable system is *truly* non-deterministic, and that the lack of predictability is not merely due to some limitation of our knowledge or of our abilities? We cannot infer indeterminism from ignorance alone. One needs other arguments.

---

[37]Laplace was expressing in words the fact that differential equations have a unique solution for given initial conditions. This was discussed in Appendix 2.A for *linear* differential equations. Here he is referring to the more general case of (2.A.1.4), where conditions have to be imposed on f for the existence and the uniqueness of the solution.

[38]Quoted, e.g., by Reichl [414, p. 3] and by Prigogine and Stengers [405, pp. 93–94] and [406, pp. 41–42].

Confusing determinism and predictability is an instance of what the physicist E.T. Jaynes calls the mind projection fallacy[39]:

> We are all under an ego-driven temptation to project our private thoughts out onto the real world, by supposing that the creations of one's own imagination are real properties of Nature, or that one's own ignorance signifies some kind of indecision on the part of Nature.
>
> Edwin T. Jaynes [283, p. 7]

This brings us to a second definition of determinism, one which tries to be independent of human abilities. Consider a physical system whose state is characterized by some numbers that change over time. Let us say that it is deterministic if there exists a function F that maps the values taken by that set of variables at a given instant, say t_1, to those obtained at a later time, say t_2; and, then, the latter to those at a later time, t_3, and so on.[40] This corresponds pretty much to Laplace's conception,[41] namely to the idea of predictability, but "in principle", i.e., putting aside limitations imposed by human abilities. The word "exist" should be taken here in a "Platonic" sense: it does not refer to our *knowledge*; the function in question may be unknown, or unknowable, or so complicated that, even if it were known, it would in practice be impossible to use it in order to make predictions.

Now, is determinism understood in this sense true? Well, let us suppose that the system happens to be in exactly the same state twice, at different times, say at times t_i and t_j, and at times t_{i+1} and t_{j+1} is in different states. Then, the function F does not exist, since, by definition, it is supposed to associate to each set of values of the variables another set in a unique way. To see how this problem may occur, consider a simple example, say the production of grain in a given country, counted yearly and in millions of tons. This is a single number and it is quite possible that it may take the same value, say in 1984 and 1993, but different values in 1985 and 1994. In that case, one should say that this system is not deterministic, according to the above definition. But suppose we were to count the amount of grain, not in millions of tons, but in grams. Or suppose we consider a larger set of data, say, the quantities of all items produced in that given country. Then, it becomes very unlikely that *this set of data* would ever take exactly the same values twice at given times *and, moreover*, two different sets of values at later times.[42] And, if that were the case, we could always include more data, say the amount of rainfall, the results of elections, etc., in order to avoid that situation. But then, how can the mere existence of the function F be refuted? In fact, its existence is trivial: take any finite sequence of sets of numbers

---

[39]Which is related to idealism, since the latter tends to identify what exists and what is in our mind.

[40]This idea is essentially the one proposed by Bertrand Russell in [426]; see [156] for a discussion.

[41]Except that he was speaking in terms of "continuous time", i.e., of differential equations, rather than "discrete time", which is chosen here because it is more intuitive.

[42]In fact, one never observes a sufficiently large set of data taking twice the same values, so that the condition that this set of data take two different sets of values at a later time is not really necessary; it is added only because of the theoretical possibility of "eternal return". But, even if that were true, it would not refute the existence of our function F, since the complete state of the world would then simply be a periodic function of time.

that has the property of never being in exactly the same state at different times, t_i and t_j, and in different states, at times t_{i+1} and t_{j+1}. One can always find one function, in fact many functions that map each set of numbers into the next one.

So let us propose the following definition: a physical system is *deterministic* if there exists a function F, as described here, whenever we describe it in sufficient detail. The important words here are "in sufficient detail", or what physicists call a fine-grained description, as opposed to a coarse-grained one (e.g., counting grains in grams is fine grained, but counting in tons is coarse grained). It is quite simple to see that any system that is deterministic can become non-deterministic if one coarse-grains it. For example, consider a deterministic system with two variables, x and y, both taking only the values 0 and 1. Suppose that the dynamics changes the pair $(0, 0)$ into $(1, 0)$ and the pair $(0, 1)$ into $(0, 0)$ (and any other rule for the two other pairs). Then if we look only at the variable x, which is a form of coarse graining, the system becomes non-deterministic, since $x = 0$ can be mapped onto $x = 0$ or $x = 1$, depending on the value of y.

But what is the point of this exercise? Simply to show that, if the idea of determinism is phrased in terms similar to those of Laplace, and not confused with the notion of predictability, then it is indeed hard to see how it could be refuted.[43]

Of course, the notion of determinism introduced here has very little to do with the goals of science, which are not simply to find a function like F. In a sense, scientists do look for such functions, but with extra properties: simplicity, explanatory power, and, of course, the possibility, using F, of making at least some predictions. So in a sense, the question of the existence of F is "metaphysical" and of no scientific interest. But so is the question of "determinism": either it is identified with predictability by us, humans, and determinism is trivially false, or it is defined as above, and it is most likely true, but uninteresting. It is difficult to see how to formulate the issue of "determinism" in a sense that makes it both interesting and decidable.

It is likely that the hostility to determinism comes from a desire to "save free will". Namely, to find a description of the physical universe that can be reconciled with our deep feeling that, at least on some occasions, "we" choose to do X and not Y. That is, that Y was possible, but did not happen because of our free choice. But, if everything is caused by anterior events, ultimately going back to the Big Bang, Y was not really possible (it only appeared to be so because of our ignorance) and free will is an illusion. Since most of our moral, legal, and political philosophies assume some kind of free will, a lot appears to be at stake.

But the problem is: what is the alternative to determinism *within physics*? Nothing has ever been proposed except *pure randomness*![44] Or, in other words, events with no

---

[43] In [214], Nicolas Gisin thinks that he has shown that quantum mechanics offers an example of true randomness. But his example, based on the nonlocal effects to be discussed in Chap. 4, does not prove this claim, as we will see in Sect. 5.2.1.

[44] In classical physics, non-deterministic processes are often modeled by Markov chains or more general stochastic processes in which an element of pure chance always enters. Of course, if one thinks of these models as describing non-isolated systems, then that chance may simply model the effect of unknown outside forces. But if we were to look for a fundamental theory using the same

cause. But that will not give us a picture of the world in which free will exists either. Our feeling of free will is not that there is some intrinsically random process at work in our minds, but that *conscious choices* are made. And that is simply something that no known physical theory accounts for. Our feeling of free will implies that there is a causal agent in the world, the "I", that is simply "above" all physical laws. It suggests a dualistic view of the world, which itself meets great difficulties. One solution is, as mentioned above, to declare that free will is an illusion. But if that is the case, it is a "necessary illusion" in the sense that we cannot live without, in some sense, believing in it, unlike, say, believing in the dogma of the Immaculate Conception. It is not clear what could constitute a solution to that problem,[45] but one should avoid using this problem to create within physics a prejudice in favor of indeterminism, since neither determinism nor indeterminism in physics can "save" free will.

As Bertrand Russell observed, scientists should look for deterministic laws like mushroom seekers should look for mushrooms. Deterministic laws are preferable to non-deterministic ones because they give both a way to control things more efficiently (at least in principle) and because they give more satisfactory explanations of why things are the way they are. Looking for deterministic laws behind the apparent disorder of things is at the heart of the scientific enterprise. Whether we succeed or not depends in a complicated way both on the structure of the world and on the structure of our minds. But the opposition to determinism tends to make people feel that the project itself is doomed to fail. And that state of mind does run counter to the scientific spirit.

3.4.2 Determinism and "Chaos Theory"

Since the 1960s there has been a renewed interest in the theory of dynamical systems, that is, systems evolving in time according to deterministic laws, but laws that may be more general than Newton's (this theory was initiated at the end of the 19th century by Henri Poincaré [400]). One basic notion that emerged from those studies is popularly known as "chaos".

Here is a simple example. Consider the set on which the dynamics takes place to be simply the interval $I = [0, 1[$. And take as (discrete time) dynamics the function $f : x \to 10x$ mod 1. This means, we take a number between 0 and 1, multiply it by 10, write the result as an integer plus a number between 0 and 1 and take the latter as the result, i.e., $f(x)$. This gives another number between 0 and 1, and we can repeat the operation. Upon iteration, we obtain the *orbit* of x; x itself is the initial condition. To describe the latter concretely, one uses the decimal expansion. Any number in I can be written as $x = 0.a_1 a_2 a_3 \ldots$, where a_i equals $0, 1, 2, \ldots, 9$. It is

---

(Footnote 44 continued)
mathematics (like the spontaneous collapse theories discussed in Sect. 6.2), then that pure chance would be postulated to be just there, with no further justification.

[45]The philosopher Colin McGinn has developed the interesting idea that the problem of "free will" may lie beyond the limits of human understanding [329].

easy to see that $f(x) = 0.a_2a_3 \ldots$. This is a perfect example of a *deterministic* but *unpredictable* system. Given the state x at some initial time, one has a rule giving the state of the system for arbitrary times. Moreover, for any fixed time, one can, in principle, find the state after that time, with any desired accuracy, given a sufficiently precise characterization of the initial state. This expresses the deterministic aspect. Unpredictability comes from the fact that, if we take two initial conditions at a distance less than 10^{-n}, which can be as small as one wants for n large, then the corresponding orbits could differ by, say, 1/2, after n steps, because the difference will be determined by the $(n + 1)$th decimal.[46]

This means that two systems obeying the same law may, at some moment in time, be in very similar (but not identical) states and yet, after some time, find themselves in very different states. This phenomenon is expressed figuratively by saying that a butterfly flapping its wings today in Brazil could provoke a hurricane three weeks from now in Texas. Of course, the butterfly by itself doesn't do much. But if one compares the two systems constituted by the Earth's atmosphere, described in every possible detail, but one system with the flap of the butterfly's wings and the other without it, the result three weeks from now may be very different (a hurricane or not). One practical consequence of this is that we do not expect to be able to predict the weather more than a few weeks ahead of time. Indeed, one would have to take into account such a vast quantity of data, and with such accuracy, that even the largest conceivable computers could not begin to cope with them.

However, the idea that this "butterfly effect" may occur is not so new. In fact, even as early as 1909, Henri Poincaré wrote:

> Why have the meteorologists such difficulty in predicting the weather with any certainty? Why do the rains, the tempests themselves seem to us to come by chance, so that many persons find it quite natural to pray for rain or shine, when they would think it ridiculous to pray for an eclipse? We see that great perturbations generally happen in regions where the atmosphere is in unstable equilibrium. The meteorologists are aware that this equilibrium is unstable, that a cyclone is arising somewhere; but where they can not tell; one-tenth of a degree more or less at any point, and the cyclone bursts here and not there, and spreads its ravages over countries it would have spared. This we could have foreseen if we had known that tenth of a degree, but the observations were neither sufficiently close nor sufficiently precise, and for this reason all seems due to the agency of chance. Here again we find the same contrast between a very slight cause, unappreciable to the observer, and important effects, which are sometimes tremendous disasters.

> Henri Poincaré [403 p. 398] (original [401])

However, the presentation given here, which is rather common in the popular literature, is not precise enough. Indeed, almost all processes have the property that "an arbitrarily small error in the initial conditions makes itself felt after a long enough time". Take a pendulum, or even a more modern clock; eventually, it will indicate the

---

[46]Write $f^{(n)}(x)$ for the nth iterate of f applied to x. Then, if $x = 0.a_1a_2a_3 \ldots a_na_{n+1}a_{n+2} \ldots$, we have $f^{(n)}(x) = 0.a_{n+1}a_{n+2} \ldots$. So if we take, say, $x = 0.a_1a_2a_3 \ldots a_na_{n+1}a_{n+2} \ldots$ and $y = 0.a_1a_2a_3 \ldots a_nb_{n+1}a_{n+2} \ldots$, then $|x - y| \leq 10^{-n}$, while $|f^{(n)}(x) - f^{(n)}(y)| = |a_{n+1} - b_{n+1}|/10$ which, by suitable choice of b_{n+1}, given a_{n+1}, can be greater than or equal to 1/2.

wrong time. In fact, for any system, whose initial state is imperfectly known (which is always the case in practice), an imprecision in the initial data will be reflected in the quality of the predictions we are able to make about the system's future state. In general, the predictions will become more inexact as time goes by. But the *manner* in which the imprecision increases differs from one system to another: in some systems it will increase slowly, in others very quickly. More precisely, in the first case the imprecision increases linearly or polynomially with time, i.e., as n^k for some k, where n is the (discrete) time, and in the second case exponentially, i.e., as $\exp(an)$ for some $a > 0$. An example of the latter is given by the function $f : x \to 10x$ mod 1 since there, at each time, the imprecision in the initial data is multiplied by 10, until it becomes of order unity.

To explain what this implies, let us imagine that we want to reach a certain specified precision in our final predictions, and let us ask ourselves how long our predictions will remain sufficiently accurate. Let us suppose, moreover, that a technical improvement has allowed us to reduce by half the imprecision of our knowledge of the initial state. For the first type of system (where the imprecision increases slowly, let us say like n^k with $k = 1$), the technical improvement will permit us to *double* the length of time during which we can predict the state of the system with the desired precision. But for the second type of system (where the imprecision increases quickly), it will allow us to increase our "window of predictability" by only a fixed amount: for example, by one additional hour or one additional week (how much depends on the circumstances).[47] Simplifying somewhat, we shall call systems of the first kind *non-chaotic* and systems of the second kind *chaotic* or *sensitive to initial conditions*. Chaotic systems are therefore characterized by the fact that their predictability is sharply limited, because even a spectacular improvement in the precision of the initial data (for example, by a factor of $1000 \sim 2^{10}$) leads only to a rather mediocre increase in the duration over which the predictions remain valid.[48]

The real discovery in the theory of dynamical systems is not that a very complex system, such as the Earth's atmosphere, is difficult to predict, but rather that a system describable by a *small* number of variables and obeying simple deterministic equations—for example, a pair of pendulums attached together—may nevertheless exhibit very complex behavior and an extreme sensitivity to initial conditions.

However, the discovery of "chaos" has repeatedly led people to consider it as an argument against determinism (in fact, this is the idea behind the Lighthill quote mentioned above); but it can only be considered as such if one confuses determinism with

---

[47]To be precise, that window increases only logarithmically with the improvement in the precision of the initial data. Indeed, if the uncertainty grows like $\exp(an)$, then, if the precision in the initial data increases by a factor of, say, $1000 \sim 2^{10}$, the additional time over which our predictability remains the same as before is given by $\exp(an) = 1000$, i.e., $n = (1/a) \ln 1000 \sim (10/a) \ln 2$.

[48]It is important to add one qualification: for some chaotic systems, the fixed amount that one gains when doubling the precision in the initial measurements can be very long, which means that in practice these systems can be predictable much longer than most non-chaotic systems. For example, one knows that the orbits of some planets have a chaotic behavior, but the "fixed amount" is here of the order of several million years [304].

predictability. Actually, the existence of chaotic dynamical systems supports universal determinism rather than contradicts it. Indeed, suppose for a moment that no classical mechanical system behaves chaotically. That is, suppose that there exists a theorem proving that any such system must eventually behave in a non-chaotic fashion. It is not completely obvious what the conclusion would be, but certainly *that* would be an embarrassment for the classical deterministic world view. Indeed, so many physical systems (like the weather) seem to exhibit sensitivity to initial conditions that one would be tempted to conclude that classical mechanics cannot adequately describe those systems. One might suggest that there must be an inherent indeterminism in the basic laws of Nature. Deterministic chaos increases the explanatory power of deterministic assumptions, and therefore, according to normal scientific reasoning, *strengthens* those assumptions.

Moreover, empirical tests have been designed to check whether a series of numbers is "random". But some deterministic chaotic systems (such as the function $f : x \rightarrow 10x \mod 1$) produce sequences of numbers that can pass all these tests. This is a very strong argument against the idea that one can ever prove that some phenomenon is "intrinsically random", in the sense that there does not exist a deterministic mechanism underlying and explaining its behavior. So the recent discoveries about chaos do not force us to change a single word of what Laplace wrote and, in fact, they support his position.

3.4.3 Probabilities in Classical Physics

Since the notion of probability appears to lie at the heart of quantum mechanics, it might be useful to explain first what probability means in classical physics.

There are at least two different meanings given to the word "probability" in the natural sciences. The first notion is the so-called "objective" or "statistical" one, i.e., the view of probability as something like a "theoretical frequency": if one says that the probability of the event E under condition X, Y, Z equals p, one means that if one reproduces the conditions X, Y, Z sufficiently often, the event E will appear with frequency p.

In the example of coin tossing, this means that, if one throws a "fair" coin in an "unbiased" way sufficiently often, it will land approximately the same number of times heads or tails. Of course, "sufficiently often" is vague and this is the source of much criticism of this notion of probability. But putting that objection aside for a moment and assuming that "sufficiently often" can be given a precise meaning in concrete circumstances, probabilistic statements are, according to this view, factual statements that can be confirmed or refuted by observations or experiments.

By contrast, the "subjective" or Bayesian use of the word probability refers to a form of reasoning, and not (at least not directly) to a factual statement. Used in that sense, assigning a probability to an event expresses a (rational) judgment on the likelihood of that single event, based on the information available at that moment. Note that one is interested here in the probability of a single event, and not in what happens when one reproduces many times the "same" event, as in the objective approach. This is of course very important in practice: when I wonder whether I need to take my umbrella because it might rain, or whether the stock market will crash next week, I am not so much interested in the frequencies with which such events occur as in what will happen here and now. Of course, these frequencies may be part of the information that is used in arriving at a rational judgment, but they are typically not the only information available.

How does one assign subjective probabilities to an event? In elementary textbooks, a probability is defined as the ratio between the number of favorable outcomes and the number of "possible" ones. While the notion of favorable outcomes is easy to define, that of possible outcomes is much harder. Indeed, for a Laplacian demon, nothing is uncertain and the only possible outcome is the actual one; hence, all probabilities are equal to 0 or 1. But *we* are not Laplacian demons[49] and it is here that ignorance enters. We try to reduce things to a series of cases about which we are "equally ignorant", i.e., the information that we do have does not allow us to favor one case over the other, and that defines the number of "possible" outcomes. The standard examples include the throwing of a dice or of a coin, where the counting is easy, but that is often not the case.

At the time of Laplace, this method was called the "principle of indifference". The modern version of that principle, which we will not discuss in detail, is the maximum entropy principle. One starts by identifying a space of states in which the system under consideration can find itself. If there are N possibilities, one assigns to each probability distribution $\vec{p} = (p_i)_{i=1}^N$ its Shannon (or Gibbs) entropy, given by

$$S(\vec{p}) = -\sum_{i=1}^{N} p_i \ln p_i. \tag{3.4.3.1}$$

One then chooses the probability distribution that has the maximum entropy, among those that satisfy certain constraints that incorporate the information that we have about the system. This probability distribution is then updated when new information, i.e., new constraints, become available.[50]

In the example of coin tossing, one might start with a probability of one half for heads and one half for tails (which just reflects our ignorance, and maximizes the entropy). But if the coin lands regularly heads more often than tails, one would suppose that the coin is biased and one would update the probability to take this bias into account, thereby making better predictions for the future.

The rationale, as for the indifference principle, is not to introduce bias in our judgments, namely information that we do not have (like people who believe in lucky numbers). And one can reasonably argue that maximizing the Shannon entropy is

---

[49]This was of course Laplace's main point, as discussed in Sect. 3.4.1.

[50]For a good exposition of Bayesian reasoning, see, e.g., [282, 284].

indeed the best way to formalize that notion. Without giving a complete argument,[51] we note (as one can easily check) that the maximum value of $S(\vec{p})$ is reached when $p_i = 1/N, \forall i = 1, \ldots, N$, and then equals $\ln N$. The minimal value is 0, which is reached for the distribution $p_i = 1$, for some i and $p_j = 0, \forall j \neq i$. So, the most uniform distribution or the most uncertain one has the highest entropy and the one where one event is certain and the others are impossible has the lowest entropy.

Note that probabilistic statements, understood subjectively, are forms of reasoning, although not deductive ones. Therefore, one cannot check them empirically. If someones says: Socrates is an angel; all angels are immortal; therefore Socrates is immortal, this is valid (deductive) reasoning. Likewise, if I say that all I know about a coin is that it has two faces and that it looks symmetric, therefore the probability of "heads" is one half, this is valid probabilistic reasoning. Throwing the coin a thousand times with a result that is always tails would not disprove the reasoning; it would only indicate that the initial assumption (of symmetry) was probably wrong (just as seeing Socrates dead would lead one to reconsider the notion that he is an angel or that the latter are immortal). The main point of Bayesian ideas is to give rules that allow one to update one's probabilistic estimates, given previous observations.

Some people think that a Bayesian or subjective view of probabilities presupposes some form of idealism, meant as a doctrine in philosophy or philosophy of science (discussed in Sect. 3.1). To make matters worse, Bayesians sometimes talk as if all of science was about "information" and never about facts or laws. Moreover, Bayesians often stress the idea that probabilities reflect our ignorance or quantify our ignorance and that makes some physicists uneasy: putting aside parapsychology, our knowledge or our ignorance do not play a causal role in the physical world, so why should they enter in a fundamental way in our physical theories?

But there is no logical connection here: a subjectivist about probabilities may very well claim that there are objective facts in the world and laws governing them, and consider probabilities as being a tool used in situations where our knowledge of those facts and those laws is incomplete. In fact, one could argue that, if there is any connection between Bayesian ideas and philosophical idealism, it goes in the opposite direction; a Bayesian should naturally think that one and only one among the "possible" states is actually realized, and that there is a difference between what really happens in the world and what we know about it. But the philosophical idealist position often starts by *confusing* the world and our knowledge of it (for example, much loose talk about everything being information often ignores the fact that "information" is ultimately information about something which itself is not information). Thus, Bayesians should not be thought of as natural fellow-travellers of philosophical idealists.

Besides, even though ignorance does enter into computations of probabilities, we will see in the next section that this does not mean that either knowledge or ignorance are assumed to play a fundamental role in physics.

A related objection raised against the Bayesian view of probability is: what is the physical meaning when we say there is a probability of one half for a single event?

---

[51] See, e.g., Jaynes [284] for detailed arguments.

Bayesian thinking may be useful in bets or in various practical situations where decisions have to be made, but what role could it have in physical theories, which are supposed to describe the world as it is and not to deal with problems of practical rationality? We will now answer that question.

3.4.4   The Law of Large Numbers and Scientific Explanations

The law of large numbers allows us to give an empirical meaning to Bayesian probability statements: the calculus of probabilities—viewed now as part of deductive reasoning—leads one to ascribe subjective probabilities close to one for certain events that are precisely those that the objective approach deals with, namely the frequencies with which some events occur when we repeat the "same" experiment many times.[52]

To explain the law of large numbers, consider the simple example of coin tossing. Let 0 denote heads and 1, tails. The "space" of results of any single flip, viz., $\{0, 1\}$, will be called the physical space while the space of all possible results of N flips, viz., $\{0, 1\}^N$, will be called the phase space. The variables N_0, N_1 that count the number of heads (0) or tails (1) will be called macroscopic (to use the terminology from statistical mechanics). Here we introduce an essential distinction between the macroscopic variables, or the macrostate, and the microstate. The *microstate*, for N flips, is the sequence of results for all the flips, while the *macrostate* simply specifies the values of N_0 and N_1.

Now fix a small number $\epsilon > 0$ and call a configuration such that $|N_0/N - 1/2| \leq \epsilon$ *typical*, for that given ϵ, and *atypical* otherwise. In other words, a configuration is typical if the coin falls heads (and tails) approximately one-half of the time. Then (a weak form of) the law of large numbers[53] states that $\forall \epsilon > 0$, $\mathcal{P}_N(T(N, \epsilon)) \to 1$ as $N \to \infty$, where $T(N, \epsilon)$ is the set of typical configurations and $\mathcal{P}_N$ the probability distribution that assigns independent probabilities 1/2 to each outcome of each flip. A more intuitive way to say the same thing is that, if we simply count the number of microstates that are typical, we find that they form a fraction of the total number of microstates close to 1, for N large.

This law allows us to make a link between probabilities and scientific explanations. A first form of scientific explanation is given by *laws*. If state A produces deterministically state B, then the occurrence of B can be explained by the occurrence of A. If A is prepared in the laboratory, this kind of explanation is rather satisfactory. Of course, if B is some natural phenomena, then A itself has to be explained, and that leads to a potentially infinite regress, or at least a regress going back to the origin of the universe.

---

[52]Of course, when we say that we repeat the "same" experiment, or that the results of different experiments are "independent" of each other, we also try to quantify the knowledge that we have, i.e., the fact that we do not see any differences or any causal connections between those experiments.

[53]See Appendix 3.A for a proof and more general statements.

However, in many situations, we do not have strict knowledge of those laws or of the initial conditions, e.g., in coin tossing, and thus we have to see what role probabilities play in our notion of explanation. Observe first that, if we toss a coin many times and we find approximately half heads and half tails, we do not feel that there is anything special to be explained. If, however, the result deviates strongly from that average, we'll look for an explanation (e.g., by saying that the coin is biased). This leads to the following suggestion. First, as discussed above, probabilities enter situations where our knowledge is incomplete and Bayesian methods allow us to make the most rational predictions in those situations. Now, suppose we want to explain some phenomenon when our knowledge *of the past* is such that this phenomenon could not have been predicted with certainty.

We suggest that our knowledge, although partial, is *sufficient* to "explain" that phenomenon if we would have predicted it using Bayesian computations and the information we had about the past. That notion of "explanation" incorporates, of course, as a special case, the notion of explanation based on laws. It also fits with our intuition concerning the coin-tossing situation discussed above: being ignorant of any properties of the coin leads us to predict a fraction of heads or tails around one-half. Hence, such a result is not surprising or, in other words, does not "need to be explained", while a deviation from it requires an explanation.

A basically similar form of explanation is used in macroscopic physics, for example, when one wants to account for the second law of thermodynamics, the law of increase of entropy.[54] We do not know all the microstates of, say, a gas; nor do we know their evolution. But we can assign, in a Bayesian way, a probability distribution on microstates, given some information that we have on the initial macrostate of the system, which may be a non-equlilibrium or low entropy state. Since for each microstate the deterministic evolution leads to a well-defined evolution of the macrostate, we can, in principle, compute the future probability, relative to our initial distribution on the microstates, of a given macrostate. If it happens that the one which is predicted on that basis coincides with the one which is observed at a later time, then one can say that the latter has indeed been accounted for by what we knew about the initial macrostate and by the above reasoning. In fact, one can generalize this kind of reasoning to establish a link between deterministic systems and statistical predictions.

3.4.5 "Randomness" and Deterministic Dynamical Systems

It will be important, when we discuss the de Broglie–Bohm theory in Chap. 5, to understand how a deterministic system can give rise to statistical predictions. Obviously, if the system is deterministic, any "randomness" must come from randomness in initial conditions. But what does that mean? Consider a set of states of a system and a deterministic dynamics on that set: each initial condition therefore gives rise to

---

[54]See, e.g., [80, 225, 307] for more details.

a trajectory, i.e., a continuous or discrete family of states evolving in time. One may then introduce a probability distribution on the initial conditions of the system.[55] Then every trajectory of the deterministic dynamics has a probability given by the probability of the initial condition that produces that trajectory. And, at a given time t, the probability of a given state is the probability of the trajectory (or the trajectories) arriving at that state at time t.

But in order for that idea to lead to statistical predictions, we need a large number of identical subsystems that can be considered independent of each other, like many coins or many dice, although we are interested in more "physical" systems, like many electrons, many atoms, many photons, etc. Then, the probability distribution on the initial conditions of those subsystems leads to a probability distribution on their states at a later time (now, for example) and, through the law of large numbers, to definite empirical predictions, that can be compared with results of experiments. This is how a deterministic dynamics may produce statistical predictions that could be interpreted (if we didn't know the laws of the dynamics) as the effect of some intrinsic randomness in the system.

Of course, the extent to which this scheme can be considered an explanation of those apparently random results depends on how natural the probability distribution on the initial conditions of the subsystems is. This question will be important when we discuss the origin of "randomness" in the de Broglie–Bohm theory (in Sect. 5.1.7).

3.4.6 Quantum Mechanics and Determinism

There are two questions that can be asked about the relationship between quantum mechanics, determinism, and probability:

1. To what extent does quantum mechanics prove indeterminism, or to what extent are probabilities in quantum mechanics irreducible, or to what extent are quantum events truly or intrinsically random? (These are reformulations of the same basic idea.)
2. How do we understand the fact that the quantum state, whose meaning is purely probabilistic, nevertheless evolves according to a physical law?

For the first question, as we said above, it is very difficult to prove that a phenomenon is genuinely random, in the sense that one cannot give any deterministic description of it. The mere fact that quantum mechanics is extremely successful and is formulated in probabilistic terms does not in the least rule out a priori the possibility of a deterministic theory underlying the apparent randomness of quantum mechanics. Of course, such a theory would require a more detailed description of physical systems, or, to use the standard terminology, the introduction of some "hidden variables".

---

[55]It may be easier to think of the set of states as being countable. Otherwise, one has to introduce a probability density.

For the second question, the problem is that probabilities, understood in Bayesian terms, do not evolve according to a physical law, but depend on our information. If we throw a coin but don't look at the result after it has fallen on one of its faces, the probabilities of heads and tails will be (1/2, 1/2), but if we look, they will jump to 0 or 1, depending on which is the case, while nothing changes in the state of the coin. But the quantum state has a probabilistic meaning *and* yet evolves, between measurements at least, according to a physical law (the Schrödinger equation).[56]

In ordinary quantum mechanics, there is no clear answer to that question because of the fundamental lack of clarity regarding the status of the quantum state, as we discussed in Sect. 2.5: is it a physical object or a mere reflection of our knowledge? One view leads to the problem of macroscopic superpositions and the other view to the difficulty posed by the no hidden variables theorem.

These questions can only be answered within a more complete theory than ordinary quantum mechanics, and we will answer them within the de Broglie–Bohm theory in Chap. 5. But before doing that, we must first consider another "mystery" of quantum mechanics, far more surprising than anything we have discussed so far.

Appendix

3.A The Law of Large Numbers

Consider first a finite set of real numbers $E = \{a_k\}_{k=1}^n$ and a random variable ω taking values in E, with probability distribution $P(\omega = a_k) = p_k, k = 1, \ldots, n$, with $p_k \geq 0$, $\sum_{k=1}^n p_k = 1$. Then, take N independent variables $\omega_i, i = 1, \ldots, N$, with the same distribution as ω, and define the frequency $\rho_N(\Omega, k)$ with which the variables $\Omega = (\omega_1, \ldots, \omega_N)$ take the value a_k by

$$\rho_N(\Omega, k) = \frac{\left|\{i = 1, \ldots, N | \omega_i = a_k\}\right|}{N} = \frac{\sum_{i=1}^N \delta(\omega_i = a_k)}{N}, \qquad (3.A.1)$$

where $|E|$ is the number of elements of the set E and δ the Kronecker delta. The set of frequencies $\{\rho_N(\Omega, k)\}_{k=1}^n$ is called the *empirical distribution* of the variables $\omega_i, i = 1, \ldots, N$. In the language of Sect. 3.4.4, and of statistical mechanics, the variables $\{\rho_N(\Omega, k)\}_{k=1}^n$ define the *macrostate* of the system, while the variables $\Omega = (\omega_1, \ldots, \omega_N)$ define its *microstate*. In the example of coin tossing in that subsection, we have $n = 2$, $a_1 = 0$, $a_2 = 1$ and $\rho_N(\Omega, 1) = N_0/N$, $\rho_N(\Omega, 2) = N_1/N$.

---

[56]There is a school of thought, called QBism, that tries to give a purely Bayesian interpretation of the quantum state. That will be discussed in Sect. 6.4.

Let $\mathcal{P}_N$ be the joint probability distribution of the variables ω_i, $i = 1, \ldots, N$:

$$\mathcal{P}_N(\omega_1 = a_1, \ldots, \omega_N = a_N) = \prod_{i=1}^{N} p_i. \qquad (3.A.2)$$

Then, the *law of large numbers* says that, $\forall \epsilon > 0$,

$$\lim_{N \to \infty} \mathcal{P}_N(\max_{k=1,\ldots,n} |\rho_N(\Omega, k) - p_k| \geq \epsilon) = 0. \qquad (3.A.3)$$

To prove (3.A.3), let $A_k = \{\Omega | |\rho_N(\Omega, k) - p_k| \geq \epsilon\}$, and write

$$\mathcal{P}_N(\max_{k=1,\ldots,n} |\rho_N(\Omega, k) - p_k| \geq \epsilon) \leq \mathcal{P}_N(\cup_{k=1}^{n} A_k) \leq \sum_{k=1}^{n} \mathcal{P}_N(|\rho_N(\Omega, k) - p_k| \geq \epsilon),$$
$$(3.A.4)$$

then use Chebyshev's inequality to get

$$\sum_{k=1}^{n} \mathcal{P}_N(|\rho_N(\Omega, k) - p_k| \geq \epsilon) \leq n \max_{k=1,\ldots,n} \mathcal{P}_N(|\rho_N(\Omega, k) - p_k| \geq \epsilon)$$

$$\leq n \frac{1}{\epsilon^2} \max_{k=1,\ldots,n} \mathcal{E}_N\left([\rho_N(\Omega, k) - p_k]^2\right), \qquad (3.A.5)$$

where $\mathcal{E}_N$ is the expectation with respect to $\mathcal{P}_N$. Using (3.A.1), we get

$$\mathcal{E}_N\left([\rho_N(\Omega, k) - p_k]^2\right) \leq \frac{1}{N^2} \sum_{i,j=1}^{N} \left[\mathcal{E}_N\left(\delta(\omega_i = a_k)\delta(\omega_j = a_k)\right) - p_k^2\right], \quad (3.A.6)$$

where we used $\mathcal{E}_N\left(\delta(\omega_i = a_k)\right) = p_k$.

Now insert $\mathcal{E}_N\left(\delta(\omega_i = a_k)^2\right) = \mathcal{E}_N\left(\delta(\omega_i = a_k)\right) = p_k$ and $\mathcal{E}_N\left(\delta(\omega_i = a_k)\delta(\omega_j = a_k)\right) = p_k^2$ for $i \neq j$ in (3.A.6). Since $p_k - p_k^2 \leq 1$, we get

$$\mathcal{E}_N\left([\rho_N(\Omega, k) - p_k]^2\right) \leq \frac{1}{N}.$$

Inserting this in (3.A.5), and combining (3.A.5) and (3.A.4), we get (3.A.3).

Thus, the law of large numbers says that, for N large, the empirical distribution $\{\rho_N(\Omega, k)\}_{k=1}^{n}$ of the variables ω_i, $i = 1, \ldots, N$ is almost equal to the numbers $(p_k)_{k=1}^{n}$, with a probability close to 1. This gives a physical meaning to the probabilities $(p_k)_{k=1}^{n}$. For a single event, the probability of an outcome can be any number between 0 and 1, since it reflects partly our knowledge and partly our ignorance. But when we have N copies of the same random event, we can make statements about empirical distributions whose probabilities are close to 1, for N large, and which thus become near certainties.

One can even prove a stronger result (the strong law of large numbers): the probability distribution (3.A.2) can be extended[57] to infinite sequences $\Omega = (\omega_1, \omega_2, \omega_3, \dots)$ and, for that probability distribution, denoted $\mathcal{P}$, there is a set of probability equal to 1 on which $\lim_{N\to\infty} \rho_N(\Omega_N, k) = p_k$, $\forall k = 1, \dots, n$, where Ω_N denotes the first N elements of Ω.

Let us define a sequence of events $\Omega = (\omega_1, \omega_2, \omega_3, \dots)$ to be *typical* if, for every $k = 1, \dots, n$, $\lim_{N\to\infty} \rho_N(\Omega_N, k) = p_k$. Then, the set of typical sequences has $\mathcal{P}$-probability 1.

Consider now a continuous variable $\omega \in \mathbb{R}$ with a probability density $p(x)$: for $B \subset \mathbb{R}$, $P(\omega \in B) = \int_B p(x)dx$. Let $\Omega = (\omega_i)_{i=1}^N$ be N independent variables with identical distribution whose density is $p(x)$. We define the empirical distribution of those N variables, for any (Borel) subset $B \subset \mathbb{R}$:

$$\rho_N(\Omega, B) = \frac{|\{i = 1, \dots, N | \omega_i \in B\}|}{N} = \frac{\sum_{i=1}^N \mathbb{1}_B(\omega_i)}{N}, \tag{3.A.7}$$

where $\mathbb{1}_B$ is the indicator function of the set B.

Let $\mathcal{P}_N$ be the joint probability distribution of the variables ω_i, $i = 1, \dots, N$, whose density in $\mathbb{R}^N$ is $\mathcal{P}_N(x_1, \dots, x_N) = \prod_{i=1}^N p(x_i)$. Then the law of large numbers says that, for any set B, and $\forall \epsilon > 0$,

$$\lim_{N\to\infty} \mathcal{P}_N \left(|\rho_N(\Omega, B) - P_B| \geq \epsilon \right) = 0, \tag{3.A.8}$$

where $P_B = \int_B p(x)dx$. To prove this result, one argues as in (3.A.4)–(3.A.6), using the fact that $\mathcal{E}_N(\mathbb{1}_B^2(\omega_i)) = \mathcal{E}_N(\mathbb{1}_B(\omega_i)) = P_B$ and $\mathcal{E}_N(\mathbb{1}_B(\omega_i)\mathbb{1}_B(\omega_j)) = P_B^2$ for $i \neq j$.

For continuous variables, it is the family of functions $\rho_N(\Omega, B)$ which is the *empirical distribution* of the variables $\Omega = (\omega_i)_{i=1}^N$ and which defines the macrostate of the system, while its microstate is given by $\Omega = (\omega_i)_{i=1}^N$.

One can again strengthen this result by extending, as in the discrete case, the probability distribution $\mathcal{P}_N$ to a probability $\mathcal{P}$ defined on infinite sequences $\Omega = (\omega_1, \omega_2, \omega_3, \dots)$ and then considering the family of sets B of the form $]-\infty, z]$, for $z \in \mathbb{R}$. There is a set of probability $\mathcal{P}$ equal to 1 on which $\lim_{N\to\infty} \rho_N(\Omega_N,]-\infty, z]) = P_{]-\infty,z]}$, $\forall z \in \mathbb{R}$, where Ω_N denotes the first N elements of Ω. The convergence is even uniform in z, by the Glivenko–Cantelli theorem (see, e.g., [490, p. 266]).

Since $P_{]-\infty,z]} = \int_{-\infty}^z p(x)dx$, one can consider $p(x)$ as the *empirical density distribution* of the variables $\Omega = (\omega_1, \omega_2, \omega_3, \dots)$. One can also define a sequence of events $\Omega = (\omega_1, \omega_2, \omega_3, \dots)$ to be *typical* if $\lim_{N\to\infty} \rho_N(\Omega_N,]-\infty, z]) = P_{]-\infty,z]}$, $\forall z \in \mathbb{R}$. Then the set of typical sequences has $\mathcal{P}$-probability 1. So, just as in the

---

[57]One assigns to "cylinder sets" of the form $\omega_1 = a_1, \dots, \omega_N = a_N$, $\omega_i \in E$, $i > N$ the probability (3.A.2) and then it is standard to extend this probability to the sigma-algebra generated by the cylinder sets.

discrete case, the empirical distribution $\rho_N(\Omega, B)$, for B an interval, is, for typical configurations of many variables, close to the probability P_B of a single variable belonging to B.

Chapter 4
The Second Mystery: Nonlocality

4.1 Introduction

As we saw in Chaps. 1 and 2, Einstein held a heterodox and minority view during the debates around the "Copenhagen" interpretation of quantum mechanics. He maintained that the quantum state is not a complete description of physical reality and that particles have properties beyond what is included in their quantum state (the values corresponding to these properties being called "hidden variables"). In other words, he thought that quantum mechanics provides a very accurate statistical description of quantum systems, but not a complete physical description of individual systems.

Einstein gave several indirect arguments in which he attempted to prove that quantum mechanics is incomplete,[1] and in 1935, together with Podolsky and Rosen [164], he produced the following reasoning: suppose there is a quantity I can measure on one system and which, thanks to conservation laws such as conservation of momentum, immediately tells me the value of the corresponding quantity for another system which is far away. Then, if we assume that the measurement on the first system does not affect the physical state of the second system, because of their spatial separation, the second system must have had that property all along. But assuming that this property exists means that there are "hidden variables" that complete the quantum description given by the quantum state.

The Einstein, Podolsky, and Rosen (or EPR) argument was based on a thought experiment, but similar experiments were actually performed later. Contrary to a popular misconception, the EPR argument was not faulty, at least if it is stated as follows: if a certain assumption of locality or of no "action at a distance" is granted, then quantum mechanics is incomplete in the sense introduced here, and there necessarily exist hidden variables[2] that are not specified by the quantum state.

---

[1] We will discuss some of them, as well as the historical controversies and misunderstandings about Einstein, Podolsky, and Rosen in Sect. 7.1.

[2] Such as the value of the momentum in the example given above, or the value of the spin in the examples discussed in Sect. 4.4. Of course, we know, because of the theorem in Sect. 2.5 that it is

© Springer International Publishing Switzerland 2016
J. Bricmont, *Making Sense of Quantum Mechanics*,
DOI 10.1007/978-3-319-25889-8_4

Einstein, Podolsky, and Rosen did not state their result like that, because for them the locality assumption was too obvious to be stated explicitly as an assumption. Moreover, the EPR argument was generally ignored or misrepresented (except mainly by Schrödinger [441]), and most physicists thought that it had been countered by Bohr (although Bohr's argument is far less clear than the EPR paper, as we will show in Sect. 7.1.4).

Almost thirty years later (in 1964), Bell showed that, in the context of the EPR experiment, if we assume the existence of the hidden variables that are necessary if we accept the locality assumption, then we obtain a contradiction with quantum mechanical predictions [35]. And these predictions were later verified experimentally.[3]

Unfortunately, Bell assumed at the beginning of his own reasoning that the EPR argument was well known.[4] But since most people had forgotten or misunderstood the EPR paper, Bell's argument was taken to be just one more proof that quantum mechanics is complete in the sense that no other variables, that would complete the quantum description, can be introduced without contradicting quantum mechanical predictions.

Of course, that was the opposite of what Bell showed, and the opposite of what he explicitly said: the conclusion of his argument, combined with the EPR argument, is rather that there are nonlocal physical effects (and not just correlations between distant events) in Nature.[5]

We will start by a little known, but very simple, thought experiment, known as Einstein's boxes. This example will allow us to raise and explain the issue of

---

(Footnote 2 continued)

not easy to introduce hidden variables. But we will re-examine here the issue of hidden variables from a different perspective to the one in Chap. 2.

[3]Our goal in this book is to present ideas in their simplest form, and we will do that also in this chapter. But there are several caveats to be made: first, while we discuss below particles with spin, actual experiments are made with photons, with their polarization playing the role of spin. Moreover, we will base our arguments in Sect. 4.4 on some perfect correlations predicted by quantum mechanics. But the actual experiments do not test those correlations directly. Instead, they are based on the Clauser–Horne–Shimony–Holt (CHSH) inequality [95]. We refer to [229] for a discussion of various forms of the Bell and CHSH theorems and the relation between them. In any case, the logic of our arguments always takes for granted the empirical correctness of quantum mechanical predictions, which have been verified to great accuracy in all experiments, even if the tests were made in somewhat different circumstances than the ones considered here. See [19–21] for some of the original experiments and [229] for references to later ones. Several possible loopholes are closed in [262].

[4]We note that Bell made no assumption whatsoever about "realism" or "determinism". We mention this here because there is an enormous amount of confusion in the literature based on the supposition that Bell makes such assumptions. The reader unfamiliar with the literature on Bell's theorem may simply ignore this remark. Besides, those confusions will be discussed later, in Sect. 7.5.

[5]For discussions of Bell's results similar to the one in this chapter, see, e.g., [10, 211, 229, 319, 327, 355–358, 459, 460, 478].

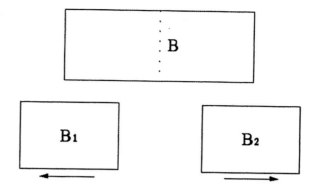

Fig. 4.1 Einstein's boxes. Reproduced with permission from T. Norsen: Einstein's boxes, *American Journal of Physics* **73**, 164–176 (2005). Copyright 2005 American Association of Physics Teachers

locality. Then we will define precisely what we mean by nonlocality and give a simple derivation of Bell's argument (due to [149]), *combined* with the EPR argument. That is the simplest and clearest way to arrive at the main conclusion, namely that the world is nonlocal in a sense made explicit in Sect. 4.3.

4.2 Einstein's Boxes

Consider the following thought experiment.[6] There is a single particle in a box B (see Fig. 4.1), and its quantum state is $|\text{state}\rangle = |B\rangle$, meaning that its quantum state is distributed[7] over B. One cuts the box into two half-boxes, B_1 and B_2, and the two half-boxes are then separated and sent as far apart as one wants.

According to ordinary quantum mechanics, the state becomes

$$\frac{1}{\sqrt{2}}\big(|B_1\rangle + |B_2\rangle\big) \, ,$$

---

[6]We base ourselves in this section on [355], where the description of the experiment is due to de Broglie [118, 119]. Einstein's original idea was expressed in a letter to Schrödinger, written on 19 June 1935, soon after the EPR paper was published [186, p. 35]. However, Einstein formulated his "boxes" argument in terms of macroscopic objects (small balls) and then gave a slightly different and genuinely quantum mechanical example, but illustrating the same point as the one made here. Figure 4.1 is taken from [355]. The "boxes" argument is also mentioned by Deltete and Guy [120] in a discussion of Einstein's objections to the quantum orthodoxy. A somewhat similar experiment to the one with the boxes was done by A. Ádám, L. Jánossy, P. Varga, in 1955 [1]; we will return to it in Sect. 7.1.5.

[7]The precise distribution does not matter, provided it is spread over B_1 and B_2 defined below; it could be distributed according to the square of the ground state wave function of a particle in the box B.

where the state $|B_i\rangle$ means that the particle "is" in box B_i, $i = 1, 2$ (and again, each state B_i is distributed in the corresponding box).[8] Here, we put scare quotes around the verb "is" because of the ambiguity inherent in the meaning of the quantum state[9]: if it reflects our knowledge of the system, then the particle *is* in one of the boxes B_i, without quotation marks. But if one thinks of the quantum state as being physical and of the position of the particle as being created or realized only when someone measures it, then the quotation marks are necessary and "is" means: "would be found in box B_i after measurement".

Now, if one opens one of the boxes (say B_1) and one does *not* find the particle in it, one *knows* that it is in B_2. Therefore, the state "collapses" instantaneously: it becomes $|B_2\rangle$ (and if one opens box B_2, one will find the particle in it!).

Since B_1 and B_2 are spatially separated (and as far apart as we wish), if we reject the notion of action at a distance, then it follows that acting on B_1, namely opening that box, cannot have any physical effect whatsoever on B_2. However, if opening box B_1 leads to the collapse of the quantum state into one where the particle is necessarily in B_2, it must be that the particle was in B_2 all along. That is, of course, the common sense view and also the one that we would reach if the particle was replaced by any large enough object.

But in the situation of the particle in the box, if we reject action at a distance, then we must admit that quantum mechanics is not complete, in the sense that Einstein gave to that word: there exist other variables than the quantum state that describe the system, since the quantum state does not tell us which box the particle is in and we just showed, assuming no action at a distance, that the particle *is* in one of the two boxes, before one opens either of them.

In any case, with this argument, Einstein had already proven the following dilemma: either there exists some action at a distance in Nature (opening box B_1 changes the physical situation in B_2) or quantum mechanics is incomplete. Since action at a distance was anathema to him (and probably to everybody else at that time), he thought that he had shown that quantum mechanics is incomplete.

There are many examples at a macroscopic level that would raise a similar dilemma and where one would side with Einstein in making assumptions, even very unnatural ones, that would preserve locality. Suppose that two people are located far apart, and each tosses coins with results that are always either heads or tails, randomly, but always the same for both throwers. Or suppose that in two casinos, far away

---

[8]This is of course a thought experiment. We assume that we can cut the box in two without affecting the particle. But similar experiments have been made with photons and were already suggested by Heisenberg in 1929 (see Sect. 7.1.5).

[9]If we refer to the four possible positions mentioned in Sect. 2.5 concerning the status of the quantum state, then viewing the quantum state as reflecting our knowledge of the system is part of the third reaction, while thinking of the quantum state as being physical and of the position of the particle as being created when one measures it is part of the second reaction. In Chap. 2, we showed that the third reaction is untenable if it applies to *all* "observables", but not necessarily if it applies only to particle positions (in Chap. 5, we will see how a position can be attributed to particles independently of measurements and in a consistent way). We also showed that the second reaction is untenable because of the linearity of Schrödinger's equation, unless one appeals to a non-physical agent.

from each other, the roulette always ends up on the red or black color, randomly, but always the same in both casinos. Or imagine twins far apart that behave exactly in the same fashion.[10] In all these examples (and in many others that are easy to imagine), one would naturally assume (even if it sounded very surprising) that the two coin throwers or the casino owners were able to manipulate their apparently random results and coordinate them in advance or, for the twins, one would appeal to a strong form of genetic determinism. Who would suppose that one coin tosser immediately affects the result of the other one, far away, or that the spinning of the ball in one casino affects the motion of the other ball, or that the action of one twin affects the behavior of the other twin? In all these cases, one would assume a locality hypothesis; denying it would sound even more surprising than whatever one would have to assume to explain those odd correlations.

But one thing should be a truism, namely that those correlations pose a dilemma: the results are either coordinated in advance or there exists a form of action at a distance. Note, however, that Einstein's assumption in the case of the boxes (incompleteness of quantum mechanics), which is similar to the assumptions we made here about coin throwers, casinos, and twins, was actually very natural.

As an aside, let us mention that the example of the boxes also raises a serious question about the transition from quantum to classical physics. Indeed, if the quantum particle is replaced by a "classical" one, meaning a large enough object, nobody denies that the particle *is* in one of the boxes before one opens either. But where is the dividing line between the quantum realm and the classical one? The transition from quantum to classical physics is usually thought of as some kind of limit, like considering large masses or large energies (compared to the ones on the atomic scale); but a limit is something that one gets closer and closer to when a parameter varies. Here, we are supposed to go from the statement "the particle is in neither of the boxes" to "the particle is in one of them, but we do not know which one". This is an ontological jump and not the sort of continuous change that can be expressed by the notion of limit.

However, Bell's arguments prove the existence of nonlocal effects in Nature, independently of whether one considers quantum mechanics complete or not. Before showing that, we must define more precisely what we mean by nonlocality.[11]

4.3 What Is Nonlocality?

Let us consider what kind of nonlocality or actions at a distance would be necessary, in the example of the boxes, in order to deny Einstein's conclusion about the incompleteness of quantum mechanics. So assume that the particle is in neither box, before

---

[10]This example is given by Bell in his interview with Jeremy Bernstein [56, p. 63].

[11]We should stress (for the experts) that our notion of locality is not the same as what is sometimes called locality or local commutativity in quantum field theory. See [48, 229] for a discussion of this point.

one opens one of them (which is what quantum mechanics being complete means). Then opening one box, say B_1, creates the particle, either in B_1 or in B_2, and consider the latter possibility here. This would nonlocally create a particle in the unopened box B_2, and that action at a distance would have the following properties[12]:

1. *That action should be instantaneous*[13]: the particle has to be entirely in box B_2, at the same time as we open box B_1.
2a. *The action extends arbitrarily far*: the fact that the particle is entirely in box B_2, once we open box B_1, does not change with the distance between the boxes.
2b. *The effect of that action does not decrease with the distance*: the effect is the creation of the particle in box B_2 and that effect is the same irrespective of the distance between the boxes.
3. *This effect is individuated*: suppose we have a thousand boxes, each containing one particle, and that we cut each of them into two half-boxes, then send both half-boxes far apart from each other. Then, opening one half-box will affect the state in the other half-box (coming from the cutting in two of the same box) but not in any other half-box.
4. *That action cannot be used to transmit messages*: if we open box B_1, we learn what the state becomes in box B_2, but we cannot use that to transmit a message from the place where B_1 is to the one where B_2 is. Indeed, if one repeats the experiment many times with several boxes, one obtains that the particles are sometimes in B_1, sometimes in B_2, in an apparently random fashion. Since we have no way, by acting on one box, to choose in which of the two boxes the particle will be, it is impossible to use that experiment to send messages.

The impossibility of sending messages is sometimes taken to mean that there is nothing nonlocal going on. But nonlocality refers here to causal interactions as described (in principle) by physical theories. Messages are far more anthropocentric than that, and require that humans be able to control these interactions in order to communicate. As remarked by Maudlin, the Big Bang and earthquakes cannot be used to send messages, but they have causal effects nevertheless [319, pp. 136–137].

Let us compare this kind of nonlocality with the nonlocality in Newtonian gravity. The latter also allows actions at a distance: since the gravitational force depends on the distribution of matter in the universe, changing that distribution, say by moving my body, instantaneously affects all other bodies in the universe. That action at a distance has properties 1 and 2a, but not the others. Of course, its effect decreases with the distance, because of the inverse square law, and it affects all bodies at a given distance equally (it is not individuated). On the other hand, it can in principle be used to transmit messages: if I decide to choose, at every minute, to wave my arm

---

[12]See Sect. 6.2 for examples of theories, different from the de Broglie–Bohm theory, in which there is indeed such a nonlocal creation of particles.

[13]Of course, instantaneity is not a relativistic notion, so let us say, instantaneous in the reference frame where both boxes are at rest. We will discuss relativity in Sect. 5.2.2 (when we speak of relativity in this book, we always refer to the special theory, unless we mention the general one). For the conflict between relativity and ordinary quantum mechanics in the example of the boxes, see [355, Note 58].

or not to wave it, then one can use that choice of movements to encode a sequence of zeros and ones and, assuming that the gravitational effect can be detected, one can therefore transmit a message instantaneously and arbitrarily far (but the further away one tries to transmit it, the harder the detection). Of course, all this refers to Newton's *theory*. There have been no experiments performed in this framework that could *prove* that gravitational forces really act instantaneously or at least at speeds faster than the speed of light (and, as we shall see, this is a major difference with the situation in quantum mechanics).

It is well known that Newton did not like this aspect of his own theory.[14] In a letter to the theologian Richard Bentley, he wrote:

> […] that one body may act upon another at a distance through a vacuum without the mediation of any thing else […] is to me so great an absurdity that I believe no man who has in philosophical matters any competent faculty of thinking can ever fall into it.
>
> Isaac Newton [330, 350]

Not surprisingly, Einstein also firmly rejected the idea of action at a distance. In his discussions with Max Born, he wrote:

> When a system in physics extends over the parts of space A *and* B, then that which exists in B should somehow exist independently of that which exists in A. That which really exists in B should therefore not depend on what kind of measurement is carried out in part of space A; it should also be independent of whether or not any measurement at all is carried out in space A.
>
> Albert Einstein [79, p. 164]

In the example of the boxes, this means that opening one half-box cannot possibly influence the physical situation in the other half-box.

The Dutch physicist Hendrik Casimir underlined the fundamental problem with nonlocal actions:

> If the results of experiments on free fall here in Amsterdam depended appreciably on the temperature of Mont Blanc, on the height of the Seine below Paris, and on the position of the planets, one would not get very far.
>
> Hendrik Casimir [93], quoted in [48]

If everything is connected with everything through nonlocal actions, then science becomes impossible, because, in order to test scientific theories, one always needs to assume that one can isolate some systems or some variables. However, luckily, the nonlocality in quantum mechanics does not go so far as to make isolated systems impossible to realize in practice.[15]

---

[14]Newton thought that gravitation was mediated by particles moving at a finite speed, so that the effect of gravitation could not be instantaneous. See [327] for more details.

[15]This is related to the phenomenon of decoherence, to be discussed in Sect. 5.1.6.

However, because of the problems linked with nonlocality, post-Newtonian physics has tried to eliminate property 1, while classical electromagnetism or the general theory of relativity have kept only property 2a and the negation of 4. And, due to special relativity, the combination of 1 and the negation of 4 allows in principle the sending of messages into one's own past,[16] so that, if 1 holds, 4 must also hold.

One may ask whether quantum mechanics proves that there are physical effects displaying properties 1–4. The example of Einstein's boxes does not allow that conclusion, because one can consistently think that the quantum description is not complete and that the particle is always in one of the boxes. Indeed, that is exactly what happens in the de Broglie–Bohm theory (explained in Chap. 5). In order to prove nonlocality in the sense introduced here, i.e., a phenomenon having properties 1–4 above, we have to turn to a more sophisticated situation.

4.4 A Simple Proof of Nonlocality

4.4.1 An Anthropomorphic Thought Experiment

Let us start with an anthropomorphic thought experiment, but which is completely analogous to what happens in real experiments and could even, in principle, be realized in the anthropomorphic form presented here.[17] Two people, A (for Alice) and B (for Bob) are together in the middle of a room and go towards two different doors, located at X and Y. At the doors, each of them is given a number, 1, 2, 3 (let's call them "questions", although they do not have any particular meaning) and has to say "Yes" or "No" (the reason why we introduce 3 questions will be clear in the next section). This experiment is repeated many times, with A and B meeting together each time in the middle of the room, and the questions and answers vary apparently at random (Fig. 4.2). When A and B are together in the room, they can decide to follow whatever "strategy" they want in order to answer the questions, but the statistics of their answers, although they look random, must nevertheless satisfy two properties. We impose these properties because they are analogous to what happens in real experiments with photons, but translating the experiments into an anthropomorphic language may help us see how paradoxical our final conclusions are.

The *first property* is that, when the same question is asked at X and Y, one always gets the same answer. How can that be realized? One obvious possibility is that A and B agree upon which answers they will give before moving toward the doors. They may decide, for example, that they will both say "Yes" if the question is 1, "No"

---

[16]At least, according to the usual understanding of relativity. We will return to this question in Sects. 5.2.1 and 5.2.2. See [48, 319] for a more detailed discussion.

[17]Readers who prefer to see the real physical situation directly can proceed to the next section.

Fig. 4.2 The
anthropomorphic experiment

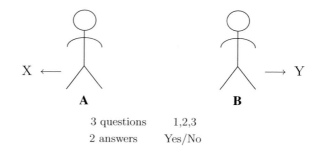

$$X \longleftarrow$$

A   **B**

3 questions   1,2,3
2 answers   Yes/No

$$\longrightarrow Y$$

if it is 2 and "Yes" if it is 3. They can choose different strategies at each repetition of the experiment and choose those strategies "at random" so that the answers will look random.

Another possibility is that, when A reaches door X, she calls B and tells him which question was asked and the answer she gave. Then, of course, B can just give the same answer as A if he is asked the same question and any other answer if the question is different.

But let us assume that the answers are given simultaneously,[18] so that the second possibility is ruled out unless there exists some superluminal action at a distance between A at X and B at Y. Maybe A and B communicate by telepathy! Of course, this is not to be taken seriously, but that is the sort of interaction that Einstein had in mind when he spoke of "spooky actions at a distance" [79, p. 158].

The question that the reader should ask at this point is whether there is *any other possibility*: either the answers are predetermined or (assuming simultaneity of the answers) there is a "spooky action at a distance", namely a communication of some sort takes place between A and B *when* they are asked the questions.[19] This is similar to the dilemma about the boxes: either the particle is in one of the boxes or there is some physical action between the two boxes.

Note that, to raise this dilemma, one question suffices instead of three: if the answers on both sides are always the same, then they must be predetermined if no

---

[18] In the reference frame in which the experiment takes place.

[19] Someone who certainly thinks that the answers are predetermined is Robert Griffiths, who offers the following analogy to illustrate the situation described here:

> Colored slips of paper, one red and one green, are placed in two opaque envelopes, which are then mailed to scientists in Atlanta and Boston. The scientist who opens the envelope in Atlanta and finds a red slip of paper can immediately infer, given the experimental protocol, the color of the slip of paper contained in the envelope in Boston, whether or not it has already been opened.

Robert B. Griffiths [249]

We will discuss Griffiths' views about quantum mechanics in Sect. 6.3.

communication is possible between the two sides.[20] This dilemma can be called the EPR part of our argument, although Einstein, Podolsky, and Rosen used variables taking continuous values (position and momentum) instead of discrete ones as here (Yes/No). The reason we need three possible questions is that there is a *second property* of the statistics of the answers: when the two questions addressed to A and B are *different*, then the answers must be the same in only a quarter of the cases. And this property, combined with the idea that the properties are predetermined, leads to a contradiction. The whole argument provides a (very simple) version of *Bell's theorem*.[21]

Theorem (Bell) *We cannot have these two properties together:*

1 *The answers are determined before the questions are asked and are the same on both sides.*
2 *The frequency of having the same answers on both sides when the questions are different is 1/4.*

Proof. There are three questions numbered 1, 2, and 3, and two answers Yes and No. If the answers are given in advance, there are $2^3 = 8$ possibilities:

1	2	3
Y	Y	Y
Y	Y	N
Y	N	Y
Y	N	N
N	Y	Y
N	Y	N
N	N	Y
N	N	N

In *each case* there are at least *two questions* with the same answer. Therefore,

---

[20]One of the weaknesses of the original EPR paper was that they considered two quantities, position and momentum, instead of one, which would have been sufficient for their argument to work: if one can predict the momentum of particle A by measuring that of particle B, far away from A, and if that measurement does not affect particle A (by assumption of locality), then particle A must have had a well defined momentum all along. The same argument holds for the position: if the two particles have opposite momenta and start from the same place, then measuring the position of one particle allows us to infer the position of the other and therefore, assuming once again no effect on A from the measurement on B, this means that particle A had a position all along. But by considering both position and momentum, Einstein, Podolsky, and Rosen may have given the impression that they were trying to prove that one could *measure* these two quantities simultaneously, which was not their point, at least not Einstein's point. We will discuss this further in Sect. 7.1.

[21]The argument given here is taken from [149]. In the original paper by Bell [35], the proof, although fundamentally the same, was more complicated. There has also been quite some discussion in the literature about the difference between *outcome independence* and *parameter independence*, but our approach bypasses that distinction (see [229, 319] for a critical discussion of these notions).

Frequency (answer to 1 = answer to 2)

    + Frequency (answer to 2 = answer to 3)

       + Frequency (answer to 3 = answer to 1) ≥ 1 .

But if

Frequency (answer to 1 = answer to 2)

    = Frequency (answer to 2 = answer to 3)

       = Frequency (answer to 3 = answer to 1) $= 1/4$,

we get $3/4 \geq 1$, which is a contradiction.   ■

The inequality above, with the sum of the frequencies greater than or equal to 1, is an example of a *Bell inequality*, i.e., an inequality which is a logical consequence of the assumption of pre-existing values, and which is violated by quantum predictions.[22] But before drawing conclusions from this theorem, let us see how the two people described here could realize these "impossible" statistics.

4.4.2   A Real Quantum Situation

Let us first describe the situation in the previous section in a non-anthropomorphic manner.[23] A and B are replaced by particles with spin $1/2$, and there are pieces of apparatus at X and Y that "measure the spin" along some direction, these being similar to the boxes discussed in Chap. 2. The "questions" 1, 2, and 3 are three possible directions for that "measurement" (we put scare quotes here because, as we shall see, no pre-existing value is measured in these experiments). The answers Yes/No correspond to results Up/Down for the spin (we will define this correspondence more precisely below).

The particles are sent towards the pieces of apparatus and the initial quantum state of the two particles is

$$|\Psi\rangle = \frac{1}{\sqrt{2}}\left(|A\ 1\ \uparrow\rangle|B\ 1\ \downarrow\rangle - |A\ 1\ \downarrow\rangle|B\ 1\ \uparrow\rangle\right)$$

$$= \frac{1}{\sqrt{2}}\left(|A\ 2\ \uparrow\rangle|B\ 2\ \downarrow\rangle - |A\ 2\ \downarrow\rangle|B\ 2\ \uparrow\rangle\right)$$

$$= \frac{1}{\sqrt{2}}\left(|A\ 3\ \uparrow\rangle|B\ 3\ \downarrow\rangle - |A\ 3\ \downarrow\rangle|B\ 3\ \uparrow\rangle\right) , \quad (4.4.2.1)$$

---

[22]In [332], Mermin provides nice idealized illustrations of possible results of experiments showing the violation of Bell's inequality, in the version given here.

[23]This formulation of the EPR argument, with spin variables instead of position and momentum, is due to Bohm [61] and was used by Bell in [35] and later. However, there exists an unpublished note by Einstein discussing the problem in terms of spin variables [432].

where we use a similar notation to the one in Chap. 2, so that $|A\ 1\ \uparrow\rangle$ is the state in which particle A has its spin up in direction 1 (meaning that a particle in that state will have its spin up with certainty after a spin measurement in direction 1, or in other words, this state is the "up" eigenstate of the spin operator in direction 1),[24] and the other symbols are defined analogously. We leave aside the issue of how to create such a state in practice and note only that it can be done. We also accept without proof the fact that this state has three similar representations[25] in each of the directions 1, 2, or 3. We also leave aside the "spatial" part of that quantum state: we assume implicitly that the state $|A\ 1\ \uparrow\rangle$ is coupled to a wave function moving towards the apparatus located at X, while the state $|B\ 1\ \downarrow\rangle$ is coupled to a wave function moving towards the apparatus located at Y, and similarly for the other parts of the state $|\Psi\rangle$.

Consider now the standard quantum mechanical description of a measurement of the spin of A at X in direction 1, without measuring anything on particle B for the moment. If one sees $\uparrow$, the state becomes $|A\ 1\ \uparrow\rangle|B\ 1\ \downarrow\rangle$ (by the collapse rule). If one sees $\downarrow$, the state becomes $|A\ 1\ \downarrow\rangle|B\ 1\ \uparrow\rangle$. And, of course, we get similar results if we measure the spin of A in directions 2 or 3.

But then the state changes *nonlocally* for particle B, because, if one sees $\uparrow$, the state "collapses", i.e., becomes $|A\ 1\ \uparrow\rangle|B\ 1\ \downarrow\rangle$, while the part $|A\ 1\ \downarrow\rangle|B\ 1\ \uparrow\rangle$ of the state is suppressed by the collapse. Another way to say this is that, after the measurement of A at X, any measurement of B at Y is guaranteed to yield the opposite result to what was found for A, while, before the measurement of A, the result for B was undetermined, since in the state (4.4.2.1), the result of measuring the spin at B could be either up or down. This gives rise to the same dilemma as for Einstein's boxes: either the measurement on A affects the physical situation of B, or the particle B had its spin determined in advance, and anticorrelated with the spin of A. Since we could measure the spin of B first and then the spin of A, the same reasoning implies that the spin at A is also predetermined. And this argument works for each of the three directions. One is therefore led to the following dilemma:

- either the spin values, up or down, were predetermined before the measurement, and in all three directions, because the same reasoning applies to each one of them,
- or there is some form of action at a distance between X and Y, even if X and Y are located arbitrarily far apart.

But the first assumption leads to a contradiction with observations made when the directions in which the spin is "measured" are *different* for A and B. To see this, denote by $v_A(\mathbf{a})$, $v_B(\mathbf{a})$ the pre-existing values of the results of measurements on A and B (assuming that they exist), where $\mathbf{a}$ denotes a unit vector in one of the directions 1, 2, 3 (to be specified later) in which the spin is measured at X or Y. We make the following conventions at X and Y : $v_A(\mathbf{a}) = +1$ means that the answer is "Yes" and $v_A(\mathbf{a}) = -1$ means that the answer is "No", but $v_B(\mathbf{a}) = +1$ means that the answer is "No" and $v_B(\mathbf{a}) = -1$ means that the answer is "Yes". With that convention, we

---

[24]See also Appendix 2.B for more details.

[25]This follows from the rotational invariance of the state (4.4.2.1) in spin space.

see that we always get the same answer when the same questions are asked on both sides. The contradiction comes from the fact that we get the same answer only 1/4 of the time when we ask different questions at X and Y, and the theorem of the previous section shows that this is impossible. We explain in Appendix 4.A how to obtain the number 1/4 for the frequency of answers to different questions, for an appropriate choice of the directions 1, 2, 3 (this is just a standard quantum mechanical calculation).

Finally, if we wish to reproduce the anthropomorphic experiment described in the previous section, we simply send the particles toward the two doors, once Alice (A) and Bob (B) have reached them. At each door, there is an instrument with which Alice and Bob "measure the spin" along the directions corresponding to the numbers given to them. They then answer the "questions" according to the results of their "measurements". In that way the "impossible" statistics mentioned in the theorem of the previous section can also be reproduced by human beings.

Of course, one has to run the experiment many times in order to get the "impossible" statistics (impossible without accepting the existence of action at a distance). But there are variants of Bell's argument (with three particles instead of two) that do not require any statistics: assuming locality implies the existence of "hidden" spin values and this leads directly to a contradiction.[26]

Before drawing conclusions from what has been proven, let us stress that the nonlocality proven by Bell does indeed have the properties 1–4 discussed in Sect. 4.3. The effect is in principle instantaneous, but one cannot check instantaneity experimentally. However, it can at least propagate at speeds far greater than the speed of light, something that can be checked experimentally. The effect does not decrease with the distance between X and Y and is individuated, since it depends on a pair of particles being in the state (4.4.2.1): acting on one particle in the pair will affect the other particle in that pair, but no other particle. Finally, this effect cannot be used to send messages from X to Y. The reason for this impossibility is similar to the one applying in the case of Einstein's boxes. Each side sees a perfectly random sequence of results "spin up" or "spin down". Since there is no mechanism that allows, given the initial quantum state (4.4.2.1), to control or affect that result by acting on one side of the experiment, there is no way to send a message from one side to the other.[27]

4.4.3   Conclusions

Taken by itself and forgetting about the EPR argument, Bell's result can be stated as a "no hidden variables theorem", similar to the one in Sect. 2.5. Indeed, Bell showed that the mere supposition that the values of the spin pre-exist their "measurement",

---

[26]This is somewhat similar, but not identical, to the contradiction derived in Appendix 2.F. See [240], [335, Sect. VIII], or [70, p. 146], for more details.

[27]For a general proof of the impossibility of using EPR-type experiments to send messages, see [37, 158, 208], [70, p. 139], or [319, Chap. 4].

combined with the perfect anticorrelation when the axes along which measurements are made are the same, and the 1/4 result for the frequencies of correlations when measurements are made along different axes, leads to a contradiction. Since the last two claims are empirical predictions of quantum mechanics that have been amply verified (in a somewhat different form), this means that these hidden variables or pre-existing values cannot exist.

But Bell, of course, always presented his result *in combination with* the EPR argument, which shows that the mere assumption of locality, combined with the perfect correlation when the directions of measurement (or questions) are the same, implies the existence of those hidden variables that are "impossible". So for Bell, his result, combined with the EPR argument, was not a "no hidden variables theorem", but a nonlocality theorem, the result about the impossibility of hidden variables being only one step in a two-step argument.

To repeat, the EPR part of the argument shows that, if there are no pre-existing values, then the perfect correlations when the directions are the same imply some action at a distance. The Bell part of the argument, i.e., the theorem of the previous section, shows that the mere assumption that there are pre-existing values leads to a contradiction when one takes into account the statistics of the results when the directions are different. That is why we used scare quotes on the word "measurement": there is no real property of the particle that is being "measured", since no spin values exist before the interaction with the "measuring" device (why this is so will become clearer in Sect. 5.1.4).

But what does this mean? In fact, it means that some action at a distance does exist in Nature, but it does not tell us what this action consists of. And we cannot answer that question without having a theory that goes beyond ordinary quantum mechanics. In ordinary quantum mechanics, what is nonlocal is the collapse of the quantum state, as we see in the transformation of (4.4.2.1) into $|A \ 1 \ \uparrow\rangle|B \ 1 \ \downarrow\rangle$ or $|A \ 1 \ \downarrow\rangle|B \ 1 \ \uparrow\rangle$, depending on the result of a measurement at A. This affects the state at B, since the second part of $|\Psi\rangle$ has now been suppressed.

Since the meaning of the quantum state and its collapse is ambiguous in ordinary quantum mechanics, it is not clear that this is a real physical effect. But, as we have emphasized, if there is no physical effect whatsoever or, if one interprets the collapse of the quantum state as a mere gain of information, then this means that we must have those predetermined values that lead to a contradiction. In Sect. 5.2.1, we will explain what nonlocality means in a concrete theory going beyond ordinary quantum mechanics, namely de Broglie–Bohm's theory.

Given the radical nature of the conclusions of the EPR–Bell argument, many attempts have been made to avoid them, i.e., to claim that the world is local after all. One such strategy is to maintain that the perfect correlations between the answers when the same questions are asked is simply a coincidence that does not need to be explained (see for example [187, 492]). In the same vein, one sometimes claims that science limits itself to predictions of empirical correlations, not to explanations.

But the whole of science can be seen as an attempt to account for correlations or empirical regularities: the theory of gravitation, for example, accounts for the regularities in the motion of planets, moons, satellites, etc. The atomic theory of matter accounts for the proportions of elements in chemical reactions. The effects of medicines account for the cure of diseases, etc. To refuse to account for correlations, without giving any particular reason for doing so, is in general a very unscientific attitude. As Bell puts it:

> You might shrug your shoulders and say 'coincidences happen all the time', or 'that's life'. Such an attitude is indeed sometimes advocated by otherwise serious people in the context of quantum philosophy. But outside that peculiar context, such an attitude would be dismissed as unscientific. The scientific attitude is that correlations cry out for explanation.

John Bell [49, p. 152]

Another variant of the "shrugging one's shoulders" argument, is to invoke a sort of "conspiracy": for example, that each person has an answer to only one question but that, each time, and no matter how many times the experiment is repeated, that happens to be the question that is being asked to him or her. If we make that assumption, then our theorem cannot be derived (for the proof of the theorem to work, we need to assume pre-existing answers for three questions).[28]

That move can be considered as an instance of the Duhem–Quine thesis; no matter what the data are, one can always save one's favorite theory (here it would be rejection of nonlocality) if one is willing to make sufficiently ad hoc assumptions. But, again, "outside that peculiar context, such an attitude would be dismissed as unscientific". As the authors of [229] observe, "if you are performing a drug versus placebo clinical trial, then you have to select some group of patients to get the drug and some group of patients to get the placebo." But for that to work, you have to assume "that the method of selection is independent of whatever characteristics those patients might have that might influence how they react to the drug" [229, Note 17]. If, by accident, the people to whom the placebo is given were exactly those that are cured spontaneously, while those to whom the drug is given are so sick that the drug has little effect on them, then of course the study would be biased. And no matter how "random" the chosen sample is, this will always remain a logical possibility.

In any case, refusing to face a problem is not the same thing as solving it.[29] In Chap. 6, we will see other attempts to "save locality", in certain interpretations of

---

[28]Another suggestion sometimes made is that the ordinary rules of probability do not apply in the EPR–Bell situation. But, as pointed out by Tumulka [478], since the reasoning here relies only on frequencies of results of experiments, and since the latter obviously do satisfy the ordinary rules of probability, this attempt to "save locality", by trying to deny the implications of the EPR–Bell argument, does not work.

[29]Another reaction which, in a sense, also avoids the problem, is the one proposed by Gisin in [214]: he does emphasize that Nature is nonlocal, but he attributes the nonlocal correlations to "pure chance" and thinks that this proves the non-deterministic nature of the Universe. But since there exists a nonlocal deterministic theory, the de Broglie–Bohm theory, that accounts for these nonlocal effects, one cannot use them to prove that a deterministic account is impossible (see Chap. 5 and [313].). One may "like" the de Broglie–Bohm theory or not, but one cannot deny its existence.

quantum mechanics. But these attempts also consist in refusing to face the dilemma that the perfect correlations pose. One thing is certain: nobody has yet proposed a local explanation for those perfect correlations, and indeed nobody could do so, since Bell has shown that it is impossible.

It is, however, important to realize that one cannot use this nonlocal effect to send messages, as we explained at the end of the last section. This contradicts all the pseudo-scientific uses of Bell's result: there is no telepathy of any sort that can be based on that result.

But if Alice and Bob tell each other which "measurements" have been made (1, 2, or 3), *without* telling the results, they both know which result has been obtained on the other side when the same measurement is made on both sides. Then they both share a common sequence of Yes/No or up/down, which is a form of "information". This information was not transmitted explicitly, by hypothesis, and it cannot come from the source (the one that has emitted the two particles), because of the non-existence of the pre-existing spin values. Thus, some nonlocal transfer of information must have taken place.[30]

Once Alice and Bob share a common sequence of Yes/No, known only to themselves, they can use that to encrypt messages. These will be transmitted at subluminal speeds, but they will be undecipherable by a third party.[31] This is one of the foundations of the field of "quantum information",[32] and in particular of "quantum cryptography", whose development will hopefully will lead to a better appreciation of the radical consequences of Bell's discovery (although, as we will see in Chap. 7, this appreciation is far from being realized at the present time).

Finally, one should emphasize that Einstein's speculations, which looked purely philosophical or even "metaphysical" to many,[33] have led to what is probably "the most profound discovery of science", to use Henri Stapp's apt phrase [458, p. 271], namely the existence of nonlocal effects in the world. And the EPR and Bell papers laid the foundation for the quantum information revolution. This should be a lesson for "pragmatists".

On the other hand, it is ironical that this result refutes Einstein's most basic assumption about Nature. As Bell said, the question raised by EPR was answered in a way that Einstein would probably "have liked least", by showing that the "obvious" assumption of locality made by EPR is actually false [36, p. 11].

---

[30] See the book on relativity and nonlocality by Maudlin [319] for a detailed discussion of the differences between messages and information and of what exactly is compatible or not with relativity.

[31] One can show that if a spy tries to intercept the particles being transmitted from the source to Alice and Bob, in order to know which results Alice and Bob will get when they do their measurements, then because of the collapse rule, that interception will necessarily have effects on those results such that Alice and Bob can detect the presence of the spy. In that sense, quantum cryptography is foolproof.

[32] See, e.g., Nielsen and Chuang [352] or Preskill [404].

[33] See, for example, the quote by Pauli in Sect. 1.5.

Appendix

4.A The Frequency of "Answers" to Different "Questions"

Here we derive the number 1/4 for the frequency of answers to different questions, for an appropriate choice of the directions 1, 2, 3. Compute first $\mathbf{E_{a,b}} \equiv \langle \Psi | \sigma_{\mathbf{a}}^A \otimes \sigma_{\mathbf{b}}^B | \Psi \rangle$, where $\mathbf{a}, \mathbf{b}$ are unit vectors in the directions (1, 2, or 3, specified below) in which the spin is measured at A or B, and $\sigma_{\mathbf{a}}^A \otimes \sigma_{\mathbf{b}}^B$ is a tensor product of matrices, each one acting on the A or B part of the quantum state (with $\sigma_{\mathbf{a}} = a_1 \sigma_1 + a_2 \sigma_2 + a_3 \sigma_3$, where, for $i = 1, 2, 3$, a_i are the components of $\mathbf{a}$ and σ_i the usual Pauli matrices introduced in Appendix 2.F). The quantity $\mathbf{E_{a,b}}$ is bilinear in $\mathbf{a}, \mathbf{b}$ and rotation invariant, so it must be of the form $\lambda \mathbf{a} \cdot \mathbf{b}$, for some $\lambda \in \mathbf{R}$.

For $\mathbf{a} = \mathbf{b}$, the result must be -1, because of the anti-correlations (if the spin is up at A, it must be down at B and vice versa). So $\lambda = -1$, and thus $\mathbf{E_{a,b}} = -\cos\theta$, where θ is the angle between the directions $\mathbf{a}$ and $\mathbf{b}$. We know that $v_A(\mathbf{a}), v_B(\mathbf{b}) = \pm 1$. Thus,

$$\mathbf{E_{a,b}} = P\big(v_A(\mathbf{a}) = v_B(\mathbf{b})\big) - P\big(v_A(\mathbf{a}) = -v_B(\mathbf{b})\big) = 1 - 2P\big(v_A(\mathbf{a}) = -v_B(\mathbf{b})\big),$$

and

$$P\big(v_A(\mathbf{a}) = -v_B(\mathbf{b})\big) = \frac{1 - \mathbf{E_{a,b}}}{2} = \frac{1 + \cos\theta}{2}.$$

One then chooses the directions

$$1 \longleftrightarrow 0 \text{ degree},$$

$$2 \longleftrightarrow 120 \text{ degree},$$

$$3 \longleftrightarrow 240 \text{ degree}.$$

Since $\cos 120 = \cos 240 = -1/2$, we get $P\big(v_A(\mathbf{a}) = -v_B(\mathbf{b})\big) = 1/4$. Thus we have perfect anticorrelations only 1/4 of the time when $\mathbf{a}$ and $\mathbf{b}$ are different. With our convention, this means that one gets the same answer when one asks different questions on both sides only 1/4 of the time.

Chapter 5
The de Broglie–Bohm Theory

In this chapter, we will outline a theory of "hidden variables" (although they are not really hidden) that accounts for all the phenomena predicted by ordinary (non-relativistic)[1] quantum mechanics, is not contradicted by the no hidden variables theorems, explains why measurements do not in general measure pre-existing properties of a system (in other words, it explains why measuring devices have an "active role"), and incorporates and to some extent explains the nonlocality implied by Bell's theorem. It would seem that, given all the claims to the effect that such a theory is impossible,[2] its mere existence should be a subject of considerable interest, but this is not the case. Although interest in the de Broglie–Bohm theory is probably increasing, it is still widely ignored or misrepresented, even by experts on foundations of quantum mechanics.

The theory was introduced at approximately the same time as the Copenhagen interpretation, in 1927, by Louis de Broglie, but it was rejected at the time by a large majority of physicists, and ignored even by critics of the Copenhagen school, like Einstein and Schrödinger. The theory was even abandoned by its founder, only to be rediscovered and completed by David Bohm in 1952, then further developed and advertised by John Bell. Our presentation will very much rely on the work of Detlef Dürr, Sheldon Goldstein, and Nino Zhanghì, and several of their collaborators [152, 153].

Since many expositions of the de Broglie–Bohm theory are now available,[3] we will simply explain the main points, relying often on simple examples. Once those

---

[1] In this chapter, we will discuss only the non-relativistic de Broglie–Bohm theory. The issue of relativity (and of quantum fields) will be addressed in Sect. 5.2.2. By "relativity", we mean the special theory of relativity, unless the general one is mentioned.

[2] We saw such claims made in Chap. 1 and we will discuss them again in Chap. 7.

[3] See, e.g., [8, 9, 474] for elementary introductions and [24, 55, 70, 152, 153, 231, 268, 359, 473] for more advanced ones. There are also pedagogical videos made by students in Munich, available at: https://cast.itunes.uni-muenchen.de/vod/playlists/URqb5J7RBr.html.

© Springer International Publishing Switzerland 2016
J. Bricmont, *Making Sense of Quantum Mechanics*,
DOI 10.1007/978-3-319-25889-8_5

are understood, the reader may look at the advanced literature for a more detailed understanding. Moreover, we will use mostly words and pictures in this chapter, leaving the mathematics to the appendices (except in the next section).[4]

5.1   The Theory

5.1.1   The Equations of the Theory

The complete state of a system with N variables[5] at time t is specified by $\big(\Psi(t), \mathbf{X}(t)\big)$, where $\Psi(t)$ is the usual quantum state, $\Psi(t) = \Psi(x_1,\dots,x_N,t)$ and $\mathbf{X}(t) = \big(X_1(t),\dots,X_N(t)\big)$ are the actual positions of the particles.[6] We use capital letters for the positions of the particles and lower case for the arguments of $\Psi(x_1,\dots,x_N,t)$. We use boldface letters $\mathbf{X}$ or $\mathbf{x}$ to denote vectors.

The theory assumes that the particles have positions at all times, and therefore trajectories, independently of whether one measures them or not. These are, by convention, what are called the "hidden variables" of the theory, because they are not included in the purely quantum description, given by Ψ. But here, the word "hidden" is silly since the positions are not hidden at all: they are the only things that are directly "seen". For example, in the double-slit experiment, one detects particle positions on the screen in Fig. 2.8. Actually, as we shall explain below, they are also the only things that are directly "seen" in *any* experiment.

The state $(\Psi, \mathbf{X})$ is a function of time and its evolution is given by two laws[7]:

1. Ψ obeys the Schrödinger equation at all times[8]:

$$i\frac{\partial}{\partial t}\Psi(x_1,\dots,x_N,t) = H\Psi(x_1,\dots,x_N,t) , \qquad (5.1.1.1)$$

with

$$H = -\sum_{j=1}^{N}\frac{1}{2m_j}\frac{\partial^2}{\partial x_j^2} + V(x_1,\dots,x_N) , \qquad (5.1.1.2)$$

---

[4]In Sects. 5.1.3, 5.1.4, and 5.2.1, we will use the presentation of the de Broglie–Bohm theory in Chap. 7 of David Albert's book *Quantum Mechanics and Experience* [8].

[5]We let N be the number of variables associated with the system. So if the system consists of M particles moving in three-dimensional space, we have $N = 3M$.

[6]We will call below the pair $\big(\Psi(t), \mathbf{X}(t)\big)$ the state of the system, to be distinguished from the quantum state $\Psi(t)$, which is only part of it.

[7]The equations here are (partial) differential equations, similar to those discussed in Appendix 2.A. The reader unfamiliar with such equations can skip them and proceed to the remarks. The consequences of these equations will be discussed and illustrated later.

[8]In this chapter, we set $\hbar = 1$. See Appendix 2.A for more details on Schrödinger's equation.

where V is the classical potential and m_j the mass of the particle of index j. We emphasize that $\Psi(x_1, \ldots, x_N, t)$ always satisfies (5.1.1.1) *whether one "measures" something or not.* The quantum state is never reduced.

2. The evolution of the positions of the particles is determined by the quantum state at time t. One writes $\Psi(x_1, \ldots, x_N, t) = R(x_1, \ldots, x_N, t) \exp i S(x_1, \ldots, x_N, t)$, and the dynamics is simply

$$\frac{d}{dt} X_k(t) = \frac{1}{m_k} \frac{\partial}{\partial x_k} S\big(X_1(t), \ldots, X_N(t), t\big) , \qquad (5.1.1.3)$$

$\forall k = 1, \ldots, N$. This can also be written as

$$\frac{d}{dt} X_k(t) = V_\Psi^k \big(X_1(t), \ldots, X_N(t), t\big) = \frac{1}{m_k} \frac{\operatorname{Im}(\Psi^* \cdot \partial \Psi / \partial x_k)}{\Psi^* \cdot \Psi} \big(X_1(t), \ldots, X_N(t), t\big) ,$$

$$(5.1.1.4)$$

$\forall k = 1, \ldots, N$. This latter version can be generalized to particles with spin,[9] by letting the dots in $\Psi^* \cdot \Psi$ and $\Psi^* \cdot \partial \Psi / \partial x_k$ stand for the scalar product between the spin components.[10]

When Bohm derived his version of the theory [62], he chose to write the dynamics as a second order equation:

$$m_k \frac{d^2}{dt^2} X_k(t) = -\frac{\partial}{\partial x_k} \Big[V\big(X_1(t), \ldots, X_N(t), t\big) + Q\big(X_1(t), \ldots, X_N(t), t\big) \Big] ,$$

$$(5.1.1.5)$$

$\forall k = 1, \ldots, N$, where V is the same potential as in (5.1.1.2) and Q is the "quantum potential" defined by

$$Q = -\sum_{j=1}^{N} \frac{1}{2m_j} \frac{1}{|\Psi|} \frac{\partial^2 |\Psi|}{\partial x_j^2} . \qquad (5.1.1.6)$$

We derive this equation from (5.1.1.1) and (5.1.1.3) in Appendix 5.B.

This equation has to be supplemented by a constraint on the initial conditions:

$$\frac{d}{dt} X_k(0) = \frac{1}{m_k} \frac{\partial}{\partial x_k} S\big(X_1(0), \ldots, X_N(0), 0\big) , \qquad (5.1.1.7)$$

$\forall k = 1, \ldots, N$. This constraint is preserved in time by the dynamics (5.1.1.1) and (5.1.1.5), meaning that

---

[9] Of course, in that case, the Schrödinger equation must be replaced by Pauli's equation, but we will not discuss that (see, e.g., [152, Sect. 8.4]).

[10] The reader familiar with quantum mechanics will recognize the right-hand side of (5.1.1.4) as the quantum probability current divided by $|\Psi|^2$. This formula is derived in Appendix 5.A.

$$\frac{d}{dt}X_k(t) = \frac{1}{m_k}\frac{\partial}{\partial x_k}S\big(X_1(t), \ldots, X_N(t), t\big) \, ,$$

for all t. Since this constraint is identical to (5.1.1.3), many authors find this reformulation of the dynamics as a second order equation unnatural. One advantage is that it is similar to Newton's equations, where the usual potential is changed by adding the "quantum potential". This is sometimes useful when comparing classical and quantum motions, and in order to obtain some intuition about the latter.

Remarks

1. Equations (5.1.1.3) and (5.1.1.4) are called the guiding equations: the quantum state guides the motion of the particles. Given those equations, initial conditions $(\Psi_0, \mathbf{X}_0)$ determine the state at later times. Mathematically, we first show that the Schrödinger equation (5.1.1.2) has a solution for all times, which is true for a large class of potentials (see, e.g., [412, 413]), and then insert the solution in (5.1.1.3)–(5.1.1.4) and prove that the latter has a solution $\mathbf{X}(t)$ for all times and a sufficiently large class of initial conditions. This was done in [54, 471]. Since particles have trajectories, they also have velocities at all times; and so, contrary to what standard quantum mechanics may suggest, particles have *both a position and a velocity* at all times. We will see in Sect. 5.1.8 why this does not contradict Heisenberg's uncertainty relations.[11]

2. The solution of the Schrödinger equation $\Psi(x_1, \ldots, x_N, t)$ is a function defined on $\mathbb{R}^N \times \mathbb{R}$. The variables $(x_1, \ldots, x_N)$ represent all the possible positions of the particles, but not the actual ones. By contrast, in Eq. (5.1.1.3), $\big(X_1(t), \ldots, X_N(t)\big)$ denote the actual positions of the particles. This is a theory of *point particles*, that people sometimes regard as unrealistic.[12] But the same idealization occurs in classical physics, including electromagnetism. Nobody claims that this version of the de Broglie–Bohm theory, which is not even relativistic, is the ultimate theory.

3. Since the system is deterministic, any randomness must be due to "randomness" in the initial conditions, as explained in Sect. 3.4.5. We will discuss this in detail below (in Sects. 5.1.3, 5.1.4, and 5.1.7).

4. The particles are always guided by the value of the wave function and its derivative at the point where the particles are, as can be seen from (5.1.1.3) or (5.1.1.4). In particular, if a wave function is a sum of two functions with disjoint supports (which means that their product always vanishes), the only part of the wave function that matters as far as the motion of the particles is concerned is the function in whose support the particles are actually located. Of course, the two component wave functions with disjoint support might recombine in the future,

---

[11] A point that is clearly misunderstood by some, e.g., Gisin [214, p. 46]: "If there are hidden positions, there must therefore also be hidden velocities. But that contradicts Heisenberg's uncertainty principle which is a key part of the quantum formalism [. . .].".

[12] See, for example, Gouesbet [237, Chap. 17], for an explicit objection to the de Broglie–Bohm theory on that basis.

as in the Mach–Zehnder interferometer. There, we have two paths for the wave function, $|2 \uparrow\rangle$ and $|2 \downarrow\rangle$, and they recombine at the arrow in Fig. 2.3 (discussed in Sect. 2.2), in which case one cannot simply forget about the other wave function forever. However, one may do so as long as no recombination takes place.

5. There is some analogy between the de Broglie–Bohm theory and electromagnetism: electric and magnetic fields are produced by moving charges (think of static charges and electric fields if you are not familiar with the magnetic ones) and they in turn act on the charges and move them (via the "Lorentz force", see, e.g., [152, Sect. 2.4] for a summary of electromagnetism). So in that sense, the electromagnetic fields guide the particles, as does the wave function in (5.1.1.3). And, like the wave function, the electromagnetic fields are defined everywhere,[13] but the way they act on the particles depends only on the value of the fields at the place where the particle actually is.

However, there are two major differences between the wave function and the electromagnetic fields. First, the wave function, unlike the electromagnetic field, has no source (the sources of the latter being the moving charges themselves). Of course, there is a relation between the wave function and the particles, e.g., the dimension of the space on which the wave function is defined depends on the number of particles, and the potential entering into (5.1.1.2) also depends on the properties of the particles, but the wave function is just *there*; it guides the particles but it is not considered to be created by them.

The second difference, which has more radical consequences, is that the electromagnetic field is always defined on the ordinary, physical space $\mathbb{R}^3$, while the wave function is not. Take, for example, an electric field: if $E_1(\mathbf{x})$, $\mathbf{x} \in \mathbb{R}^3$, is the field created by charge e_1 and $E_2(\mathbf{x})$ is the field created by charge e_2, then the field produced by both charges is simply $E(\mathbf{x}) = E_1(\mathbf{x}) + E_2(\mathbf{x})$, and it is still a function of $\mathbf{x} \in \mathbb{R}^3$.

However, as mentioned at the end of Appendix 2.A, the wave function of two particles $\Psi(\mathbf{x}_1, \mathbf{x}_2)$, for example, is defined on $\mathbb{R}^3 \times \mathbb{R}^3$. For M particles, it is defined on $\mathbb{R}^{3M}$, which is called *configuration space*, the space of all the possible configurations of particles.

6. Another analogy one can make is between the wave function and the Hamiltonian in classical mechanics. The latter also guides the motion of the particles, via Hamilton's equations, is also defined everywhere, and its action depends only on the value of the derivatives of the Hamiltonian at the location of the particles. And, unlike the electromagnetic field, it is defined on the "abstract" *phase space*, $\mathbb{R}^{6M}$ for M particles in $\mathbb{R}^3$, the space of all the positions and the momenta of the particles. However, if there are no forces acting directly between the particles, the Hamiltonian becomes a sum of terms, one per particle, and each term guides the motion of the corresponding particle. As we will see in Sect. 5.2, in the de Broglie–Bohm theory, the situation is more subtle.

We will now first illustrate, via simple examples, what sort of trajectories the de Broglie–Bohm dynamics produces. And then, we will explain how that dynamical

---

[13] Setting aside the problems coming from the self-interaction of the particle with its own field.

system reproduces the quantum mechanical predictions, first for measurements of
positions, then for measurements of other quantities, such as spin, momentum, etc.

5.1.2   How Does the de Broglie–Bohm Dynamic Work?

The reader who wants to see an explicit solution of (5.1.1.3), for the Gaussian wave
function (2.A.2.23), can find it in Appendix 5.D. But here we will consider the double-
slit experiment, described in Appendix 2.E, which can be analyzed without formulas.
It is interesting to see a numerical solution of that experiment within the de Broglie–
Bohm theory, as shown in Fig. 5.1. Each (wavy) line in the figure represents the
trajectory of a single particle (since particles are sent one by one, there is no interaction
between the particles). Different lines correspond to different initial conditions. One
assumes that the initial wave function Ψ is symmetric: $\Psi(z) = \Psi(-z)$, where z is
the coordinate of the vertical axis, with $z = 0$ half-way between the slits.

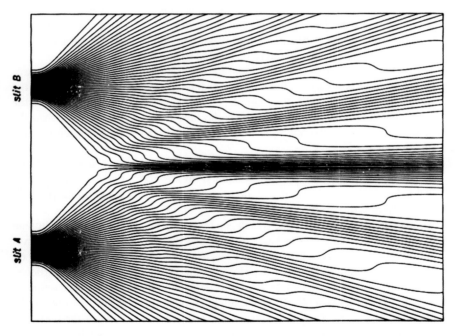

Fig. 5.1 De Broglie–Bohm trajectories computed for the double-slit experiment. Reproduced with
the kind permission of *Società Italiana di Fisica* and the authors, from C. Philippidis, C. Dewdney,
B.J. Hiley: Quantum interference and the quantum potential, *Il Nuovo Cimento B* **52**, 15–28 (1979)

Each particle goes through only one slit, but the wave function evolves differently
when both slits are open than when only one of them is, and this in turn affects the

motion of the particles via (5.1.1.3). This is rather easy to understand intuitively: the function Ψ propagates like a wave through (5.1.1.1) and spreads out in space (which is not surprising for a wave) [see (2.A.2.23) and (2.A.2.24)]. Obviously, a wave beyond the slits will be different if it has two sources (one for each slit) or one. Thus, the behavior of the particle (depicted here for both slits open) is affected by the fact that the slit *through which it does not go* is open or not. As an analogy, imagine a water wave coming through the slits, and a small, light object being carried by that water wave; evidently, the form of the wave will depend on whether one slit is open or both, and that will affect the motion of the object, even though the latter goes through only one slit (of course, this is just an analogy, because the equations here are not the same as for water waves).

Note that the motion here happens *in vacuum* (there is no force or potential acting directly on the particles). The de Broglie–Bohm theory is highly *non-classical*. It does not satisfy Newton's first law of motion, namely propagation along straight lines in the absence of forces.[14] But here, of course, the motion "in vacuum" is still affected by the wave function. In the Bohm version of the theory (5.1.1.5), the ordinary potential $V(x)$ vanishes beyond the slits, but the quantum potential Q, defined by (5.1.1.6) with $N = 2$, which depends solely on Ψ, does not; and neither is it constant.[15] Note also the presence of a nodal line: by the symmetry of Ψ, i.e., $\Psi(z) = \Psi(-z)$, we have $d\Psi(0)/dz = 0$, so by (5.1.1.3) the speed vanishes on that line and the particles cannot cross it. As a consequence, one can determine the slit the particle went through a posteriori by detecting the particle on the screen.

It is also interesting to compare this numerical solution with an experimental result for the two slit experiment, as shown in Fig. 5.2. This is the result of a so-called weak measurement of velocities, from which trajectories are reconstructed. The idea of weak measurement[16] was introduced by Aharonov et al. [3]. Roughly speaking, to obtain a weak measurement of velocities, one first measures the position of the particle "weakly", which means without disturbing the wave function very much. But then, one does not get a precise location of the particle either. Since that measurement is weak, one can do a "strong", i.e., ordinary, measurement, a little later and obtain the precise location then. By repeating that operation many times, one can produce a statistical distribution of positions, and by taking its average, obtain a position at the first location. But then, one has two consecutive positions and a time interval between them, so one can compute from that a velocity which one may associate

---

[14]However, if the wave function is a *plane wave*, viz., $\Psi(x) = \exp(ikx)$, a rather idealized situation, then (5.1.1.3) yields straight line motion with velocity k/m. We will see in the next two subsections that this violation of Newton's first law does not contradict conservation of momentum, which also holds in quantum mechanics. Indeed, this conservation law concerns results of measurements and the de Broglie–Bohm theory will predict, in a subtle but natural way, the same results of measurements as ordinary quantum mechanics.

[15]For a picture of that potential, see [70, p. 34].

[16]See [151, 362, 522] for a discussion of weak measurements in the context of the de Broglie–Bohm theory.

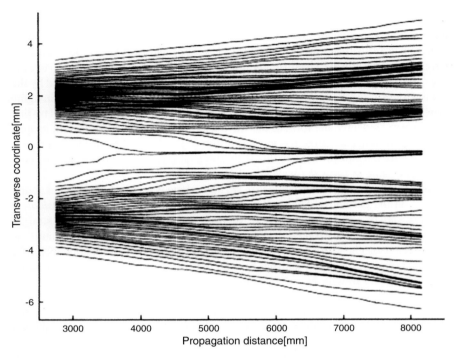

Fig. 5.2 Weak measurement of trajectories in the double-slit experiment. From S. Kocsis, B. Braverman, S. Ravets, M.J. Stevens, R.P. Mirin, L.K. Shalm, and A.M. Steinberg: Observing the average trajectories of single photons in a two-slit interferometer, *Science* **332**, 1170–1173 (2011). Reprinted with permission from AAAS

with the first position.[17] It can be shown that this yields the true velocities in the de Broglie–Bohm theory [151, 522].

By repeating that "weak measurement" in different places, one obtains a velocity field (i.e., an assignment of velocities at each point in space) and the trajectories can be reconstructed from this field by drawing the lines whose tangents are given by the velocity field.

This is not meant to "prove" that the de Broglie–Bohm theory is correct, because other theories will make the same empirical predictions,[18] but the result is neverthe-less suggestive, because the predictions made here by the de Broglie–Bohm theory is very natural within that theory, and is confirmed by these observations. Moreover, it should be said that the experiments reported in Fig. 5.2 are done with photons, not

---

[17]This does not contradict Heisenberg's uncertainty relations, because we need to do many opera-tions and take an average to get this result. Heisenberg's uncertainty relations apply to the results of strong, i.e., ordinary, measurements.

[18]Those theories will be discussed in Sect. 5.4.1.

electrons, unlike the trajectories shown in Fig. 5.1. An experiment with electrons has been proposed (but not performed) in [317].

Here is how John Bell summarized the de Broglie–Bohm theory in the case of the double-slit experiment:

> Is it not clear from the smallness of the scintillation on the screen that we have to do with a particle? And is it not clear, from the diffraction and interference patterns, that the motion of the particle is directed by a wave? De Broglie showed in detail how the motion of a particle, passing through just one of two holes in the screen, could be influenced by waves propagating through both holes. And so influenced that the particle does not go where the waves cancel out, but is attracted to where they cooperate. This idea seems to me so natural and simple, to resolve the wave–particle dilemma in such a clear and ordinary way, that it is a great mystery to me that it was so generally ignored.

> John Bell [49, p. 191]

It is interesting to compare this statement to the ones by Feynman and Landau and Lifshitz, quoted in Appendix 2.E, with regard to the double-slit experiment:

> Nobody knows any machinery. Nobody can give you a deeper explanation of this phenomenon than I have given; that is, a description of it.

> Richard Feynman [185, p. 145]

> It is clear that [the results of the double-slit experiment] can in no way be reconciled with the idea that electrons move in paths. [...] In quantum mechanics there is no such concept as the path of a particle.

> Lev Landau and Evgeny Lifshitz [302, p. 2]

What is surprising is the dogmatic assurance of the latter statements: how does one know that *nobody* can do something that has not yet been done? Or that an experiment *can in no way* be reconciled with a concept? Moreover, the de Broglie version of the theory already explained that experiment back in 1927 (if anyone had bothered to look at it), and Feynman was quite close to Bohm and seems to have held in him great esteem.[19]

5.1.3 What About the Statistical Predictions of Quantum Mechanics?

So far, we have been discussing the way various individual trajectories behave, but what about the statistics of the results? As we said in Sect. 3.4.5, this must come from statistical assumptions regarding the initial conditions, since the dynamics is completely deterministic.

Let us consider, for simplicity, systems composed of a single particle (like in the double-slit experiment or the Mach–Zehnder interferometer discussed in the next

---

[19]See [386, pp. 272–273].

subsection); the extension to systems with many particles is straightforward. Let us also first discuss the quantum mechanical predictions for measurements of positions.

Consider a large number of particles (meaning a large number of copies of single particle systems). In Appendix 3.A, we introduced the notion of the *empirical density distribution* of those particles, which is a positive function $\rho(\mathbf{x})$, with $\int_{\mathbb{R}^3} \rho(\mathbf{x})d\mathbf{x} = 1$, such that, for every[20] subset $B \subset \mathbb{R}^3$, the fraction of particles in B is approximately equal to $\int_B \rho(\mathbf{x})d\mathbf{x}$. Of course, this becomes precise only in the limit where the number of particles tends to infinity, hence the use of the word "approximately", but this is the case with all applications of the law of large numbers in probability. In practice, the statistical predictions will be sufficiently precise to be compared with observations for all sets B such that $\int_B \rho(\mathbf{x})d\mathbf{x}$ is not too small. We always assume below that the number of particles is large enough, so that the approximation we make by "taking it to infinity" does not matter.

Now suppose we have a large set of particles, with empirical density distribution $\rho(\mathbf{x}, 0)$ at some initial time $t = 0$ and that those particles obey some deterministic dynamical law (any such law, not necessarily the one specified by the de Broglie–Bohm theory): $\mathbf{X}(0) \rightarrow \mathbf{X}(t)$. Let $\phi^t(\mathbf{x})$ denote the solution of (5.C.1.3) at time t when the initial condition is $\mathbf{X}(0) = \mathbf{x}$ at time 0. Then, at a later time $t > 0$, the particles will have a well-defined empirical density distribution $\rho(\mathbf{x}, t)$ defined by[21]

$$\int_B \rho(\mathbf{x}, t)d\mathbf{x} = \int \mathbb{1}_B(\phi^t(\mathbf{x}))\rho(\mathbf{x}, 0)d\mathbf{x} \,, \qquad (5.1.3.1)$$

where $\mathbb{1}_B$ is the indicator function of the set B. The identity (5.1.3.1) means that the fraction of particles in B at time t is simply defined by the fraction of particles at time $t = 0$ whose trajectories are in B at time t.

Now, an elementary property of the de Broglie–Bohm dynamics is that it is *equivariant*, to use an important notion introduced in [141]. This means that, if we assume, for a system in $\mathbb{R}^N$ and some initial time 0, that a set of particles is distributed with an empirical density distribution $\rho(\mathbf{x}, 0) = |\Psi(\mathbf{x}, 0)|^2$, where $\Psi(\mathbf{x}, 0)$ is the quantum state of a single particle in $\mathbb{R}^3$ at time 0, then at any later time $t > 0$, we will have

$$\rho(\mathbf{x}, t) = |\Psi(\mathbf{x}, t)|^2 \,, \qquad (5.1.3.2)$$

where $\Psi(\mathbf{x}, t)$ comes from Schrödinger's equation (5.1.1.1), and $\phi^t(\mathbf{x})$ in the definition (5.1.3.1) of $\rho(\mathbf{x}, t)$ comes from the deterministic de Broglie–Bohm dynamics through (5.1.1.4). This in proven in Appendix 5.C.[22] We also prove the property of equivariance explicitly in Appendix 5.C, using the solution of the Schrödinger

---

[20]To be mathematically precise, one should add "measurable" or "Borel" here.

[21]Here we use the fact that a function can be defined by giving the value of its integral on every (Borel) subset, if the measure defined by those integrals is absolutely continuous, and we assume here that the map $\phi^t(\mathbf{x})$ is such that this measure has this property.

[22]To be precise, this is rather a consequence of what is called equivariance in [141]. See the second part of that appendix.

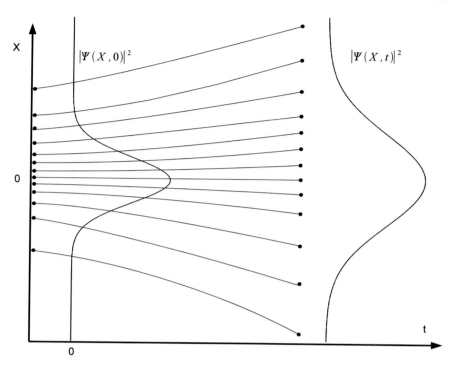

Fig. 5.3 Illustration of the property of equivariance of the $|\Psi(\mathbf{x}, t)|^2$ distribution, in one dimension, for a Gaussian Ψ. Each dot represents the position of a particle, both at time 0 and at time t, connected by trajectories. The density of particles is (approximately) given by $|\Psi(\mathbf{x}, 0)|^2$ on the left of the picture and by $|\Psi(\mathbf{x}, t)|^2$ on the right. An exact computation of the de Broglie–Bohm dynamics in this situation is given in Appendix 5.D.1. Note that the point with initial condition $X(0) = 0$ does not move

equation (5.1.1.1) and of the guidance equation (5.1.1.3), for the special case of the Gaussian wave function (2.A.2.23).

The property of equivariance (for a one-dimensional system) is illustrated in Fig. 5.3, where each dot denotes the position of a particle and each line its trajectory. We see that the density of particles, which is a function of $x \in \mathbb{R}$ here, is more or less equal to the square of the wave function, both at time 0 and at time t.

Equivariance can also be illustrated by Fig. 5.1. There, one assumes a distribution of particles (i.e., of initial positions just behind the slits) given by $|\Psi(\mathbf{x}, 0)|^2$, 0 denoting the time of passage through the slits. Then, the right side of the figure indicates the place where the particle lands on the screen and these dots are distributed according to $|\Psi(\mathbf{x}, t)|^2$, t being the time of arrival on the screen and $\Psi(\mathbf{x}, t)$ being the solution of the Schrödinger equation with initial data $\Psi(\mathbf{x}, 0)$. Since the trajectories of the particles are numerically computed from the guidance equation (5.1.1.3), the identity $\rho(\mathbf{x}, t) = |\Psi(\mathbf{x}, t)|^2$ holds here too.

So if we assume that the empirical density distribution of particles at some initial time is given by $\rho(\mathbf{x}, 0) = |\Psi(\mathbf{x}, 0)|^2$, this will be true at all later times. In Sect. 5.1.7, we will discuss the naturalness of this assumption, but let us first explain why this implies the agreement between the de Broglie–Bohm theory and *all* quantum predictions.

5.1.4   Measurements of "Observables" in the de Broglie–Bohm Theory

We have seen in Sect. 2.5 that the idea that quantum "measurements" actually measure, in general, intrinsic properties of quantum systems is untenable: the no hidden variables theorems forbid us to think of measurements as revealing pre-existing values of the quantities associated with quantum "observables", such as, for example, spin components or momentum.

How does the de Broglie–Bohm theory deal with that? And how does one obtain the results of these measurements if the only hidden variables of the theory are the positions of the particles? Besides, how can one reconcile the fact that particles *have* a position and a velocity at all times with the Heisenberg uncertainty relations, which prohibit a joint measurement of those quantities beyond a certain accuracy?[23] Even more seriously, how can one reconcile this fact with the second part of the no hidden variables theorem (see Sect. 2.5.2), which says that one cannot consistently view the numbers resulting from a measurement of position and of momentum as revealing pre-existing properties of the particles?

The fact that the de Broglie–Bohm theory can escape these dilemmas, or even apparent dead ends, and does so in a extremely natural way, is probably its main virtue.

First, let us explain why all measurements can in the end be reduced to position measurements. Consider a "spin measurement", as described in Sect. 2.1. What we directly see are particles exiting the boxes through one hole or the other in Fig. 2.1 or 2.2, i.e., we detect their *positions*. We never directly see their spin (assuming this expression means something). Or, to take a more concrete example, consider the setup in Fig. 5.4, to be analyzed below. A magnetic field deflects the particles, upwards for those with spin up, downward for those with spin down. Again, what we see are particle positions, never the spin.

Alternatively, suppose we try to measure a momentum. One way to do it, as we saw in Appendix 2.C, would be to measure the distance travelled by a particle in a certain amount of time, which again brings us back to a measurement of positions.

One can convince oneself that the same is true for other measurements of quantum observables. But one can also use a more abstract argument: in the end, all these measurements are "seen" by traces on macroscopic devices: the position of a pointer on a screen, or the positions of ink dots on a computer printout. The upshot is that

---

[23]See Appendix 2.C.

any theory that makes the correct predictions for the (statistics) of the positions of all particles will automatically get the correct predictions for the results of all the other "observations".

One might object that the results of some measurements could be signaled by a color, say a red light if the particle in Fig. 5.4 goes up and a green one if it goes down. But in that situation, one would simply correlate a color signal with the detection of a position. Once we get the right predictions for the positions, we get them also for the colors. We shall now analyze two special cases.[24]

Spin Measurements

We start by describing how the strange effects associated with the spin measurements described in Chap. 2 are explained within the de Broglie–Bohm theory.[25] First of all, the boxes "measuring" the spin in Chap. 2 are in practice Stern–Gerlach devices. This means that a charged particle with spin (an electron for example) goes through a device where an inhomogeneous magnetic field is present.

Suppose we start with a quantum state $\Psi(x, z)|1 \uparrow\rangle$, and the field is oriented in direction 1 (by assumption, this is the direction indicated in Fig. 5.4), $\Psi(x, z)$ denotes the spatial part of the quantum state, and $|1 \uparrow\rangle$ its (normalized) spin part. We consider here a two-dimensional motion, where z denotes the "vertical" direction in which the spin is "measured" by the Stern–Gerlach apparatus and x the transverse direction in which the particle moves towards the apparatus.

Assume that $\Psi(x, z)$ is a function whose support is concentrated around zero along both axes. As far as the free motion of the particle is concerned, we have a rectilinear motion along the x axis (there is also some spreading of the wave function, but we will neglect that, assuming that the duration of the experiment is short enough). We will furthermore neglect that motion and replace $\Psi(x, z)$ by $\Psi(z)$.

If one includes an inhomogeneous magnetic field in the Hamiltonian at some time, say 0, the solution of the time evolution is then, with much simplification[26]:

$$\Psi(z - t)|1 \uparrow\rangle , \qquad (5.1.4.1)$$

---

[24]We will take for granted in this subsection that one can detect the positions of particles, and we will see how the de Broglie–Bohm theory accounts for the usual quantum predictions, given that assumption. However, this detection does of course necessitate the coupling of the particle with a macroscopic device, and that coupling, described in conventional terms in Appendix 2.D, will be explained in the context of the de Broglie–Bohm theory in Sect. 5.1.6 and Appendix 5.E.

[25]It should be stressed that all the "experiments" are only meant to illustrate the theory, not to explain how real experiments are performed.

[26]The time evolution is given by Pauli's equation rather than Schrödinger's equation. See [152, Sect. 8.4] or [360] for a detailed discussion. The solution given here is similar to the one in Appendix 2.D, except that here we consider the evolution of the wave function of the particle, while in that appendix we considered the evolution of the wave function of the pointer.

where t means vt, with the velocity set equal to 1 for simplicity (thus t here denotes a length), which means that the particle will always go in the direction of the field. If we start with a quantum state $\Psi(x, z)|1 \downarrow\rangle$ the solution will be

$$\Psi(z + t)|1 \downarrow\rangle , \tag{5.1.4.2}$$

and the particle will always go in the direction opposite to the field.

A more interesting situation is when we start with a quantum state

$$\Psi(z)\big(|1 \uparrow\rangle + |1 \downarrow\rangle\big) . \tag{5.1.4.3}$$

The solution of the time evolution is then

$$\Psi(z - t)|1 \uparrow\rangle + \Psi(z + t)|1 \downarrow\rangle . \tag{5.1.4.4}$$

This means that the particle has a quantum state which is composed of two parts, one localized near t, the other near $-t$, along the z axis. If t is not too small, those two regions are far apart, and by detecting the particle, we can see in which region it is. If it is near t, the spin is up and if it is near $-t$, it is down. This is illustrated in Fig. 5.4.

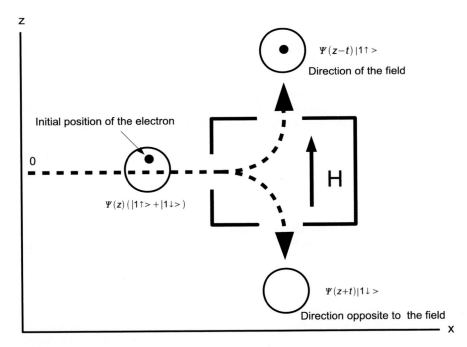

Fig. 5.4 An idealized spin measurement

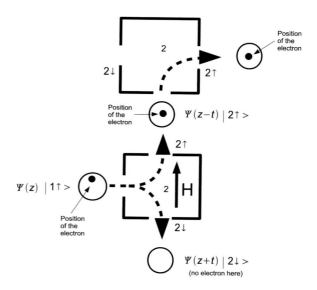

Fig. 5.5 Two successive spin measurements in the same direction

As already mentioned, all we see in that "measurement" is the position of the particle. In the de Broglie–Bohm theory, the particle has a position and therefore a trajectory, so it will go either up or down.

Now, suppose that Ψ is symmetric in z: $\Psi(z) = \Psi(-z)$. Then the derivative $\partial\Psi(z)/\partial z$ vanishes at $z = 0$ and, by (5.1.1.4), the particle velocity is zero for $z = 0$. Therefore, the particle never crosses the line $z = 0$ (which is said to be *nodal*). From this, we conclude that the particle will go up if its initial condition satisfies $z > 0$ and down otherwise.[27]

Let us now see how the de Broglie–Bohm theory accounts for the phenomena described in Sects. 2.1 and 2.2. First, suppose that we start with a quantum state $\Psi(x, z)|1 \uparrow\rangle$, and that the field is oriented in direction 2 (by assumption, this is the direction indicated at the bottom of Fig. 5.5). With the same conventions and simplifications as above, the solution of the time evolution is then

$$\Psi(z - t)|2 \uparrow\rangle + \Psi(z + t)|2 \downarrow\rangle . \qquad (5.1.4.5)$$

If one starts with a particle position as in Fig. 5.5 and the initial wave function is symmetric, i.e., $\Psi(x, z) = \Psi(x, -z)$, the particle will go upward, as indicated in the figure. But then, since the particle is guided only by the part of the wave function in the support of which it lies, we only have to consider the quantum state $\Psi(z - t)|2 \uparrow\rangle$. Thus, if we perform a second measurement[28] in the direction 2, indicated by the box

---

[27]One can check that the presence of a magnetic field does not destroy the symmetry $\Psi(z) = \Psi(-z)$.

[28]In that part of the figure, as well as in Fig. 5.6, the holes are not put in a natural way (top and bottom), but that is just a convention, chosen for graphical convenience.

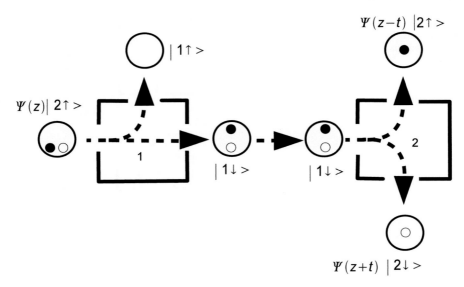

Fig. 5.6 Two successive spin measurements in different directions

at the top of Fig. 5.5, we will necessarily get the result $|2 \uparrow\rangle$, since the quantum state of the particle is now effectively $\Psi(z-t)|2 \uparrow\rangle$.

This explains the rather unsurprising fact that, once we obtain a given result for a measurement of the spin in direction 2, we will get the same result with certainty if we again measure the spin in direction 2, even if the original state, before the first measurement, was a superposition like (5.1.4.5) (and, of course, the statistical results of the first measurement are 50 % up, 50 % down, as we will show at the end of this subsection).

Now consider Fig. 5.6, which shows what happens if we first measure the spin in direction 1 and then in direction 2, starting with a particle for which the spin part of the quantum state is $|2 \uparrow\rangle$ and thus a superposition of $|1 \uparrow\rangle$ and $|1 \downarrow\rangle$ (this experiment was described in Fig. 2.2).

Again, the quantum state becomes a sum like (5.1.4.5) after the first measurement (with $|1 \uparrow\rangle$ and $|1 \downarrow\rangle$ instead of $|2 \uparrow\rangle$ and $|2 \downarrow\rangle$ and different spatial parts of the wave function, corresponding to different conventions for the holes in Figs. 5.5 and 5.6). If we consider two initial particle positions, as in Fig. 5.6, and if they are both below the nodal line where $\partial\Psi(z)/\partial z = 0$, they will both exit the first box through the "down" hole. But each particle position will be modified by the time evolution, and we may get the situation before the measurement of the spin in direction 2, on the right of Fig. 5.6, where the quantum state, being of the form $|1 \downarrow\rangle$ (for its spin part) is a superposition of $|2 \uparrow\rangle$ and $|2 \downarrow\rangle$. Then, the introduction of a magnetic field in direction 2 will produce a quantum state of the form $\Psi(z-t)|2 \uparrow\rangle + \Psi(z+t)|2 \downarrow\rangle$. As for the particle positions, if they are as in Fig. 5.6 before entering the box measuring the spin in direction 2, one particle will go up and the other down, if we assume that there is again an axis of symmetry and thus that the particles cannot cross the horizontal line in the middle of that box.

This explains what we saw in Fig. 2.2, namely that, if we measure the spin in direction 2 after measuring it in direction 1, we do not always obtain the up result, even if we started with spin state $|2 \uparrow\rangle$ before the measurement in direction 1. In fact, we will get 50 % of the particles exiting with spin $|2 \uparrow\rangle$ and 50 % with spin $|2 \downarrow\rangle$.

Consider finally the interference phenomena illustrated by Figs. 2.3 and 2.4. What happens in the de Broglie–Bohm theory is illustrated in Fig. 5.7. The two parts of the wave functions follow different paths, one for the $|2 \uparrow\rangle$ part and one for the $|2 \downarrow\rangle$ part. We assume as before that the wave function is symmetric so that there is a nodal line. Two possible initial conditions of the particles are indicated in Fig. 5.7: one particle, starting above the nodal line will go up and its wave function will follow the path $|2 \uparrow\rangle$, whereas the other particle, starting below the nodal line, will go through the $|2 \downarrow\rangle$ hole and its wave function will follow the path $|2 \downarrow\rangle$. In each case, the particle follows a unique path and is guided by the part of the wave function that follows that path. If both paths are open, the two parts of the wave function recombine at the arrow, and the particle is guided by a pure $|1 \downarrow\rangle$ quantum state, independently of the path that it followed. Thus, upon measurement in direction 1 after the arrow, the result "down" will occur for both particles. What happens here is like the behaviour described by (5.1.4.2).

If on the other hand the wall is inserted along the path followed by the $|2 \downarrow\rangle$ part of the state, as in Fig. 2.4, then the wall effectively blocks that part and it will not contribute to the recombination of the quantum state at the arrow.[29] Thus, after the arrow, the quantum state will be purely $|2 \uparrow\rangle$ and the result of a measurement of the spin in direction 1 will give 25 % up, 25 % down. The same is true for each of the boxes measuring the spin in direction 1, as in Fig. 2.5: we have a quantum state which is either $|2 \uparrow\rangle$ or $|2 \downarrow\rangle$ and in each case the result of a measurement of the spin in direction 1 will give 25 % up, 25 % down.

There is no problem either, in the de Broglie–Bohm theory, with the "delayed choice" experiment discussed in Sect. 2.2. The particle is guided at all times by the wave function in whose support it is located. Whether one inserts a wall or not or whether one inserts or removes the arrow, will affect the behavior of the particle when it hits the wall or when it goes through the arrow (if they are there), but before that, it follows one path, while its associated wave function goes through both paths. In the de Broglie–Bohm theory, there is no sense in which our present choices affect the past.[30]

The Elitzur–Vaidman bomb testing mechanism,[31] although counterintuitive, is equally unsurprising.[32] The bombs that are duds do not affect the wave function in

---

[29]To be more precise, "blocking" here has to be understood as a collapse of the quantum state, whose meaning, in the de Broglie–Bohm theory, will be explained in the next subsection.

[30]See [38, 69] for further discussion of the "delayed choice" experiment in the de Broglie–Bohm theory.

[31]See Sect. 2.2 and Appendix 2.D.

[32]Englert, Scully, Sussman, and Walther have suggested that the de Broglie–Bohm trajectories are "surrealistic" rather than realistic [175], because those trajectories can have counterintuitive

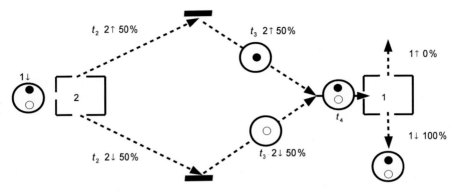

Fig. 5.7 Interference viewed in the de Broglie–Bohm theory

any way (putting a dud on the path is like putting nothing), while those that are active block the part of the wave function along the path that the particle does not take, without exploding (they would explode if they detected the particle). Thus, they act then like a wall and the particle is guided later only by the quantum state associated with the other path alone |path 2 ↑⟩. And, if the bomb is on the path of the particle, it detects it and explodes. In that way, one can identify 25 % of the initial stock of active bombs as being active without exploding them, since half of the particles that do not take the path of the active bombs (that is, a quarter of the total number of particles) will have their spin equal to |1 ↑⟩ if one measures the spin in direction 1 after the arrow. Since one would always have gotten the result |1 ↓⟩ if there had been no active bomb on the path |path 2 ↓⟩, we know that, whenever one obtains the result |1 ↑⟩ after the arrow, that the bomb on the path |path 2 ↓⟩ is active.

However, one can find statements, similar to those of Feynman and Landau and Lifshitz about the double-slit experiment quoted in Sect. 5.1.2, and claiming that the de Broglie–Bohm theory explanation of the Mach–Zehnder interferometer is simply impossible: in a caption to a figure similar to our Fig. 2.3, one reads:

> Any explanation which assumes that the photon takes exactly one path through the interferometer leads to the conclusion that the two detectors should on average each fire on half the occasions when the experiment is performed.[33] But experiment shows otherwise!

> David Deutsch, Artur Ekert, Rosella Lupacchini [129]

(Footnote 32 continued)
aspects, like the ones discussed here. Originally, the situation described in [175] was presented as an argument against the existence of de Broglie–Bohm trajectories, but in fact those "surrealistic" trajectories can be perfectly understood, in a natural way, within the de Broglie–Bohm theory. See [28, 130, 143, 176, 265] for further discussion of these trajectories.

[33]In our language, where we speak of electrons rather than photons, it would mean that the particles are detected after the arrow with spin up and spin down in the direction 1 half of the time. (Note by J.B.).

But this confuses the fact that there is a difference between the photon *following* a path and that same photon *being detected* along that path. The detector along one path changes the quantum state (by "collapsing" it), even if it does *not* detect a photon, and therefore modifies the behavior of the particle following the other path when it is detected after the arrow.

Let us now go back to the situation illustrated in Fig. 5.4 and show something apparently surprising about it, but which is fundamental if one wants to understand the de Broglie–Bohm theory. Suppose that we reverse the direction of the magnetic field. Then, the solution of the time evolution becomes (again, with much simplification)

$$\Psi(z + t)|1 \uparrow\rangle + \Psi(z - t)|1 \downarrow\rangle . \qquad (5.1.4.6)$$

So now, if the particle is near $-t$, the spin is up and if it is near t, it is down.

However, if we do the experiment starting with *exactly* the same quantum state and *exactly* the same particle position, the particle will go in the upward direction again if its initial condition satisfies $z > 0$ and downward otherwise, since it cannot cross the line $z = 0$. But now, as we just saw, it will be in the support of the wave function with spin down if it goes in the positive z direction, and in the support of the wave function with spin up, if the particle goes in the negative z direction. This is illustrated in Fig. 5.8.

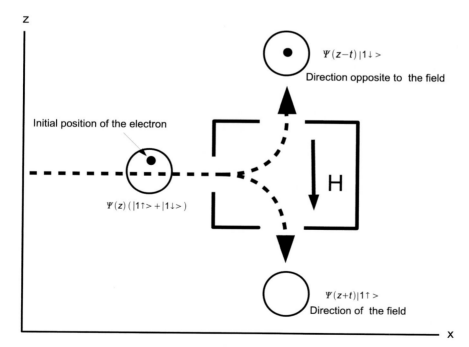

Fig. 5.8 An idealized spin measurement with the field reversed relative to Fig. 5.4

In other words, the value up or down of the spin that actually results from the measurement is "contextual": that value does not depend only on the quantum state and the original particle position but on the concrete arrangement of the "measuring" device.[34] Here the scare quotes are used because we finally see the truth of something that we have emphasized several times: there is no intrinsic property of the particle that is being "measured", in general, in a "measurement".

Of course, since the system is deterministic, once we fix the full initial state (the quantum state and the position) of the particle *and* the experimental device, the result of the experiment is predetermined. But that does not mean that the spin value that we "observe" is predetermined, because, as we saw, we can measure the spin along a given axis by orienting the magnetic field in one direction or the opposite one along that same axis. So the value of the "spin" of the particle that results from a measurement depends on our conventions, which means that it does not exist as an intrinsic property of the particle.

So not only is the word "hidden variable" misleading, but so are the words "observable" or "measurement", since, except when one measures positions, one does not observe or measure any intrinsic property of the system being observed or measured. This vindicates in some sense Bohr's emphasis on:

> [...] the *impossibility of any sharp distinction between the behavior of atomic objects and the interaction with the measuring instruments which serve to define the conditions under which the phenomena appear.*
>
> Niels Bohr [73, p. 210], quoted in [49, p. 2] (italics in the original)

But in the de Broglie–Bohm theory, this follows from the equations of the theory and not from some more or less a priori notion. Besides, this is good news, since it explains how the de Broglie–Bohm theory can introduce hidden variables (the positions of the particles) without having to introduce the hidden variables (the spin values) that would lead to a contradiction, because of the no hidden variables theorem, at least if one assumes that those hidden spin variables are distributed according to the quantum mechanical predictions for the results of their "measurements".[35]

Finally, one may ask how the de Broglie–Bohm theory can reproduce the quantum statistics in the case of spin measurements. Since the theory is deterministic, all "randomness" must be in the initial conditions, and the answer again relies on the notion of equivariance.

---

[34] The word "contextual" sometimes has a different meaning, as we saw in Appendix 2.F, namely that the result of a measurement may depend on which other measurement is performed simultaneously with the first one. But since one sees here that the result of the "measurement" depends on the details of the apparatus, one can understand that it may also depend on whether another measurement is performed simultaneously. We will come back to this notion of contextuality in Sects. 5.1.5 and 5.3.4.

[35] Within the de Broglie–Bohm theory, one can introduce, if one wants, "hidden" spin variables, but their value is a function of the quantum state and the positions of the particles (hence, they are redundant) and the actually measured spin values will be affected by the measuring devices and thus not pre-exist their "measurement". See [268, Chap. 9] or [70, Chap. 10] and references therein for more details on these spin variables.

Consider, instead of (5.1.4.3), a more general quantum state of the form

$$\Psi(z)\big(c_1|1\uparrow\rangle + c_2|1\downarrow\rangle\big),\qquad(5.1.4.7)$$

with $|c_1|^2 + |c_2|^2 = 1$ and $\int |\Psi(z)|^2 dz = 1$ (we suppress here the x variable). Consider an ensemble of particles distributed at the initial time according to the empirical density distribution $\rho(z) = |\Psi(z)|^2$. The solution of the Schrödinger equation (in the situation described by Fig. 5.4) will be

$$\Psi(z - t)c_1|1\uparrow\rangle + \Psi(z + t)c_2|1\downarrow\rangle,\qquad(5.1.4.8)$$

and, by equivariance (5.1.3.2), the empirical density distribution of the particle positions will be given by $|\Psi(z - t)c_1|^2 + |\Psi(z + t)c_2|^2$. But if t is not too small, the supports of $\Psi(z - t)$ and $\Psi(z + t)$ will be approximately disjoint.[36] Let $B_\pm$ denote the respective supports of $\Psi(z \pm t)$, assumed to be strictly disjoint for simplicity. Then we get

$$\int_{B_-} \big|\Psi(z - t)c_1\big|^2 + \big|\Psi(z + t)c_2\big|^2 dz = \int_{B_-} \big|\Psi(z - t)c_1\big|^2 = |c_1|^2,\qquad(5.1.4.9)$$

because of the disjointness of the supports B_- and B_+ and the normalization $\int_{B_-} |\Psi(z - t)|^2 dz \approx \int |\Psi(z)|^2 dz = 1$. We get the same result for B_+, with $|c_2|^2$ instead of $|c_1|^2$.

Thus, equivariance (5.1.3.2) implies that, in the situation depicted in Fig. 5.4, a fraction approximately equal to $|c_1|^2$ of the particles will go upwards and a fraction approximately equal to $|c_2|^2$ of them will go downwards, so that the usual quantum predictions ("Born's rule") are recovered in the de Broglie–Bohm theory.

The same *statistics* hold also if we reverse the direction of the field, as in Fig. 5.8. Indeed, instead of (5.1.4.8), one then gets

$$\Psi(z + t)c_1|1\uparrow\rangle + \Psi(z - t)c_2|1\downarrow\rangle,\qquad(5.1.4.10)$$

whence a fraction approximately equal to $|c_1|^2$ of the particles will go downwards (i.e., now with spin up) and a fraction approximately equal to $|c_2|^2$ of them will go upwards. This is why we only need to know the "observable" being "measured" and not the details of the measurement setup if we are only interested in the statistics of the measurement. But, as we saw by contrasting Figs. 5.4 and 5.8, the result for *individual* measurements depends on the details of this setup.

---

[36]This is only approximate (hence the use of the symbol $\approx$ below), because of the spreading of the wave functions under free evolution, which implies that, even if the initial wave function has a bounded support, this will not be true at later times. One says that the time evolution produces "tails" of the wave function, namely regions where it is small but nonzero. However, in practice, the probability of finding the particle in one of the tails of the wave function is exceedingly small.

Now that we have seen how a measurement of something else than position emerges naturally from an analysis of measurements of positions within the de Broglie–Bohm theory, and before discussing further the implications of this observation, we will analyze another example.

Momentum Measurements

Let us consider a wave function that is real $\Psi(x_1, \ldots, x_N, t)$, as many wave functions are.[37] Then, by (5.1.1.3), nothing moves! That sounds paradoxical, but is not a priori contradictory. However, it means that the particle has zero momentum, while the probability density for the results of measurements of the momentum[38] is given by $|\hat{\Psi}(p, t)|^2$, where $\hat{\Psi}(p, t)$ is the Fourier transform of $\Psi(x, t)$, and this distribution is never exactly concentrated at zero. This does seem to lead to a contradiction.

But in order to "measure" that momentum, we cannot look at the particle as it is, with "God's eye" so to speak. One has to interact with the particle in a definite way and observe, at the end of that interaction, its position. It is by analyzing that interaction that one can see that there is no contradiction.[39]

Let us illustrate this with a simple example, discussed mathematically in Appendix 5.D, which has some historical significance, since it is based on an objection that Einstein raised against the de Broglie–Bohm theory in his contribution to the book published in honor of Max Born when he retired [172]. Consider a particle in a box of side L (which we consider in one spatial dimension, to simplify the notation). Suppose it is free to move in that box. There is a family of wave functions $\Psi_n(x, t)$ for that system whose associated probability distributions are stationary, i.e., $|\Psi_n(x, t)|^2$ is independent of time, and whose phase is constant in x, so that nothing moves, but for which quantum mechanics predicts a nonzero distribution of velocities (or momenta), if the latter are measured.

Einstein used this example to raise two objections. One was that, according to standard quantum mechanics, the quantum behavior should approach, in some limit (corresponding here to the index n of Ψ_n being large), the classical behavior, and a classical particle should be somewhere in the box, while the wave functions $\Psi_n(x, t)$ only assign a probability of being observed at a given place in the box. This is a familiar objection of Einstein against the completeness of ordinary quantum mechanics. The other objection was directed at de Broglie and Bohm. Einstein thought that it was unacceptable to have a zero velocity, since in the classical large n limit, one would expect the particle to move back and forth between the two walls of the box, as classical particles are expected to do.

---

[37]Or, which amounts to the same, with a phase S that is independent of $(x_1, \ldots, x_N)$. Ground state wave functions are, in general, real.

[38]See Appendix 2.C.

[39]In [70, Sect. 6.3], Bohm and Hiley give a general analysis of such an interaction. We will discuss only a concrete example.

Bohm's reply to Einstein [66] was that his theory was not contradicted by any experimental facts and that Einstein's demand was therefore arbitrary.[40] But it is nevertheless interesting to show that, even in the situation considered by Einstein, we obtain the usual quantum mechanical predictions if we "measure" the momentum of the particle, even though its original momentum is zero.

Let us first see how one can measure the momentum of the particle. One way to do it is to remove the walls of the box in which the particle is trapped and to detect the position x of the particle after a certain time t. Then x/t gives a "measurement of momentum" (we have set the mass $m = 1$), as we saw in Appendix 2.C. But removing the walls of the box modifies the wave function of the particle, and gives it a non-constant phase $S(x)$, and thus, because of (5.1.1.3), it also implies that the particle starts moving. This is explained mathematically in Appendix 5.D. We also show in that appendix, by explicit computation, that one recovers two kinds of motions in the limit of large n, one with a given velocity along the positive x axis and one with the same absolute velocity, but with opposite sign. So one does recover Einstein's expectation in that "classical" limit, but only after interacting with the particle (by removing the walls of the box).

One could go on and discuss more examples (see for example [70, 268] for a number of them), but we prefer here to draw a more general conclusion.

5.1.5   *"Contextuality" and Naive Realism About Operators*

The statement that the measurement of an observable depends on the concrete experimental arrangement used to "measure" it is true, not only for spin and momentum, as we saw, but also for all quantum mechanical "observables", other than position[41] [105]. John Bell repeatedly emphasized the misleading character of the word "measurement", because it gives the impression that some real property of the system is being measured. He raised the following "charge" against the use of this word:

> [...] the word comes loaded with meaning from everyday life, meaning which is entirely inappropriate in the quantum context. When it is said that something is 'measured' it is difficult not to think of the result as referring to some pre-existing property of the object in question. This is to disregard Bohr's insistence that in quantum phenomena the apparatus as well as the system is essentially involved.

John Bell [46, p. 34]

Moreover, it is because of "measurements" that we have this dual nature of the time evolution in quantum mechanics: Schrödinger's equation between measurements and collapse when they occur. Would it not be more natural to have a theory in which

---

[40]For another reply, see Gondran and Hoblos [234, 235] who, following Born [77], consider a non-stationary initial wave function and then obtain a de Broglie–Bohm motion close to the classical one.

[41]However, one can even have measurements of the position operator that do not measure positions. See [147, Sect. 7.5] for an example of this situation.

measurements were just ordinary physical processes, governed by the same physical laws as those acting outside of measurements? But, as we just saw, that is exactly what the de Broglie–Bohm theory does for us. The price to pay, so to speak, is that measurements do not in general measure anything pre-existing to them.

But that is no price to pay at all! Indeed, the logical consequence of the no hidden variables theorems is that measurements are "contextual", namely that one cannot think that a pre-existing value $v(A)$ is revealed when one "measures" the "observable" A, at least not for each A in a natural class of such observables. But this contextuality is naturally explained by the de Broglie–Bohm theory, since the result of measurement may depend, in each individual case, on the precise experimental setup of the measuring device (although the statistical results will not depend on this setup).

It is usually the orthodox discourse that falls into a sort of "naive realism with respect to operators" [105], by giving the impression that one measures the values of observables represented by matrices (or operators). Dürr, Goldstein, and Zanghì have clearly expressed what is wrong about this sort of realism:

> "Properties" that are merely contextual are not properties at all; they do not exist, and their failure to do so is in the strongest sense possible!
>
> Detlef Dürr, Sheldon Goldstein, Nino Zanghì [147]

The only thing we ever "see" are particle positions, and the calculus with operators allows us to compute the statistics of those positions in certain types of interactions, called, misleadingly, measurements.[42] But that does not mean that one has "measured" a physical quantity intrinsic to the quantum system.

5.1.6   What About the Collapse of the Quantum State?

The short answer, which we will elaborate in this subsection, is that there is never any collapse of the quantum state in the de Broglie–Bohm theory, but there is an effective collapse or a collapse "in practice", which coincides with the one in ordinary quantum mechanics, at least when the latter is unambiguous.[43]

As we saw, if a quantum state is a superposed state, namely a sum of two (or more) terms corresponding to different physical situations, for example spin up and spin down, or going through one slit or the other (in the double-slit experiment), then one has to keep both terms in order to predict the future behavior of the system correctly, although when the terms have disjoint supports, the particle will be guided by only one of them, the one whose support it happens to lie in. This is because the

---

[42]See [147] for a more detailed discussion.

[43]The last sentence is needed because in ordinary quantum mechanics the exact time when the collapse occurs is not well defined. It is defined by the "observation" but, as we discussed before, that notion itself is imprecise.

two terms may recombine later, in the sense that their supports overlap again, and then both terms will affect the behavior of the particle. This is what happens both in the double-slit experiment and in the Mach–Zehnder interferometer.[44]

On the other hand, we are supposed to keep only one term when we observe a quantum system. What is the difference? In order to observe something, we need the particle to interact with a macroscopic system, because that is the only sort of thing we can perceive. A macroscopic system means that N in (5.1.1.1) and (5.1.1.4) is large, of the order of Avogadro's number $N \sim 10^{23}$. Such a system could be any detector in a laboratory, a pointer pointing up or down, a cat that can be alive or dead, etc. In Appendix 5.E, following Bohm and Hiley [70, Sect. 6.1], we consider an ion chamber.

Suppose we have two terms, each of which corresponds to macroscopically disjoint situations. Consider again the example (2.5.1.1) of Chap. 2:

$$c_1 \begin{pmatrix} 1 \\ 0 \end{pmatrix} \varphi^{\uparrow}(z) + c_2 \begin{pmatrix} 0 \\ 1 \end{pmatrix} \varphi^{\downarrow}(z) , \qquad (5.1.6.1)$$

where $\varphi^{\uparrow}(z)$ and $\varphi^{\downarrow}(z)$ correspond to the last two pictures in Fig. 2.6. It is important to realize that the variable z is a function of a large number of variables $\sim 10^{23}$. For example, it could be defined as the center of mass of all the variables in the pointer.

The particle will be in the support of one of the terms. However, in principle we must keep both terms because they may recombine later. But is such a recombination possible in practice for *macroscopic* systems? The answer is no, and the reason is that, as we illustrate in Appendix 5.E, one would have to get the support of the quantum state to overlap again for *each* of the N variables. And while this is possible for small N, it is quite a different matter to do it for $N \sim 10^{23}$. Thus, if we can be sure that no overlap will occur in the future between the two terms, we can simply keep the term in the support of which the particle happens to lie (and we know which one it is because of the coupling between the particle and the macroscopic device, by simply looking at the latter), as far as the predictions for the future behavior of the system are concerned.[45]

So, in some sense, we do "collapse" the quantum state when we look at the result of an experiment. But this is only a practical matter. We can still consider that the true quantum state is and remains forever given by the time evolution of the full quantum state (5.1.6.1). It is simply that one of the terms of the quantum state no longer guides the motion of the particle, either now or at any time in the future.

The measuring process here is an entirely physical process, with no role whatsoever left to the observer. And the latter only uses the reduction of the quantum

---

[44]See Bell [38] for a discussion, similar to the one given here, of the effective collapse of the quantum state in the de Broglie–Bohm theory, but in the example of the double-slit experiment.

[45]It is sometimes thought that, for such an effective collapse to occur, one needs the measuring device to interact with an environment, such as the air molecules surrounding it, and ultimately the entire universe. But that is not true: any sufficiently macroscopic device suffices, even if the latter were perfectly isolated from the rest of the universe (which is never the case, but that is not relevant since what we say would be true even if perfect isolation were possible).

state as a practical tool for further calculations on the system. We also see that this reduction is a matter of degree: as the number N increases it becomes more and more difficult to make the two terms of the quantum state interfere in the future, but there is no N for which there would be a sharp jump from a non-reduced quantum state to a reduced one.

This effect, namely the fact that the two terms in (2.5.1.1) will not overlap or interfere in the future, is called *decoherence* and has been the subject of an extensive literature. But, for our purposes, the basic idea, already given by David Bohm [62, Part1, Sect.7] in 1952 and outlined here, is already sufficient.

This is also very similar to what happens in classical statistical mechanics. It is well known that Newton's equations are "time-reversal invariant". This means that if we let any system of N particles, whose positions are $\mathbf{X}(t) = (X_1(t), \ldots, X_N(t))$, evolve from time 0 to time T and, at time T, reverse the velocities of all the particles, then the system will retrace its path backwards until it reaches, at time $2T$, its initial positions, with all the velocities reversed.

To give a concrete example of this behavior, consider a box divided in two halves by a wall, with all the particles in one half of the box, and let us remove the wall at time $t = 0$. Then the particles will disperse themselves throughout the whole box. We never see them going back to the half box from which they started, although we just gave a mechanism (reverse the velocities of all the particles) that will, in principle, produce such a result. But what is possible in principle becomes impossible in practice once N is large, although it would be possible if N were small.[46] We say that the behavior is irreversible for large N. But, just like the reduction of the quantum state, this is a practical consideration, not a matter of principle, and it is a question of degree: there is no sharp jump from microscopic and reversible to macroscopic and irreversible as N increases. Reversal of the velocities just becomes more and more difficult, in practice, as N increases. If one accepts that there is no mystery in the irreversible behavior of classical statistical mechanical systems, one should likewise accept that there is no mystery in the reduction of the quantum state within the de Broglie–Bohm theory.

Moreover, once the particles are more or less homogeneously distributed throughout the whole box, we can take this homogeneous distribution as a new "initial" state for all future behavior of the system, i.e., we can forget the fact that all the particles came from one half box in the past. For example, suppose that our box is part of a second, larger box, and that we let the gas expand into that larger box once it is homogeneously distributed in the first box. Then, there will be no noticeable difference between the behavior of the gas if we assume that it starts from a uniform distribution in the first box or if we remember that it came earlier from one half of that box.

This is again very similar to the situation in quantum mechanics, where one can, in practice, use the reduced state, after the measurement, as the "real" state of the system, as far as predicting its future behavior is concerned, and forget about the other

---

[46]For more details, see, e.g., [80, 225, 307].

"branches" of the quantum state, where the branches refer to the different terms in (2.5.1.1) or in any other macroscopic superposition, such as (5.1.6.2).

Finally, let us stress that there is a rather common misconception about decoherence, namely that this phenomenon not only allows us to make sense of the effective collapse (hence to solve the "measurement problem") within the de Broglie–Bohm theory, but that it is also, on its own, sufficient to account for the collapse. The idea is roughly that, if the different terms in the sum involving a macroscopic object like (5.1.6.1) do not overlap, then we just pick up the one we see at the end of the experiment in order to predict the future behavior of the system. The problem with this view is that, if there is nothing in the "outside world" to distinguish that term from the others in the sum, we get back to putting the observations in the very formulation of our physical theories, which is what the de Broglie–Bohm theory manages to avoid.[47]

There are more sophisticated ways to try to make decoherence the cornerstone of a solution to the measurement problem, such as the many-worlds interpretation and the decoherent histories approach, but these will be discussed in Chap. 6.

We note finally that this analysis can be extended without change to the general measurements described in Appendix 2.D, resulting in states of the form

$$\sum_n c_n \varphi_0(z - \lambda_n t) |e_n(x)\rangle . \tag{5.1.6.2}$$

For t not too small, in each term of the sum, $\varphi_0(z - \lambda_n t)$ will correspond to macroscopically different situations (for example, different positions of a pointer) and the terms will in practice "decohere", i.e., it will be impossible to make them overlap again.

5.1.7 What Are the Meaning and the Origin of the Statistical Assumptions on the Initial Conditions in the de Broglie–Bohm Theory?

As we saw, we recover the statistical predictions of quantum mechanics if we assume that the initial conditions in a given experiment are distributed with the probability density $|\Psi(x, 0)|^2$, where $\Psi(x, 0)$ is the initial wave function of the system, and, for simplicity, $x \in \mathbb{R}$. This assumption is called the *quantum equilibrium* assumption. But what does this assumption mean exactly and how can it be justified?

As regards its meaning, we already sketched it in Sect. 5.1.3 and Fig. 5.3. If we have N particles in one spatial dimension, each of which have the same wave function $\Psi(x, 0)$, then we assume that the positions of the particles are independent random variables with identical distribution $|\Psi(x, 0)|^2$. By the law of large numbers, this

---

[47]There is also the crucial question, which we put aside for now, of what it could mean for the world to *be* a wave function (see Sect. 6.1).

means that the empirical density distribution of the particles (in the sense explained in Sect. 5.1.3 and in Appendix 3.A) will be given by the density $\rho(x) = |\Psi(x, 0)|^2$.

But this raises the question: why should the initial positions be distributed according to $|\Psi(x, 0)|^2$? By the equivariance (5.1.3.2) of the $|\Psi(x, t)|^2$ distribution, we know that any system that has the distribution $|\Psi(x, 0)|^2$ at the initial time 0 will have the corresponding distribution $|\Psi(x, t)|^2$ at a later time. So, ultimately, we just have to assume that the particles were distributed according to the $|\Psi(x, 0)|^2$ density at the beginning of the universe. Having to make assumptions about the initial conditions of the universe is something that always makes some people uneasy. But in a deterministic world, assumptions about initial conditions always ultimately refer to those of the universe.

In fact, if one wants to account for the increase of entropy (even from the point of view of classical statistical mechanics), one has to assume that the universe started in a very low entropy state, namely a very special or "improbable" one,[48] and that assumption is less easy to justify than the one we have to make in the de Broglie–Bohm theory. But since there is only one universe, and not an ensemble of universes, what does it mean to make statistical assumptions about it (whether in a classical or quantum context)? Obviously, the empirical significance of such an assumption can only come from empirical distributions of a large number of identical subsystems of the "universe". But to make sense of that, we must first explain what the wave function of a subsystem means.

In ordinary quantum mechanics, one assumes that a system is "prepared" in a given quantum state, which means that, by performing a measurement on a system and looking at the result, one knows, because of the collapse rule, which quantum state the system is in after that measurement.[49] So let us consider what happens in a measurement. We write the coordinates of the particles in the system and the apparatus as (X, Y), X for the system, Y for the apparatus (or the environment). After the measurement, the wave function can be written as

$$\Psi(x, y) = \psi(x)\Phi(y) + \Psi^{\perp}(x, y) , \tag{5.1.7.1}$$

where the lower case x, y denote the generic arguments of the functions, as opposed to (X, Y), which denote the actual positions.[50] We assume that the actual coordinate Y is in the support of Φ and that $\Psi^{\perp}$ denotes the "empty" part of the wave function which, as we saw when we analyzed the measurement process in the previous subsection, has a support in the y variables that is disjoint from the support of $\Phi(y)$ and remains so in the future.

---

[48] See [185, Chap. 5] and [389, Chap. 7] for pedagogical discussions of this idea.

[49] Assuming that the eigenvalue that is observed is non-degenerate.

[50] We follow the notation of [141] and do not use boldface letters, even though x, y, X, Y generally belong to a space of more than one dimension. This is because we will use boldface letters below for coordinates of several copies of the same system.

For example, in the quantum state (5.1.6.1), if the pointer is in the support of $\varphi^\uparrow(z)$, then $\psi(x)\Phi(y)$ in (5.1.7.1) will correspond to $\begin{pmatrix} 1 \\ 0 \end{pmatrix} \varphi^\uparrow(z)$ and $\Psi^\perp(x, y)$ to $\begin{pmatrix} 0 \\ 1 \end{pmatrix} \varphi^\downarrow(z)$. In the more general situation (5.1.6.2), if the pointer is in the support of $\varphi_0(z - \lambda_k t)$ for some k in the sum, then $\psi(x) = |e_k(x)\rangle$, and $\Phi(z) = \varphi_0(z - \lambda_k t)$, while $\Psi^\perp(x, y) = \sum_{n \neq k} c_n \varphi_0(y - \lambda_n t)|e_n(x)\rangle$.

The function $\psi(x)$ is then called in [141] the *effective wave function* of the system.[51] This is simply the usual quantum mechanical wave function of a system that has been "prepared" in a given quantum state by a previous measurement.

If we assume a quantum equilibrium distribution for the "universe" composed of the system and its environment for which the probability density is

$$P(X = x, Y = y) = |\Psi(x, y)|^2 , \tag{5.1.7.2}$$

and if $y \in \mathrm{supp}\, \Phi$, then we get the conditional probability density

$$P(X = x | Y = y) = |\psi(x)|^2 , \tag{5.1.7.3}$$

where $\psi(x)$ is the effective wave function of the system.[52] We did not include the time variable here, but this holds for any given time t.

In order to understand the empirical significance of (5.1.7.3), we need to consider a system composed of a large number N of identical systems, with position variables $\mathbf{x} = (x_1, \ldots, x_N)$, and the same effective wave function $\psi(x_i)$, $i = 1, \ldots, N$. We can think of several measurements of similarly prepared systems, not necessarily in the same place.[53] We assume these systems to be non-interacting with one another. Note that *all* experiments testing the statistical predictions of quantum mechanics actually have this form. Even if one considers sets of particles that are interacting, all the statistical tests are made with many non-mutually-interacting copies of those sets of particles. Then, by (5.1.7.1), for each $i = 1, \ldots, N$, we have

$$\Psi(x_i, y_i) = \psi(x_i)\Phi_i(y_i) + \Psi_i^\perp(x_i, y_i) , \tag{5.1.7.4}$$

---

[51] In the situations where the decomposition (5.1.7.1) does not hold, there is a more general notion, the *conditional wave function* [141]:

$$\Psi(x) = \Psi(x, Y) ,$$

where x are the generic variables of the system and Y is the actual position of the environment. Of course, $\Psi(x, Y) = \psi(x)\Phi(Y)$ when the decomposition (5.1.7.1) holds.

[52] We have $P(X = x | Y = y) = P(X = x, Y = y)/P(Y = y)$, and $P(Y = y) = \int P(X = x, Y = y)dx = \int |\Psi(x, y)|^2 dx$. By (5.1.7.1), if $y \in \mathrm{supp}\, \Phi$, the latter is equal to $\int |\psi(x)\Phi(y)|^2 dx = |\Phi(y)|^2$, since $\int |\psi(x)|^2 dx = 1$. So finally, $P(X = x, Y = y)/P(Y = y) = |\psi(x)\Phi(y)|^2/|\Phi(y)|^2 = |\psi(x)|^2$.

[53] Nor necessarily at the same time either, although the treatment of measurements at different times is more subtle, see [141, Sects. 8–10].

where we write only the arguments x_i, y_i, the other variables being considered as parameters.

One can show that, if Y_i is in the support of Φ_i for each i, this yields the *factorization formula*, proven in Appendix 5.F:

$$\Psi(x_1, \ldots, x_N, \mathbf{Y}) = \psi(x_1) \cdots \psi(x_N) \Phi(\mathbf{Y}) , \qquad (5.1.7.5)$$

where $\mathbf{Y} = (Y_1, \ldots, Y_N)$, and Φ is a function only of $\mathbf{Y}$. So here $\psi(x_1) \cdots \psi(x_N)$ is the effective wave function of the system composed of N identical subsystems. This implies, as in (5.1.7.3), that the probability density of $X = (x_1, \ldots, x_N)$, given $\mathbf{Y} = (Y_1, \ldots, Y_N) = \mathbf{y}$, with y_i in the support of Φ_i for each i, is

$$P(X_1 = x_1, \ldots, X_N = x_N | \mathbf{Y} = \mathbf{y}) = |\psi(x_1)|^2 \cdots |\psi(x_N)|^2 , \qquad (5.1.7.6)$$

which is the probability density of independent random variables $(x_1, \ldots, x_N)$, with identical distribution $|\psi(x_i)|^2$, $\forall i = 1, \ldots, N$ (see Appendix 3.A).

By the law of large numbers, this implies that the empirical distribution of the results of position measurements of a large number of identical systems with wave function ψ will satisfy Born's rule. But once that rule is satisfied for the position measurements, it is satisfied for all other "observables", as we saw in the examples of the spin and momentum measurements (see [141, 147] for an analysis of the general case).

So once the laws of physics are given and once the initial quantum state $\Psi(\mathbf{x})$ of the universe is chosen, with $\mathbf{x} = (x_1, \ldots, x_N)$ being the variables associated with all the particles in that universe, the only thing the physicists' metaphorical god had to choose was a set of positions for all the particles of that universe. How did he do it?

Let us consider, for simplicity, a "universe" composed of N copies of the same microscopic system, each system having a wave function $\psi(x)$. In order to account for the Born rule, all we have to assume is that this set of positions was chosen to be *typical* with respect to the distribution $|\Psi(\mathbf{x})|^2 = |\psi(x_1)|^2 \cdots |\psi(x_N)|^2$, meaning that the empirical density distribution of those particles[54] agrees to a very good approximation with $|\psi(x)|^2$. And once this holds at the initial time, it will hold for all later times, because of the equivariance of that distribution.

How should one regard that last assumption, i.e., the *quantum equilibrium* assumption? There are at least two ways to try to justify it. The first [141] is to say that, because of the equivariance of $|\Psi(\mathbf{x})|^2$, this is a natural assumption, with no need for further justification, since, once it is postulated to hold at a given time, it will hold at all times. That initial time can then be taken to be the origin of the Universe. The quantum equilibrium assumption is then viewed as analogous to the equilibrium distribution in statistical mechanics (hence the name): in statistical mechanics, it is any non-equilibrium distribution that has to be justified, since the equilibrium distribution is the most probable one. And the existence of a non-equilibrium distribution at the beginning of the universe is simply postulated, in order to account for the

---

[54]See Sects. 3.4.4 and 3.4.5 and Appendix 3.A for a definition of those notions.

increase in entropy, with no deeper justification. Therefore, the quantum equilibrium assumption is certainly more natural than the usual one made in statistical mechanics in order to account for the law of increase of entropy.

However, it should be emphasized that this justification is not natural from a Bayesian point of view. For example, if we consider particles in a one-dimensional box $[0, L]$, with the ground state wave function

$$\psi(x) = \sqrt{\frac{2}{L}} \sin \frac{\pi x}{L} ,$$

then the quantum equilibrium distribution is

$$\frac{2}{L} \left| \sin \frac{\pi x}{L} \right|^2 ,$$

while from a Bayesian or "indifference" point of view, one would rather take a uniform distribution, with density $1/L$. If we adopt that approach, the quantum equilibrium assumption has to be considered as an independent principle.

Another line of thought, in order to justify the quantum equilibrium assumption, is to argue that, even if the universe started in a non-quantum-equilibrium state, it would quickly evolve towards quantum equilibrium under the de Broglie–Bohm dynamics. This is far from being proven in general, but there is some good numerical evidence for it in special cases [100, 487].[55]

Of course, if the universe was in a non-quantum-equilibrium state in the past, the Born rule would not have been satisfied then. Moreover, as we will see in Sect. 5.2.1, a non-equilibrium distribution generally allows instantaneous transmission of messages. But provided that such an initial non-quantum-equilibrium distribution has disappeared and that we are now in quantum equilibrium, no contradiction arises from the assumption of a non-quantum-equilibrium initial state of the universe. On the other hand, it is not clear what one gains by assuming an initial non-equilibrium distribution, unless one finds evidence that such a non-equilibrium distribution has existed in the past.[56]

5.1.8 Heisenberg's Relations and Absolute Uncertainty

The Heisenberg uncertainty relations[57] do not conflict with the de Broglie–Bohm theory since the statistical predictions of the latter are the same as those of ordinary

---

[55]This convergence to equilibrium cannot be true in full generality since, for any real wave function $\Psi(\mathbf{x})$ nothing moves, hence all distributions are stationary in that situation and do not converge to anything. There have also been some attempts to prove convergence to equilibrium theoretically [67, 481–484].

[56]See [484, 485, 488] for a discussion of a possible non-equilibrium distribution in the early universe.

[57]See Appendix 2.C.

quantum mechanics, and the Heisenberg uncertainty relations are only relations between the variances of those statistical predictions. But since, according to the de Broglie–Bohm theory, particles do have positions and velocities at all times, why can't we measure them with more accuracy than is allowed by the quantum formalism?

The answer lies in Eq. (5.1.7.3) or (5.1.7.6). Everything we can know about a system, if we assume that the universe is in quantum equilibrium, is encoded in the values of the **Y** variables, which are those of the environment or the measuring devices. But even if we knew everything, down to the position of every particle in the environment, which is obviously far more than we can ever know (even if the world obeyed classical laws), the probability density, conditioned on that knowledge, is given by the right-hand side of (5.1.7.3) or (5.1.7.6). So whichever measurement we perform, we will never obtain a better knowledge of the system than the one encoded in that quantum equilibrium distribution.

Moreover, since ψ has a dynamical role and not only a probabilistic one, that distribution will tend to become more spread out under the time evolution [as we saw, for the case of a free evolution in (2.A.2.24)], so that the initial uncertainty can only grow. This is what Dürr, Goldstein, and Zanghì call *absolute uncertainty* [141, Sect. 11]. It goes beyond the Heisenberg uncertainty relations, the latter being a consequence of that fundamental uncertainty.[58]

An intuitive way to understand this is by reflecting on the fact that all our knowledge of the world depends on the latter being in a non-equilibrium situation. Indeed, if the world was in equilibrium, there would be no brains to start with and no recording devices either: think for example of a homogeneous gas or fluid as examples of equilibrium states; they are too structureless to record information. In other words, the world is assumed, in the de Broglie–Bohm theory, to be in quantum equilibrium, relative to the universal wave function. But the universal wave function itself is in a situation similar to the positions and momenta of all the particles of the universe in classical statistical mechanics and therefore has to be far from a thermal equilibrium state for anything interesting (like us, humans) to be around.[59]

Finally, let us remark that if one measures the spin of a single particle in a given direction, noted 1, then in another direction noted 2 and one repeats the operation (measuring again in direction 1, then 2 etc.) an arbitrary number of times, after each measurement, the conditional probability density will always be given by the

---

[58]Indeed, if we perform, say, a position measurement and, as a result we obtain a wave function with a relatively narrow support $\psi(x)$, then, of course, a subsequent measurement of momentum will start from that wave function and will have a variance Var(p) related by (2.C.1.4) to the variance Var(x) of the original x measurement (see Appendix 2.C for the computation of those variances for Gaussian wave functions). On the other hand, if we consider the example of a particle in a box, discussed in Sect. 5.1.4 and Appendix 5.D, the initial momentum is zero and therefore has a variance equal to zero. But, since what we "measure" is not that initial momentum but the asymptotic position of the particle after removing the walls of the box, and thus setting the particle in motion, the result of that "measurement" will coincide with the ordinary quantum predictions (see Appendix 5.D), and will therefore also satisfy the Heisenberg uncertainty relations.

[59]See [141, Sects. 12–14] for more details.

right-hand side of (5.1.7.3) for the ψ that is being observed (either up or down). So that, no matter how many measurements we perform, we can never predict the future better than what is allowed by the wave function at any given time. This sounds counterintuitive, because, after all, if we fix the initial wave function and the initial position of the particle, as well as the measuring procedure for each measurement, the entire sequence of results is determined.

Actually, this was also the basis of an objection, by von Neumann,[60] to the idea of hidden variables: if one makes those alternating measurements, in direction 1, then 2, etc., sufficiently often, then eventually one should be able to narrow down the range of the values of those hidden variables so that the future would become predictable with more precision than is allowed by quantum mechanics. The above analysis shows that this is not true. The system behaves in a way which is similar to the map $f : x \to 10x \mod 1$ on $I = [0, 1[$ encountered in Sect. 3.4.2. Here, too, the system is deterministic but if our "observations" consist in recording which is the first digit a_1 of $x = 0.a_1a_2a_3 \ldots$, then a_2 of $f(x) = 0.a_2a_3 \ldots$, a_3 of $f(f(x)) = 0.a_3a_4 \ldots$, etc., then no matter how many such observations we make, there will always be an equal probability of finding any $a_i = 0, 1, 2, \ldots, 9$ at the next step.[61]

5.1.9 What Is the Relationship Between the de Broglie–Bohm Theory and Ordinary Quantum Mechanics?

The answer to this question may shock certain people, but *it is the same theory!* Or, more precisely, the de Broglie–Bohm theory *is* a theory, while ordinary quantum mechanics is not. Indeed, quantum mechanics doesn't even pretend to be a theory, but rather claims to be an algorithm allowing us to compute "results of measurements". Another way to say this is that ordinary quantum mechanics is the algorithm used to compute results of measurements that is *derived* from the de Broglie–Bohm theory.

One might also say that ordinary quantum mechanics is simply a truncated version of the de Broglie–Bohm theory: in ordinary quantum mechanics, one ignores the particle trajectories, but since the empirical predictions of the de Broglie–Bohm theory are statistical, and are the same as those of ordinary quantum mechanics, there are no practical consequences of that omission.

Thus, ordinary quantum mechanics is sufficient "for all practical purposes" (FAPP) to use Bell's expression [46]. But it is the de Broglie–Bohm theory that *explains* why ordinary quantum mechanics is sufficient FAPP, something that is true but mysterious without de Broglie–Bohm. In particular, both the collapse rule and the statistical nature of the predictions find a natural explanation within the

---

[60] At least according to Wigner, who told the story as though von Neumann was his "friend", see [516, Note 1, p. 1009].

[61] Here, the probability refers to the Lebesgue measure on the interval $[0, 1[$, which amounts to considering all the symbols a_i to be independently distributed and giving equal probability to each possible value of $a_i = 0, \ldots, 9$.

de Broglie–Bohm theory, due firstly to the analysis of measurement as a purely physical mechanism, always governed by the Schrödinger equation, and secondly to the hypothesis of quantum equilibrium.

5.2   Some Natural Questions About the de Broglie–Bohm Theory

5.2.1   How Does the de Broglie–Bohm Theory Account for Nonlocality?

In Chap. 4, we gave an argument combining those of EPR and Bell which proved that there exist nonlocal effects in Nature. But we did not say what these effects were due to, nor how we could explain them. Indeed, in order to do that, we have to go beyond ordinary quantum mechanics, where nonlocality only manifests itself through the "collapse rule", whose status is unclear.

On the other hand, the de Broglie–Bohm theory "explains" the nonlocality, insofar as it can be explained. At least, one can see what is going on in those nonlocal effects. We will explain this in three steps: first in general terms, then in a simple example, and finally, in the EPR–Bell situation.

The wave function is defined on configuration space, not on ordinary space. As we saw in Sect. 5.1.1, this is a major difference with respect to electromagnetic fields, which are always defined on $\mathbb{R}^3$, independently of the number of particles producing them. The fact that the wave function is defined on configuration space is the source of nonlocality, because it implies that the motion of each particle depends on the positions of all the other particles. This in turn happens because, in (5.1.1.3), $S(X_1(t), \ldots, X_N(t), t)$ is evaluated at all the actual positions of all the particles. This means, say for $N = 2$, that if we change the position $X_1(t)$ of the first particle, we will in general change the value of $S(X_1(t), X_2(t))$ and of its derivatives, and therefore the velocity of $X_2(t)$, no matter how far apart $X_1(t)$ and $X_2(t)$ happen to be.

There is, however, an important exception, namely when the quantum state factorizes into a product of states, one for each particle: $\Psi(x_1, \ldots, x_N) = \prod_{j=1}^{N} \Psi_k(x_j)$. Then the phase

$$S(X_1(t), \ldots, X_N(t), t) = \sum_{j=1}^{N} S_j(X_j(t), t) ,$$

and taking derivatives, we get, for each $k = 1, \ldots, N$,

$$\frac{d}{dt} X_k(t) = \frac{1}{m_k} \frac{\partial}{\partial x_k} S_k(X_k(t), t) , \qquad (5.2.1.1)$$

so that each particle then follows its own law, independently of the other particles.

This is the situation which is analogous to the Hamiltonian being a sum of terms in classical mechanics, implying that there is no direct interaction between the particles. But, and this is the main difference with the classical situation, in the de Broglie–Bohm theory, it is not enough for there to be no direct interaction between the particles for them to move independently of each other: one must also have factorization of the quantum state into a product of states in order to obtain an equation of motion of the form (5.2.1.1). The example of the quantum state (4.4.2.1) in Chap. 4 shows that one can have non-factorization, even when there is no direct interaction between the particles.

Nevertheless, if the quantum state factorizes approximately, then the particles do behave approximately independently of each other. One expects this to occur typically for macroscopic systems, and as a result, the world is more or less local. This then explains why Casimir's worry [93], mentioned in Chap. 4, about "the results of experiments on free fall here in Amsterdam" depending appreciably "on the temperature of Mont Blanc, on the height of the Seine below Paris and on the position of the planets", will never really give cause for concern.

To illustrate the nonlocality of de Broglie–Bohm theory, consider two particles in one dimension (to simplify matters), with (X_1, X_2) denoting the positions of the particles, and let V be a potential localized around $x_1 = a$ and acting on particle 1 (Fig. 5.9). This affects $\Psi(x_1, x_2)$ via the Schrödinger equation:

$$i \frac{\partial}{\partial t} \Psi(x_1, x_2, t) = H \Psi(x_1, x_2, t) , \qquad (5.2.1.2)$$

with

$$H = -\frac{1}{2} \left(\frac{\partial^2}{\partial x_1^2} + \frac{\partial^2}{\partial x_2^2} \right) + V(x_1 - a) , \qquad (5.2.1.3)$$

where $a \in \mathbb{R}$ and we set the masses equal to 1. We assume that $V(\cdot)$ is a function localized around 0. The presence of V will affect the value of $\Psi(x_1, x_2)$, at least when x_1 is close to a (in fact, in general, the effect propagates in time even to values of x_1 far away from a, but we leave that aside).

Fig. 5.9 An illustration of nonlocality in the de Broglie–Bohm theory

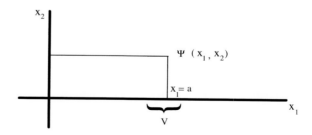

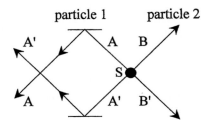

Fig. 5.10 Paths A, A', B, and B' with two mirrors. Reproduced with permission from D.A. Rice: *A geometric approach to nonlocality in the Bohm model of quantum mechanics*, American Journal of Physics **65**, 144–147 (1997). Copyright 1997, American Association of Physics Teachers

This, in turn, affects the behavior of the second particle via the guiding equation

$$\dot{X}_2 = \frac{\partial}{\partial x_2} S(X_1, X_2, t) \,, \qquad (5.2.1.4)$$

where $\Psi(x_1, x_2, t) = R(x_1, x_2, t)e^{i S(x_1, x_2, t)}$. The state Ψ can be viewed as the graph of a surface above the (x_1, x_2) plane (which is complex in general). Changing that surface around some value $x_1 = a$ affects it right along the line in the plane where the coordinate $x_1 = a$, and this may affect the motion of a particle located at X_2, if X_1 is close to a, even if X_2 is arbitrarily far away from the value a where the wave function is affected, in its x_1 variable.

To understand all this more intuitively, let us consider the following example, due to Dien Rice [416]. Figure 5.10 shows a two-particle quantum state, emitted from the source S, and which is a superposition of "particle 1 following path A and particle 2 following path B" and of "particle 1 following path A' and particle 2 following path B'":

$$|\Psi\rangle = \frac{1}{\sqrt{2}}\big(|1A\rangle|2B\rangle - |1A'\rangle|2B'\rangle\big) \,. \qquad (5.2.1.5)$$

Of course, in ordinary quantum mechanics, the particles do not follow any path, so the paths in Fig. 5.10 are just paths along which the wave function propagates. On the other hand, in the de Broglie–Bohm theory, the particles do follow paths, determined by the part of the wave function $|1A\rangle|2B\rangle$ or $|1A'\rangle|2B'\rangle$ whose support they happen to lie in. In Fig. 5.10, there are mirrors along the possible paths followed by the wave function 1 that reflect it, so that the paths A and A' cross. Now, if we detect particle 2 along path B (at the top of Fig. 5.10), this means that the support of the particles is in the part $|1A\rangle|2B\rangle$ of the wave function, and therefore that particle 1 will be found along path A, if detected (that is, at the bottom of Fig. 5.10), and vice versa for paths B' and A'.

Let us now introduce more mirrors, this time along the possible paths followed by wave function 2, as in Fig. 5.11. To proceed with our argument, we need an important but elementary property of the de Broglie–Bohm evolution (5.1.1.3): the paths of the particles can never cross each other, at the same time, in the space $\mathbb{R}^4$

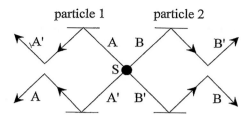

Fig. 5.11 Paths A, A', B, and B' with four mirrors. Reproduced with permission from D.A. Rice: *A geometric approach to nonlocality in the Bohm model of quantum mechanics*, American Journal of Physics **65**, 144–147 (1997). Copyright 1997, American Association of Physics Teachers

(each particle's motion takes place in a plane $\mathbb{R}^2$). This is because (5.1.1.3) is a first order equation, and the velocities are uniquely determined by the positions, at any given time. Crossing of trajectories at the same time is therefore impossible, since it would mean that the same positions (at the point where they cross) could correspond to two different velocities (each pointing in the direction of one of the crossing paths).

But then, the only possibility is that the paths *followed by the particles* bounce off each other,[62] as in Fig. 5.11. What the arrows in that figure now indicate are the possible particle paths. Note, however, that the paths A, A', B, B' of the *wave functions* cross each other, as in Fig. 5.10 for A, A' and now symmetrically for B, B', contrary to the particle paths.[63]

Thus, now, if particle 1 starts out in the support of the part A of the wave function, it will end up in the support of part A' and vice versa; and similarly for particle 2 and the support of parts B, B' of the wave function. When the paths of the wave functions cross each other, the particles will go from being in the support of the part $|1A\rangle|2B\rangle$ of the quantum state (5.2.1.5) to being in the support of the part $|1A'\rangle|2B'\rangle$. The particles switch horses so to speak, i.e., they go from being guided by one of the wave functions to being guided by the other. So if the particle 2 is detected at the top of Fig. 5.11, particle 1 will also be detected at the top of that figure (unlike what happened without the mirrors on the possible paths of particle 2).

This illustrates how nonlocality works in the de Broglie–Bohm theory: by putting mirrors along the possible paths of particle 2, one instantaneously modifies the path followed by particle 1 since, without those mirrors, if particle 1 starts in the support of the part $|1A\rangle|2B\rangle$ of the quantum state (5.2.1.5), it will end at the bottom of Fig. 5.10, while if the mirrors are inserted, it will end at the top.

---

[62]Of course, this assumes a perfect symmetry between the two parts of Fig. 5.11.

[63]The Schrödinger equation (5.1.1.1) is also first order in time, but that does not imply that the paths followed by the wave functions cannot cross each other. Indeed, in order for that crossing to be forbidden, one would need the right-hand side of (5.1.1.2) to coincide for the two wave functions when they cross, i.e., the wave functions and their second derivatives would have to coincide, and there is no reason why that should happen.

Of course, since the situation is symmetrical, one could insert or not mirrors along the possible paths of particle 1 (assuming that there are already mirrors along the possible paths of particle 2) and instantaneously affect the path taken by particle 2.

However, since this does not affect the statistics of what one detects, it cannot be used to send messages instantaneously: whatever is done to particle 2, one will detect particle 1 half of the time on each of the paths A or A'.

But this last result depends on the fact that we are in quantum equilibrium. Suppose that we had an ensemble of particles such that, even though the wave function is still given by (5.2.1.5), all the particles are in the support of the first term in (5.2.1.5) (which would of course constitute an initial state that was not in quantum equilibrium). Thus, all the particles 1 will initially follow path A and all the particles 2 will follow path B. Then by inserting or not inserting a mirror along the paths of particle 2, we can ensure that, after the trajectories cross or bounce off one another, the detection of particle 1 will be either at the top (if the mirrors are inserted) or at the bottom of the picture (if the mirrors are not inserted). By repeating the experiment several times, this would of course allow the person who controls the mirror on the paths of particle 2 to send a message to the person who detects particle 1. It is enough to associate a symbol 0 with the absence of the mirrors and 1 to their presence. Then any sequence of 0 and 1 s can be transmitted by repeating the experiment several times.[64] Here we see why the hypothesis of quantum equilibrium is crucial: without it, one could send messages instantaneously.[65]

Finally, let us analyze the EPR–Bell experiments in the framework of the de Broglie–Bohm theory. This is illustrated in Figs. 5.12 and 5.13. We start with a quantum state like (4.4.2.1), but consider only the spins in direction 1 and reintroduce the spatial part of the state:

$$|\Psi\rangle = \frac{1}{\sqrt{2}}\Big[|A\,1\,\uparrow\rangle|B\,1\,\downarrow\rangle\Psi_A(x_A, z_A)\Psi_B(x_B, z_B) - |A\,1\,\downarrow\rangle|B\,1\,\uparrow\rangle\Psi_A(x_A, z_A)\Psi_B(x_B, z_B)\Big],$$
$$(5.2.1.6)$$

where x_A, x_B, z_A, z_B are the x and z coordinates of particles A and B, which are as in Fig. 5.4: the z coordinates denote the vertical direction in which the particle may be deflected, and the x coordinates denote the direction in which the particle moves towards the measuring devices (we will neglect the x coordinates below).

Consider Fig. 5.12. If we measure first[66] the spin of particle A in direction 1, we will get the up result, given the initial particle position, assuming that things are symmetric, as in Fig. 5.4. But this means that the quantum state (5.2.1.6) becomes

---

[64]The fact that a non-equilibrium distribution would allow us to send messages instantaneously is in fact quite general [486].

[65]And, because of the relativity of simultaneity, discussed in the next subsection, one could in principle send them into one's own past, at least when one does not introduce an additional structure, such as a preferred foliation into spacetime. See [319, Chap. 4] and [48] for a discussion of the relationship between nonlocality and the sending of messages.

[66]Readers may wonder what "first" and "later" mean in a relativistic framework. This will be discussed in the next subsection.

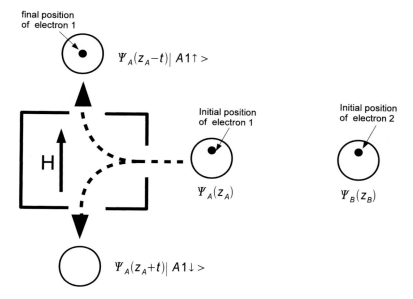

Fig. 5.12 Measurement of the spin on the left first

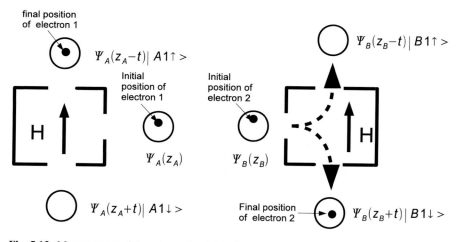

Fig. 5.13 Measurement of the spin on the right, after a measurement on the left

effectively $|A\ 1\ \uparrow\rangle|B\ 1\ \downarrow\rangle\Psi_A(z_A - t)\Psi_B(z_B)$, since the particle A is in the support of that part of the state, and the other part $|A\ 1\ \downarrow\rangle|B\ 1\ \uparrow\rangle\Psi_A(z_A + t)\Psi_B(z_B)$ now has a support disjoint from $|A\ 1\ \uparrow\rangle|B\ 1\ \downarrow\rangle\Psi_A(z_A - t)\Psi_B(z_B)$.

This implies that particle B must also be guided only by that part of the wave function. Since the spin part of the quantum state is $|B\ 1\ \downarrow\rangle$, it will go down in all cases, irrespective of its initial position, if one measures its spin in direction 1, after

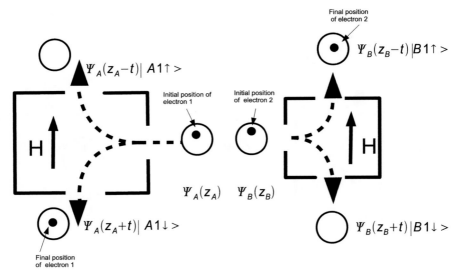

Fig. 5.14 Measurement of the spin on the right first

the measurement at A, as in Fig. 5.13. The final quantum state for the two particles will be $|A\ 1\ \uparrow\rangle|B\ 1\ \downarrow\rangle\Psi_A(z_A - t)\Psi_B(z_B + t)$.

Consider now Fig. 5.14. There one measures the spin of particle B first (in direction 1). Since the initial position is above the symmetry axis, as in Fig. 5.13, particle B goes up and the quantum state of the two particles becomes $|A\ 1\ \downarrow\rangle|B\ 1\ \uparrow\rangle\Psi_A(z_A)\Psi_B(z_B - t)$. But then a later measurement of the spin of particle A in direction 1 yields the result down, again irrespective of the initial position of that particle, and the final quantum state for the two particles is $|A\ 1\ \downarrow\rangle|B\ 1\ \uparrow\rangle\Psi_A(z_A + t)\Psi_B(z_B - t)$.

Finally, if the orientation of the magnetic field is reversed (as in Fig. 5.8), if one measures first the spin of particle A, one obtains the quantum state $|A\ 1\ \downarrow\rangle|B\ 1\ \uparrow\rangle\Psi_A(z_A - t)\Psi_B(z_B)$ and a later measurement at B produces the up result, irrespective of the initial position of the B particle (see Fig. 5.15). So the final quantum state for the two particles is $|A\ 1\ \downarrow\rangle|B\ 1\ \uparrow\rangle\Psi_A(z_A - t)\Psi_B(z_B + t)$, if one reverses also the direction of the field acting on particle B, as in Fig. 5.15, while it will be $|A\ 1\ \downarrow\rangle|B\ 1\ \uparrow\rangle\Psi_A(z_A - t)\Psi_B(z_B - t)$ if one does not reverse it.

All this illustrates how nonlocality works in the de Broglie–Bohm theory. There is a genuine action at a distance here, since choosing the orientation of the magnetic field that measures the spin of particle A will affect the motion of that particle, but also of particle B, no matter how distant the two particles are from one another. Another way to see this nonlocal action is to note that, if we measure the spin at A first, then with the initial particle positions as in Figs. 5.12 and 5.13, particle B will go down, whereas if we measure the spin at B first, then particle B will go up.

However, in the de Broglie–Bohm theory, no message can be sent in the EPR–Bell type situations if we assume quantum equilibrium, because then each side will see

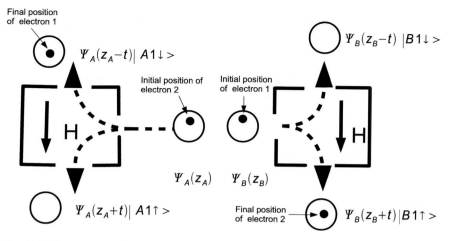

Fig. 5.15 Measurement of the spin on the left first, with the field reversed on both sides

a "random" sequence of up and down results, whichever measurement one chooses to make, and there is no way to manipulate those measurements so as to transmit a message.[67] But if one were in a non-quantum-equilibrium situation, one could send messages, just as in the situation with the mirrors described above.

   The fact that the de Broglie–Bohm theory is nonlocal is a quality rather than a defect, of course, since Bell showed that any correct theory must be nonlocal. Moreover, the nonlocality is of the right type (assuming that we are in quantum equilibrium), i.e., just what is needed to reproduce Bell's results, but not more, where "more" might be a nonlocal theory allowing the transmission of messages.

5.2.2   *What About Relativity?*

This question, which is probably the most frequent objection made against the de Broglie–Bohm theory (apart from some silly ones to be discussed later), has to be divided into several sub-questions. From the 1930s, physicists tried to extend quantum mechanics to classical field theories, such as electromagnetism, which is the simplest example of a relativistic theory. A field is (to give a very simple idea) just a function $\phi(x, t)$ depending on $x \in \mathbb{R}^3$ and on time. The idea was to "quantize" them, which in our language means that, instead of being defined as a function on $\mathbb{R}^{3N}$ (for N particles), the quantum state would have to be defined on a space of functions: $\Psi(\phi)$, with $\phi = \phi(x, t)$ being a field configuration.[68] This extension led

---

[67]See [37, 158, 208], [70, p. 139], and [319, Chap. 4] for a general argument.
[68]This is not the way quantum field theory is usually presented, but it allows us to make a straightforward connection with ordinary quantum mechanics.

to extremely successful predictions in quantum electrodynamics, and later in the theory of elementary particles.

So one may ask a first question: is there a de Broglie–Bohm version of those theories? In particular, since the usual theory deals with particles, can it deal with fields? And since the number of particles is fixed in the de Broglie–Bohm theory, how does it deal with the creation and annihilation of pairs of particles (which occurs in many elementary particle processes)?

To answer such questions, one would have to define a dynamics for the fields rather than for the particles, and the guidance equation would apply to the dynamics of field configurations. That is certainly doable and was partly achieved by Bohm back in 1952 [62]. One can also propose other Bohm-like quantum field theories, including theories of particles and their pair creation.[69] However, this issue is too technical to be discussed here. We only emphasize that the upshot is that all the predictions of the usual quantum field theories are also obtained in those Bohmian-type models and, to the extent that those models are rather ill-defined mathematically, the same thing is true for ordinary quantum field theories,[70] which is not the case for non-relativistic quantum mechanics or the corresponding de Broglie–Bohm theory for particles.

But there is another question, concerning the compatibility of the de Broglie–Bohm theory and relativity theory, which brings in the notion of Lorentz invariance. To explain briefly what is at issue here, let us start with Galilean invariance, a notion that arises in classical mechanics. To formulate the laws of physics, we need a system of coordinates, that is, we have to fix an origin of space and choose three perpendicular axes, in the directions x, y, and z. We also have to choose an origin for the time axis, and we have to choose length units such as the meter and time units such as the second. Obviously these choices are conventional and the way we formulate the laws of physics cannot depend on our conventions. A given set of such conventions (the origin of space and time and the perpendicular axes) is called a *frame of reference*.

But there is another, less obvious property of these laws. To explain it, we need the notion of *inertial frames of reference*, which can be defined as a frame of reference such that a particle on which no forces act will move on a straight line at constant speed (possibly equal to zero). Here, one defines the fact that no force acts by saying that the particle is very far away from everything else.[71] Obviously, if one described the world by attaching coordinates to a merry-go-round (that is, by defining a frame of reference in which the merry-go-round is at rest), everything would look as if it were rotating, and at increasing speed as one moved away from the merry-go-round. But this would just be an artifact of our choice of coordinates. In practice, one can find approximate inertial frames of reference, like the Earth, which do rotate, but whose rotation can be neglected, at least in most circumstances.

---

[69]See [145, 146, 148, 150]. For reviews and references to the original papers, see [464, 465, 477].

[70]See, for example, [216] for an introduction to a rigorous approach to quantum field theories. One can always put a cutoff on the range of momenta to which the theory applies in order to make it well defined, although that trick does not answer the real question, namely, what happens if one removes the cutoff.

[71]This is, of course, an idealization, but it is good enough to allow us to identify approximate inertial frames of reference.

Since in an inertial frame of reference, a particle on which no forces act moves on a straight line at constant speed, different inertial frames also move, one relative to another, on a straight line at constant speed (i.e., there is no "absolute" state of rest for an inertial frame of reference). Now, suppose that one describes the motion of a particle in two different inertial frames of reference. Is one frame preferable to the other or is the choice between them also conventional? The answer is that there is no preferred frame, and this is expressed in classical mechanics by saying that the laws of physics are invariant under certain transformations of our systems of coordinates, called Galilean transformations. These transformations are rather obvious: one simply adds to the velocity of an object in a first frame of reference the velocity of that first frame relative to a second frame, and one obtains the velocity of the object in the second frame. Newton's laws of motion are invariant under Galilean transformations, which means that they have the same form in all inertial frames of reference, and no physical effect distinguishes one inertial frame from another.

However, it turns out that the laws of electromagnetism are not invariant under these Galilean transformations![72] This is one of the discoveries that gave rise to the theory of relativity,[73] which still postulates that the laws of physics should be invariant under transformations between different inertial frames of reference, but under the Lorentz transformations rather than the Galilean ones.[74]

The main consequence, for our purposes, of the replacement of Galilean transformations by Lorentz transformations is that it implies the *relativity of simultaneity*. This means that, while we naively think that there is a "now" that applies to the entire universe (it seems to make sense to say that an event here and on the moon happen at the same time), and while this is true in classical mechanics, it is not so in relativity. Whether two events happen at the same time depends on the inertial frame of reference one uses. If someone passes my present location, but in a moving rocket, his present and my present will be different for distant objects. Thus, if the laws of physics are invariant under Lorentz transformations, they should not make any reference to the notion of "absolute simultaneity" (meaning simultaneity independent of any frame of reference).

But this poses a priori a problem for a relativistic formulation of the de Broglie–Bohm theory, since in Eqs. (5.1.1.3) and (5.1.1.4), the right-hand side depends on the positions of all the particles, but evaluated *at the same time*, which means that the very formulation of the theory seems to require a notion of absolute simultaneity.

However, this problem is in reality the expression, in the de Broglie–Bohm theory, of the existence of nonlocal causal connections proven by Bell, the latter being

---

[72]This is because the speed of light enters into those equations. But then the equations cannot be invariant under Galilean transformations, since those transformations change the speed of any object when one passes from one frame to another.

[73]For a good introduction to the theory of relativity, see [469], and for a careful conceptual discussion, see [326].

[74]Under the Lorentz transformations, the speed of light is constant and the laws of electromagnetism are invariant.

instantaneous.[75] But instantaneous in which frame? If the theory is Lorentz invariant, it should not matter. But suppose we describe an EPR experiment, like those discussed in Sect. 4.2 [see the quantum state (4.4.2.1)], using the language of the collapse of the quantum state. If the measurement at X is done first on particle A, then that also collapses the quantum state of particle B at Y, and vice versa. But if the events are spatially separated and if the collapse is instantaneous, then which comes first, the measurement on A or on B, is frame-dependent. On the other hand, if there is a cause and effect relation, we would like the cause to precede the effect in *all* frames.

But that is simply impossible! The only way to avoid the instantaneous collapse causing such a problem would be if the collapse were purely epistemic and not in any sense a physical occurrence. In the language of Chap. 4, this would mean that the answers to the "questions" are predetermined. Then there would indeed be no problem in learning about one answer by obtaining the correlated answer far way. But Bell has shown that this idea simply does not work.

In fact, if one looks at books that purport to unify quantum mechanics and relativity, specifically books on quantum field theory, one never finds a discussion of the collapse rule in relativistic terms, although this rule is essential to ordinary quantum mechanics. Why? Because quantum field theory essentially allows us to compute "scattering cross-sections", that is, the results of what happens when particles collide with one another. The results are, it is true, invariant under Lorentz transformations, but so are the predictions of Bohm-like quantum field theories, since they are the same as those of the usual theories.

The predictions of quantum field theory are extremely impressive, but contrary to received opinion, it is not true that there exists a fully relativistic quantum theory. After all, such a theory should be a generalization of non-relativistic quantum mechanics and should include (as one of its approximations) the results of EPR–Bell experiments (or any similar experiments involving a collapse of the quantum state when the latter is an entanglement of states whose parts are spatially separated). But that would require a relativistic treatment of the collapse as a physical process, and such a treatment simply does not exist. Moreover, one would encounter enormous difficulties if one tried to formulate it, as we just explained.

The reason why few people worry about this sort of question is probably just that the implications of EPR and Bell concerning the nonlocality of the world have not yet been fully grasped. We will discuss this further in Chap. 7.

So ordinary quantum mechanics, including quantum field theories, and the de Broglie–Bohm theory are on an equal footing here, except that the latter faces a problem that the former simply ignores. And within the de Broglie–Bohm framework, one can introduce a sort of simultaneity in a Lorentz invariant way [154].[76] Once such a notion of simultaneity is introduced, one can write down equations such as (5.1.1.3) and (5.1.1.4), or their analogue for relativistic quantum field theories, since they depend on the positions of all the particles at the same time. The trick that allows

---

[75] At least, since one cannot check experimentally that this simultaneity is *absolute*, much faster than the speed of light.

[76] Technically, this is called a foliation of spacetime.

one to do that in a Lorentz invariant way is to relate this notion of simultaneity to the wave function of the universe. Of course, it can be objected that introducing a global nonlocal object, such as the wave function of the universe goes against the common understanding of relativity, and it certainly goes against the way Einstein thought about relativity.

That common understanding supposes that Lorentz transformations can be applied, not only to the entire universe, but also to subsystems of that universe, which can in practice be considered isolated from the rest: for example, one often considers the Earth as an isolated system, which of course is not strictly true (e.g., because of the attraction of the Sun). If the Lorentz transformations were applicable only to the entire universe, then they would be pretty useless, since all the applications of the theory of relativity concern subsystems much smaller than the universe.[77] But the nonlocality proven by Bell shows that systems can be non-isolated, even though they do not interact directly with anything, because their quantum state may change due to a measurement done far away that collapses their state, as shown in the example of the quantum state (4.4.2.1). So because of Bell's result, we need to introduce some nonlocal structure in our theories, and the nonlocal structure introduced in [154] may therefore not be so objectionable after all.

In any case, the problem of a genuine Lorentz invariance of our most basic physical theory, namely quantum mechanics, in the face of EPR–Bell experiments is probably the biggest problem that theoretical physics faces today, even though it is not even recognized as such by most theoretical physicists. Here is how John Bell summarized the situation:

> [The usual quantum] paradoxes are simply disposed of by the 1952 theory of Bohm, leaving as *the* question, the question of Lorentz invariance. So one of my missions in life is to get people to see that if they want to talk about the problems of quantum mechanics — the real problems of quantum mechanics — they must be talking about Lorentz invariance.
>
> John Bell [47]

5.2.3 What About the Classical Limit?

In ordinary quantum mechanics, one expects the theory to approximate classical mechanics in certain limits, e.g., if objects have large masses or large energies. One expects that the wave function will not then spread too much, and its motion will be more or less classical.[78] But that does not answer the question of the status of the wave function: if it only means that it determines the probability of finding a particle

---

[77]For example, when one derives conservation laws, such as conservation of energy, momentum, and angular momentum, one always assumes that the systems to which those laws apply are isolated, otherwise there would be no way to derive those laws. But if the conservation laws held only for the universe, they would have no practical applications.

[78]In (2.A.2.24), one sees that the larger the mass, the slower the spreading, since time enters in (2.A.2.24) only through the factor t/m.

somewhere after a measurement, then even in a "classical limit", the particle is not actually in any specific location before being measured, no matter how narrowly concentrated in space the wave function is.

Actually, it is easy to give examples where the wave function is not concentrated in space, even for large objects. Consider Einstein's boxes. In that situation, as already mentioned in Chap. 4, the wave function is extended over both boxes, no matter how big the particle is. If the particle is in neither of the boxes before being observed, what happens if we replace the particle by larger and larger groups of particles? When do we make an ontological jump from "not being in either box" to "being in one box or the other, but we don't know which one"? That cannot happen just because we take a limit in some equation!

In the de Broglie–Bohm theory, there is no such problem, because particles have trajectories, whether we look at them or not and whether the particles are big or small. The only issue is whether their motion, which can be highly non-classical, as we see in Fig. 5.1, becomes approximately classical for large particles.[79] The simplest way to see when and why this happens is probably by looking at the second order equation (5.1.1.5): the equation is just like Newton's equation of motion, but with an extra term in the potential, the quantum potential. Thus, the motion will be approximately classical if the term in (5.1.1.5) equal to the derivative of the quantum potential, is so small as to be negligible. Many examples of this situation are given by Bohm and Hiley [70] and Holland [268]. A general discussion and an explicit conjecture about the validity of the classical limit is given in [11].

5.3   Other Objections

As Sheldon Goldstein observed in his Encyclopedia of Philosophy article on "Bohmian mechanics" (called the de Broglie–Bohm theory here)[80]:

> A great many objections have been and continue to be raised against Bohmian mechanics. Here are some of them: Bohmian mechanics makes predictions about results of experiments different from those of orthodox quantum theory so it is wrong. Bohmian mechanics makes the same predictions about results of experiments as orthodox quantum theory so it is untestable and therefore meaningless. Bohmian mechanics is mathematically equivalent to orthodox quantum theory so it is not really an alternative at all. Bohmian mechanics is more complicated than orthodox quantum theory, since it involves an extra equation. (This objection is based on the surprisingly common misconception that orthodox quantum theory is defined solely by Schrödinger's equation, and does not actually need as part of

---

[79]For particles in the box, this corresponds to large index n in the wave functions Ψ_n (see Appendix 5.D).

[80]Here are some remarks that may help to understand the quote. The many-worlds theory will be discussed in Sect. 6.1. The von Neumann theorem will be discussed in Sect. 7.4. The Kochen–Specker theorem is a version of the theorem in Sect. 2.5. The curved trajectories in the absence of forces are illustrated in Fig. 5.1. The stationary quantum states referred to here are those were the wave function is real, whence the particles do not move, according to (5.1.1.3). In Appendix 5.D, we give examples of such states, even with high energies.

its formulation any of the measurement postulates found in textbook quantum theory. It is only within a many-worlds framework that this view could begin to make sense, but I strongly doubt that it makes sense even there.) Bohmian mechanics requires the postulation of a mysterious and undetectable quantum potential. Bohmian mechanics requires the addition to quantum theory of a mysterious pilot wave. Bohmian mechanics, as von Neumann has shown, can't possibly work. Bohmian mechanics, as Kochen and Specker have shown, can't possibly work. Bohmian mechanics, as Bell has shown, can't possibly work. Bohmian mechanics is a childish regression to discredited classical modes of thought. Bohmian trajectories are crazy, since they may be curved even when no classical forces are present. Bohmian trajectories are crazy, since a Bohmian particle may be at rest in stationary quantum states. Bohmian trajectories are crazy, since a Bohmian particle may be at rest in stationary quantum states, even when these are large-energy eigenstates. Bohmian trajectories are surrealistic. Bohmian mechanics, since it is deterministic, is incompatible with quantum randomness. Bohmian mechanics is nonlocal. Bohmian mechanics is unintuitive. Bohmian mechanics is the many-worlds interpretation in disguise. (For a bit of discussion of some of these objections, see the exchange of letters on Quantum Theory Without Observers, in the February 1999 issue of Physics Today, particularly the last four of the eight letters [223].)

Sheldon Goldstein [231] (in the 2001 edition)

Although our discussion so far should allow the reader to answer many of those objections, we shall deal with some of them here.

5.3.1 Isn't This a Return to Classical Mechanics?

An American president might reply to that question by saying: "It depends what you mean by classical." The theory is "classical" in the sense that particles have definite properties (but only their positions[81]) without being measured or observed. But then the special and general theories of relativity are also classical in that sense. If the particles do not have any property whatsoever, then we do not know what the theory says about the world, outside of measurements.

On the other hand, the motion of the particles is highly non-classical, if "classical" refers to intuitions based on classical mechanics, since one can have particles that do not move[82] or move in a "strange" fashion, as in the double-slit experiment (see Fig. 5.1). So the de Broglie–Bohm theory should not be blamed for being a return to classical mechanics, in the same way that nobody would raise that objection against the special and general theories of relativity.

Besides, the theory introduces a minimal set of "classical" variables, only the positions, and it accounts for all the other "variables" as emerging from an analysis of the motion of the particles.

---

[81] Since their positions vary with time, they also have velocities but, as we discussed in Sect. 5.1.4, when one "measures" the "momentum", one does not get the actual value of those velocities. So in the de Broglie–Bohm theory, there are no "hidden variables" for the momenta. We will discuss this important property of the de Broglie–Bohm theory again in Sect. 5.3.3.

[82] Like the particles in the box whose states are discussed in Appendix 5.D, or certain electrons, whose wave function is also real, for example, those in an atomic "ground state".

Moreover, what could be *less* classical than nonlocal dynamics, such as we find in the de Broglie–Bohm theory? And once we understand, thanks to the EPR–Bell argument, that nonlocality is unavoidable, what could be a serious defect is turned into a major quality: making the nature of nonlocality in the world thoroughly explicit.

In fact, when discussing the de Broglie–Bohm theory, one often hears contradictory objections: first that the theory is "a return to classical mechanics", then that the motion of the particles is counterintuitive or that it does not put all observables on an equal footing (since only the positions have values independently of measurements). But one cannot have it both ways: if particles have no positions and no trajectories (as the orthodoxy maintains), then what could be strange about the trajectories that are proposed by the de Broglie–Bohm theory? People who raise that objection must have in mind some picture of what is going on outside of measurements, for example, that particles in a box bounce back and forth or that electrons move around the nucleus. But that is *not* what the orthodoxy says: in orthodox quantum mechanics, there are no trajectories and so one should not speak of "particles bouncing back and forth" or of "electrons moving around the nucleus". Similarly, when people object that positions have a "privileged status" relative to other observables in the de Broglie–Bohm theory, either they think that all observables have definite values before measurements (which leads to contradictions because of the no hidden variables theorems) or they think that nothing "exists" independently of measurements, which leads one to wonder what measuring devices are made of.

5.3.2   Isn't the Theory Too Complicated?

The theory is not very complicated, since the equation of motion (5.1.1.3) is the simplest one which could be associated with the wave function and which is equivariant (5.1.3.2). If we want to introduce particles that move, then it is natural to have their motion determined by that law. As noted by the philosopher and logician Hilary Putnam, responding to the charge made in Wikipedia that "Bohm's theory is mathematically inelegant":

> The formula for the velocity field is extremely simple: you have the probability current in the theory anyway, and you take the velocity vector to be proportional to the current.[83] There is nothing particularly inelegant about that; if anything, it is remarkably elegant!
>
> Hilary Putnam [410]

Moreover, all the quantum phenomena, including all the strange interference effects, experiments with spin, etc., can be derived from that assumption alone, together with the quantum equilibrium assumption on the initial conditions. What could be simpler?

Finally, there is a big difference between being complicated and being mysterious or incomprehensible. Ordinary quantum mechanics may seem simple, but that is true

---

[83]See Appendix 5.A for the definition of the probability current. (Note by J.B.).

only as long as one does not ask where the collapse rule comes from or what the wave function means physically. It is only by overlooking the collapse rule (which is indispensable for the quantum predictions to agree with observations) that one can pretend that ordinary quantum mechanics has only one equation (the Schrödinger equation) while de Broglie–Bohm has two equations, and hence is more complicated.

Of course, the story we have been telling about how measurements work and the assumption of quantum equilibrium is more complicated than the simple collapse rule. But this is like saying that statistical mechanics is more complicated than thermodynamics. Of course it is, because the former explains the latter. And the long story we have been telling explains why the collapse rule works, without making it into a *deus ex machina*. If someone wants to have a "simple" theory, they can just use ordinary quantum mechanics, with the dual laws of the Schrödinger evolution and the collapse rule, but knowing that there is a deeper explanation that unifies both of these rules, just as they can use thermodynamical formulas for the entropy without worrying about their justifications, but knowing that they exist.

5.3.3 What About the Symmetry Between Position and Momentum?

Heisenberg [259, p. 118] and Pauli [379, p. 33], [381, p. 12] have raised the following objection to the de Broglie–Bohm theory[84]: in quantum mechanics, there is a "perfect symmetry" between positions and momenta in the sense that the usual wave function $\Psi(x)$ is related to the probability density of positions (when they are measured!), which is given by $|\Psi(x)|^2$, while its Fourier transform $\hat{\Psi}(p)$ is related in a similar way to the probability density of momenta (also when they are measured!), which is given by $|\hat{\Psi}(p)|^2$, as we saw in Appendix 2.C.[85] The de Broglie–Bohm theory, by giving an ontological status to positions, is supposed to break that symmetry and should thus be rejected, if only on esthetic grounds.

But what exactly is the Heisenberg–Pauli argument? This symmetry suggests that the particles have positions and momenta, even when we do not measure them, and the probability densities of their distributions are given by $|\Psi(x)|^2$ and $|\hat{\Psi}(p)|^2$, respectively. That of course would contradict de Broglie–Bohm because that theory does not assign positions and momenta (when they are not measured) according to those distributions, at least not for the momenta, as we saw in the example of the particle in the box. But that is also not what ordinary quantum mechanics says, because of its emphasis on measurements. It does not say that particles have positions and momenta distributed according to $|\Psi(x)|^2$ and $|\hat{\Psi}(p)|^2$; it only says that we will find those distributions as results of our measurements. Moreover, the no hidden variables theorem discussed in Chap. 2 makes it *impossible* to assume that particles have both positions and momenta that are simply revealed by their "measurements".

---

[84]See [344] for a detailed discussion of this objection, and see also [89].

[85]This holds for any given time t, which we suppress here in the arguments of $\Psi(x)$ and $\hat{\Psi}(p)$.

The de Broglie–Bohm theory solves this problem in the most elegant way: it does assume that particles have positions and momenta, but not of course distributed according to the impossible statistics of $|\Psi(x)|^2$ and $|\hat{\Psi}(p)|^2$, and yet it explains, by taking into account the active role of measurements, why the results of measurements for both positions and momenta agree with the usual quantum mechanical predictions.

So where is the problem? Of course, the no hidden variables theorems preventing us from assigning both a position and a momentum to particles (outside of observations) whose distributions would be given by $|\Psi(x)|^2$ and $|\hat{\Psi}(p)|^2$ were not known to Heisenberg and Pauli, but it is fair to say that we can see now why their objection is devoid of substance.

5.3.4   What About the No Hidden Variables Theorems?

We have already answered that objection. The no hidden variables theorems show that certain variables, such as the spin values, cannot be introduced as existing prior to their measurements, if those values are such that measurements merely reveal them. But the de Broglie–Bohm theory accounts for those measurements precisely by *not* assuming that those measured values pre-exist and by showing how the results of measurements emerge naturally from a theory that assumes only that particles have positions.

Bohm and Hiley [70, p. 121] give the following concrete illustration of the way the de Broglie–Bohm theory avoids the impossibility implied by the no hidden variables theorems. We define the following operators:

$$A = \sigma_x^1 , \quad B = \sigma_y^2 , \quad A' = \sigma_z^1 \sigma_z^2 , \quad B' = \sigma_y^1 \sigma_x^2 , \quad C = \sigma_x^1 \sigma_y^2 ,$$

where the sigma matrices are the Pauli matrices, as in Appendix 2.F, and the superscripts 1, 2 refer to two different particles with spin (but the operators A, B, etc., are different here and in that appendix). From the relations (2.F.4)–(2.F.7) between these matrices, we get

$$C = AB = A'B' , \tag{5.3.4.1}$$

where the last identity follows from $A'B' = \sigma_z^1 \sigma_z^2 \sigma_y^1 \sigma_x^2 = \sigma_z^1 \sigma_y^1 \sigma_z^2 \sigma_x^2$, since σ_z^2 and σ_y^1 commute (2.F.7), and then one uses the commutation and anticommutation relations (2.F.5, 2.F.6) to get $\sigma_z^1 \sigma_y^1 = -i\sigma_x^1$ and $\sigma_z^2 \sigma_x^2 = i\sigma_y^2$, which yields $A'B' = C$.

The matrices A, B, C commute among themselves, and so do A', B', C, because σ_z^1 anticommutes with σ_y^1 and σ_z^2 anticommutes with σ_x^2 by (2.F.5), but A and B do not commute with A', B', because σ_x^1 does not commute with σ_z^1 and σ_y^1, and σ_y^2 does not commute with σ_z^2 and σ_x^2. So we can measure C in two different ways. The first would be to measure A and then B, and finally C : if we get the values a and b for the first two measurements, we necessarily get the value ab for the measurement of C (the quantum state resulting from the first two measurements being an eigenvector of C).[86] But we could also measure first A' and then B', and finally C : in that case, if the results of the measurements of A' and B' are a' and b', we will get the value $a'b'$ for the measurement of C.

But there is no reason why the result $a'b'$ of the second measurement should coincide with the result ab of the first measurement. Indeed, if we assumed that, we would be guilty of "naive realism with respect to operators". The two experimental arrangements corresponding to the sequence of measurements, A, B, C and A', B', C are different and since, as we saw in the example of the spin measurements in Sect. 5.1.4, the result of a measurement depends on the detailed experimental arrangement of the measuring device. Thus, there is no reason to expect these two different experimental arrangements for the measurement of C to give the same results.

In this way, the de Broglie–Bohm theory avoids the contradictions that would otherwise be implied by the no hidden variables theorems if it assumed that there were values pre-existing the measurements and that the latter revealed them. For we know that, if the theory made that assumption, and if $v(A)$ denoted this value for a given A, we would have $v(AB) = v(A)v(B)$, whenever A and B commute and, in the example above, we would then have $v(C) = v(A)v(B) = v(A')v(B')$. But as we saw in the no hidden variables theorem in Chap. 2, assuming this sort of relation leads directly to a contradiction.

5.3.5 If the Predictions of the de Broglie–Bohm Theory Are the Same as Those of Quantum Mechanics, What Is the Point of the de Broglie–Bohm Theory?

We have already answered this question: if ordinary quantum mechanics is a theory, what is it a theory of? If it is a theory about results of measurements, then it is not a physical theory, which is supposed to deal with the world outside of our laboratories and to be checked by experiments, but not be limited to them. If it is a theory about the world outside of our laboratories, then what does it say? As we saw in detail in Sect. 2.5, the answer to that crucial question is often ambiguous or even, when made precise, false (because no quantum state ever collapses under the ordinary quantum evolution, or because of the no hidden variables theorems). So the point of the de Broglie–Bohm theory is simply to be the missing theory behind the quantum algorithm.

5.3.6 Why Isn't there an Action–Reaction Principle in the de Broglie–Bohm Theory?

In classical mechanics, there is a well-known action–reaction principle: if body 1 exerts a force on body 2, then body 2 exerts a force on body 1, of the same strength,

---

[86] See Appendix 2.C.2.

but in the opposite direction to the first force. In classical field theories, such as electromagnetism or general relativity, the matter or energy content of the universe both generates fields and is affected or guided by them. But there is no relationship of this kind between particle positions in the de Broglie–Bohm theory and the quantum state. The latter guides the motion of the particles, but is not affected by them. The quantum state is just *there*. This is seen as a problem for the de Broglie–Bohm theory by some authors, for example by Anandan and Brown [17].

Dürr, Goldstein, and Zanghì have stressed the analogy between the quantum state and the Hamiltonian in classical physics, since this is also an object that is defined on an abstract space, guides the particles, and is not affected by them [144].[87] The double role of the quantum state, both dynamical and statistical (via the quantum equilibrium hypothesis), is also true for the Hamiltonian, since the probability distribution in equilibrium is given by $\exp(-\beta H)$. So the analogy is that H is similar to $-\log \Psi$, with $\beta = 2$, since the equilibrium distribution is given by[88] $|\exp(2 \log \Psi)| = |\Psi|^2$, and the equations of motion (5.1.1.3) involve the derivative of $\log \Psi$, just as they involve the derivatives of the Hamiltonian in classical mechanics.

An obvious objection to this analogy is that, in general, the quantum state changes with time, even for an isolated system, while the Hamiltonian does not. But Dürr, Goldstein, and Zanghì point out that one may have models of a quantum state for the universe that is static, yet such that one can also define quantum states for subsystems (of the universe) that evolve in time (see [144, Sect. 13] for a simple example).

Even putting aside the merits of this latter argument, the analogy with the Hamiltonian helps us think of the quantum state as being an aspect of the physical laws rather than a physical object. In the de Broglie–Bohm theory, what is fundamentally "physical" are the trajectories, and the quantum state that guides them can be considered as somewhat similar to a law of Nature.

5.4   Conclusions

In this section, we will first review the problems faced by the de Broglie–Bohm approach,[89] then summarize its merits.

5.4.1   *Trouble in Paradise?*

There are two sorts of problems that the de Broglie–Bohm theory faces. One is the issue of relativity. But as we saw, because of nonlocality, a genuine merger between

---

[87]Of course, both the Hamiltonian and the quantum state depend on the existence and the properties of the particles via the potential term in the Hamiltonian or in the Schrödinger equation.

[88]Writing $\Psi = Re^{iS}$, we get $\log \Psi = \log R + iS$ and $|\exp(2 \log \Psi)| = R^2 = |\Psi|^2$.

[89]The title of the next subsection was inspired by [89].

quantum mechanics and relativity is still an open problem *for everyone*, not just for defenders of the de Broglie–Bohm theory. Indeed, it may be the deepest current problem in theoretical physics, and certainly the deepest unrecognized problem in theoretical physics.

The other problem, which we have not discussed yet, is underdetermination. Indeed, the guiding equation (5.1.1.4) is far from being the only one that will reproduce the quantum predictions.

First of all, there exists a replacement of (5.1.1.4) by a stochastic equation, due to Imre Fényes and developed by Edward Nelson (see [182, 217, 348, 349]) that will reproduce the quantum predictions. In that theory, the trajectories will be random and will certainly not coincide with those of the de Broglie–Bohm theory. Moreover, there exists a one-parameter family of such stochastic theories.

But even restricting oneself to deterministic theories, Deotto and Ghirardi have shown [122] that there is an infinity of equations, similar to (5.1.1.4) but with a modified right-hand side, that give rise to an equivariant dynamics relative to the $|\Psi(\mathbf{x}, t)|^2$ distribution.[90] And one still has an infinity of equations, even after imposing restrictions of "naturalness" on these modified equations. All these dynamics will predict the same results for the weak measurements of Fig. 5.2, even though it is only in the usual de Broglie–Bohm theory that those measurements reveal the actual velocities (in that sense, the usual de Broglie–Bohm theory is more natural than all these alternative theories).

The situation seems to be somewhat worse when considering quantum field theories: there the problem is not that one cannot formulate a de Broglie–Bohm type of quantum field theory, but that there is no unique way to do it[91]: one can formulate field theories for bosons[92] in a rather standard way, analogous to the de Broglie–Bohm theory for particles, with a dynamics for field configurations instead of (5.1.1.4). Here what exists are only the fields, without localized particles (of course, as in the case of particles, the velocity law for fields is not unique).

However, for fermions, formulating a de Broglie–Bohm quantum field theory is more problematic. If one introduces instead a theory based on particles rather than fields, there are several possibilities, but a discussion of these would go beyond the scope of this book. However, the problem is certainly not that one cannot reproduce the usual quantum field theory predictions within a de Broglie–Bohm like theory.[93]

---

[90]This follows from the observation that, if one adds any term of the form $\mathbf{j}(\mathbf{x}, t)/|\Psi(\mathbf{x}, t))|^2$ to the right-hand side of (5.1.1.4), where $\nabla \cdot \mathbf{j} = 0$, then the $|\Psi(\mathbf{x}, t)|^2$ distribution density remains equivariant, because the left-hand side of (5.A.1) in Appendix 5.A is not modified by the addition of a current of the form (5.A.6) with $\nabla \cdot \mathbf{j} = 0$.

[91]See [464, 465] for a summary of the different possibilities and for references to the original papers.

[92]Here we introduce a distinction between bosons and fermions which is essential in quantum field theory, but which is not discussed elsewhere in this book. Without entering into a detailed discussion, one can think of bosons, e.g., photons, as mediating interactions between particles, and fermions, e.g., electrons, protons, and neutrons, as constituents of ordinary matter.

[93]In [145, 146, 148, 150], one introduces a stochastic process with creation of pairs of particles at random times and a deterministic evolution in-between. In [99], one introduces a deterministic model with a "Dirac sea".

In [226], Goldstein, Taylor, Tumulka, and Zanghì have shown that there exists an even more radical form of underdetermination: suppose that we have a theory with a certain number of types of particle (say, quarks, electrons, protons, etc.). Then, if one keeps the Schrödinger equation and the quantum state unchanged, one can modify (5.1.1.4) so as to restrict it to a subset of all the types of particles, in such a way that all our macroscopic records will be unchanged, so that there is no way to test whether all particles are real or whether only some of them are (for example, quarks could be real, but not electrons).[94] One can go even further and give "quite ridiculous" versions of that argument, to use the expression of Schrödinger about his cat being both alive and dead, and claim that nothing exists outside the solar system, or that women don't exist (for men and, of course, vice versa) and yet, all the observable predictions remain the same. We will not explain how one constructs such implausible theories, which are nevertheless empirically equivalent to the usual one [226], but the fact that these constructions are possible is related to the absolute uncertainty implied by quantum equilibrium. Because of quantum equilibrium, we cannot observe the trajectories, nor manipulate the initial conditions so as to obtain results that go beyond what the quantum formalism predicts, but as shown in [226], the uncertainty goes much further than this.

Before drawing the conclusion that this shows the unreliability of the de Broglie–Bohm theory and to use it as an argument in favor of ordinary quantum mechanics, one should observe that, from the point of view of [226], in ordinary quantum mechanics, *nothing* exists. Indeed, what is done in [226] is to restrict the set of species of particles that are assumed to exist. But in ordinary quantum mechanics, nothing at all exists (outside of measurements). Put another way, one "chooses" the empty set as subset of all the types of particles!

How should one think about these various forms of underdetermination? For the last one, one has to remember the Duhem–Quine thesis discussed in Chap. 3: one can always introduce "crazy" theories in an ad hoc way. Here it is quite natural to put all the particles (or, in some cases, the fields) that enter the quantum state and the Hamiltonian on the same footing, i.e., assume that they exist, and that a guiding equation like (5.1.1.4) holds for all of them.

For the theories exhibited by Deotto and Ghirardi or for the "choice" between stochastic (Nelson) and deterministic (de Broglie–Bohm) theories, one might a priori hope that some unique version arises as a limit of deeper, relativistic theories.[95] But then, what about the underdetermination in quantum field theories? Again, one may hope that quantum field theories are not the last word and that they arise from a unique deeper theory (strings for example?).

However, such hopes are not likely to be realized. One might also argue that some theories are more natural or elegant than others (and that argument can be made for the version of the de Broglie–Bohm theory given in this book [152, Sect. 8.1]). But

---

[94]One could also introduce only fermionic variables and not bosonic ones. This is actually Bell's approach in [42].

[95]For a discussion of the situation for relativistic de Broglie–Bohm theories, see [269, 270].

we have no good reason to believe that the world is kind enough to conform itself to what we, humans, consider natural or elegant.

Another, more modest, viewpoint is that we just have to live with underdetermination and to consider it as part of the human condition, i.e., of the limitations of our knowledge, just as the absolute uncertainty, discussed in Sect. 5.1.8 and following from quantum equilibrium, limits our ability to know a system with more precision than is allowed by the quantum formalism. But in no way does this uncertainty suggest that we should give up the project underlying the de Broglie–Bohm theory, viz., a rational understanding of quantum mechanics. Indeed, it is only because of this deeper understanding that the issue of underdetermination can be raised. In an ill-defined theory or in a "theory" restricted to predicting "measurements", everything is underdetermined in a sense, because nothing is being said about the world itself.

5.4.2   The Merits of the de Broglie–Bohm Theory

We want to emphasize two of the main merits of the de Broglie–Bohm theory. The first one is that it exists. And since it is a truism that a single counterexample is enough to refute a general claim, it is a counterexample to three claims that have been almost universally accepted by physicists, commented by philosophers, taught in classes, and sold to the general public:

1. That quantum mechanics signals the end of determinism in physics.
2. That quantum mechanics assigns a special role, in its very formulation, to the "observer". There has been quite some debate as to whether this "observer" is a set of laboratory instruments or a human consciousness, but the debate would never have got under way if the central role of observations in quantum mechanics had not been accepted to start with.
3. That quantum mechanics is something that "nobody understands", to quote Richard Feynman [185]; that quantum mechanics is mysterious and requires a far more drastic revision in our ways of thinking than any previous scientific revolution; or, even more radically, as Murray Gell-Mann reportedly said: "[T]here's much more difference [...] between a human being who knows quantum mechanics and one that doesn't than between one that doesn't and the other great apes." [203].

By simply existing, being deterministic, and describing the "measurements" as purely physical processes, the de Broglie–Bohm theory constitutes a refutation of claims 1 and 2.

A further quality of de Broglie–Bohm theory is its perfect clarity, which refutes claim 3. The philosopher of science David Albert gave a nice formulation of that idea:

This is a kind of theory whereby you can tell an absolutely low-brow story about the world, the kind of story (that is) that's about *the motions of material bodies*, the kind of story that contains nothing cryptic and nothing metaphysically novel and nothing ambiguous and nothing inexplicit and nothing evasive and nothing unintelligible and nothing inexact and nothing subtle and in which no questions ever fail to make sense and in which no questions ever fail to have answers and in which no two physical properties of anything are ever "incompatible" with one another and in which the whole universe always evolves *deterministically* [...]

David Albert [8, p.169]

Of course, some effort is needed to understand this theory, but not more than to understand, say, the theories of special and general relativity (at least not for someone who knows the mathematics of standard quantum mechanics). The de Broglie–Bohm theory contains paradoxes galore, but so do the theories of relativity. However, in neither case do the paradoxes imply any contradiction: they are simply due to the fact that we are misled by our intuitions.

Since the de Broglie–Bohm theory is nothing but a rational version of quantum mechanics, it makes no new predictions. But it has several concrete applications, such as numerical simulations of the wave function in quantum chemistry [524], many-body simulations based on Bohmian ideas [399], and semi-classical approximations (see [466] and references therein).

However, its main virtue is to clarify our ideas. Given that there are endless bookshelves of confused talk about the role of the "observer" in physics or about the death of determinism, this is no small feat, especially given the number of times that this accomplishment has been declared impossible. Moreover, the following quote by John Bell shows the essential role of Bohm's work in his own thinking and in his discovery of nonlocality:

When I first realized that [Bohm's theory is nonlocal], I asked: "Is that inevitable or could somebody smarter than Bohm have done it differently and avoided this nonlocality?" That is the problem that [Bell's] theorem is addressed to. The theorem says: "No! Even if you are smarter than Bohm, you will not get rid of nonlocality," that any sharp mathematical formulation of what is going on will have that nonlocality ... In my opinion the picture which Bohm proposed then completely disposes of all the arguments that you will find among the great founding fathers of the subject—that in some way, quantum mechanics was a new departure of human thought which necessitated the introduction of the observer, which necessitated speculation about the role of consciousness, and so on. All those are simply refuted by Bohm's 1952 theory.

John Bell [47]

The objections made by Einstein seemed purely philosophical or even "metaphysical" to many physicists. Recall the quote in Sect. 1.5 in which Pauli compared Einstein's queries to the question of "of how many angels are able to sit on the point of a needle" [79, p. 223].

But such speculations have led, through the work of Bell, to the discovery of nonlocality. Furthermore, because they lie at the basis of quantum information theory, they have also led to various technological applications. And as we have just seen, Bohm's papers were a major source of inspiration for Bell.

So let us leave the last word to Bell, hoping that other people will appreciate the "impossible" when it is done. After recalling the von Neumann argument, reported by Born,[96] against the impossibility of hidden variables, Bell "saw the impossible done" in Bohm's 1952 papers:

> Bohm's 1952 papers on quantum mechanics were for me a revelation. The elimination of indeterminism was very striking. But more important, it seemed to me, was the elimination of any need for a vague division of the world into "system" on the one hand, and "apparatus" or "observer" on the other. I have always felt since that people who have not grasped the ideas of those papers ... and unfortunately they remain the majority ... are handicapped in any discussion of the meaning of quantum mechanics.
>
> [...]
>
> Why is the pilot wave picture ignored in textbooks? Should it not be taught, not as the only way, but as an antidote to the prevailing complacency? To show that vagueness, subjectivity, and indeterminism are not forced on us by experimental facts, but by deliberate theoretical choice?

John Bell [49, pp.173, 160]

One possible answer to that question could be that "the pilot wave picture" is just one "interpretation" of quantum mechanics among many, so why pay attention to that one alone? The next chapter will deal with that objection.

Appendices

5.A The Continuity Equation for the $|\Psi|^2$ Distribution

Let us consider Schrödinger's equation (5.1.1.1) with H given by (5.1.1.2). We will first prove the following continuity equation:

$$\frac{\partial |\Psi(\mathbf{x}, t)|^2}{\partial t} + \nabla \cdot \mathbf{j}(\mathbf{x}, t) = 0 , \qquad (5.A.1)$$

where $\nabla = (\partial/\partial x_1, \ldots, \partial/\partial x_N)$ is the gradient, the dot denotes the scalar product, and $\mathbf{j}(\mathbf{x}, t) = (j_k(\mathbf{x}, t))_{k=1}^N$ (we index the components here by k instead of j) is the "probability current" or the "quantum flux" defined by

$$
\begin{aligned}
j_k(\mathbf{x}, t) &= \frac{1}{2im_k}\left[\Psi^*(\mathbf{x}, t)\frac{\partial}{\partial x_k}\Psi(\mathbf{x}, t) - \Psi(\mathbf{x}, t)\frac{\partial}{\partial x_k}\Psi^*(\mathbf{x}, t)\right] \qquad (5.A.2)\\
&= \frac{1}{m_k}\text{Im}\left[\Psi^*(\mathbf{x}, t)\frac{\partial}{\partial x_k}\Psi(\mathbf{x}, t)\right] .
\end{aligned}
$$

---

[96]This was in Born's book *Natural Philosophy of Cause and Chance* [76], since Bell did not read German and von Neumann's book [496] was only translated into English in 1955. We will discuss Born and von Neumann's views in Sects. 7.2 and 7.4.

To prove (5.A.1), write

$$\frac{\partial |\Psi(\mathbf{x}, t)|^2}{\partial t} = \Psi^*(\mathbf{x}, t)\frac{\partial \Psi(\mathbf{x}, t)}{\partial t} + \Psi(\mathbf{x}, t)\frac{\partial \Psi^*(\mathbf{x}, t)}{\partial t} , \qquad (5.A.3)$$

take the complex conjugate of (5.1.1.1), viz.,

$$-i\frac{\partial \Psi^*}{\partial t}(\mathbf{x}, t) = H\Psi^*(\mathbf{x}, t) , \qquad (5.A.4)$$

and insert this and (5.1.1.1) into (5.A.3). The potential term cancels out and we get

$$\frac{\partial |\Psi(\mathbf{x}, t)|^2}{\partial t} = -\frac{1}{i} \sum_{k=1}^{N} \frac{1}{2m_k} \left[\Psi^*(\mathbf{x}, t)\frac{\partial^2}{\partial x_k^2}\Psi(\mathbf{x}, t) - \Psi(\mathbf{x}, t)\frac{\partial^2}{\partial x_k^2}\Psi^*(\mathbf{x}, t) \right] .$$
$$(5.A.5)$$

It is enough to observe that

$$\frac{\partial}{\partial x_k} \left[\Psi^*(\mathbf{x}, t)\frac{\partial}{\partial x_k}\Psi(\mathbf{x}, t) - \Psi(\mathbf{x}, t)\frac{\partial}{\partial x_k}\Psi^*(\mathbf{x}, t) \right] = \Psi^*(\mathbf{x}, t)\frac{\partial^2}{\partial x_k^2}\Psi(\mathbf{x}, t)$$
$$- \Psi(\mathbf{x}, t)\frac{\partial^2}{\partial x_k^2}\Psi^*(\mathbf{x}, t)$$

to obtain (5.A.1).

Now, if we consider (5.1.1.4), we see that the function $\mathbf{j}(\mathbf{x}, t) = (j_k(\mathbf{x}, t))_{k=1}^N$ [whose arguments are the generic variables $\mathbf{x}$ rather than the actual positions of the particles as in (5.1.1.4)] is of the form

$$\mathbf{j}(\mathbf{x}) = \mathbf{V}_\Psi(\mathbf{x}, t) |\Psi(\mathbf{x}, t)|^2 , \qquad (5.A.6)$$

and we can write (5.A.1) as

$$\frac{\partial |\Psi(\mathbf{x}, t)|^2}{\partial t} + \nabla \cdot \left[\mathbf{V}_\Psi(\mathbf{x}, t)|\Psi(\mathbf{x}, t)|^2 \right] = 0 , \qquad (5.A.7)$$

which is the continuity equation[97] one would obtain for a "fluid" of density $\rho(\mathbf{x}, t) = |\Psi(\mathbf{x}, t)|^2$, with local velocity $\mathbf{V}_\Psi(\mathbf{x}, t)$. This will be useful when we come to prove the equivariance of the $|\Psi(\mathbf{x}, t)|^2$ distribution in Appendix 5.C.

---

[97]This equation was first written by Erwin Madelung, who gave it a "hydrodynamic" interpretation [315, 316].

5.B A Second Order Dynamics and the Quantum Potential

Here we derive (5.1.1.5) from (5.1.1.1)–(5.1.1.3). We insert $\Psi(\mathbf{x}, t) = R(\mathbf{x}, t)e^{iS(\mathbf{x},t)}$ into Schrödinger's equation (5.1.1.1), where $\mathbf{x} = (x_1, \ldots, x_N)$. Suppressing here the arguments $(\mathbf{x}, t)$, we get

$$\frac{i\partial R}{\partial t}e^{iS} - \frac{\partial S}{\partial t}\Psi = -\sum_{j=1}^{N}\frac{1}{2m_j}\left[\frac{\partial^2 R}{\partial x_j^2} - R\left(\frac{\partial S}{\partial x_j}\right)^2 + 2i\frac{\partial R}{\partial x_j}\frac{\partial S}{\partial x_j} + iR\frac{\partial^2 S}{\partial x_j^2}\right]e^{iS} + V\Psi \, .$$

(5.B.1)

Dividing both sides by e^{iS} and separating the real and imaginary parts give rise to two equations:

$$\frac{R\partial S}{\partial t} = \sum_{j=1}^{N}\frac{1}{2m_j}\left[\frac{\partial^2 R}{\partial x_j^2} - R\left(\frac{\partial S}{\partial x_j}\right)^2\right] - VR \, , \tag{5.B.2}$$

$$\frac{\partial R}{\partial t} + \sum_{j=1}^{N}\frac{1}{2m_j}\left[2\frac{\partial R}{\partial x_j}\frac{\partial S}{\partial x_j} + R\frac{\partial^2 S}{\partial x_j^2}\right] = 0 \, . \tag{5.B.3}$$

Dividing (5.B.2) by R gives

$$\frac{\partial S}{\partial t} = \sum_{j=1}^{N}\frac{1}{2m_j}\left[\frac{1}{R}\frac{\partial^2 R}{\partial x_j^2} - \left(\frac{\partial S}{\partial x_j}\right)^2\right] - V \, . \tag{5.B.4}$$

Multiplying (5.B.3) by $2R$ gives

$$\frac{\partial R^2}{\partial t} + \sum_{j=1}^{N}\frac{1}{m_j}\frac{\partial}{\partial x_j}\left(R^2\frac{\partial S}{\partial x_j}\right) = 0 \, ,$$

which, with $R^2 = |\Psi|^2$ and using (5.A.6), is the same as (5.A.1).

Now, if we consider (5.1.1.3) and take its time derivative, remembering that S depends on time through $(X_1(t), \ldots, X_N(t))$ and t, we get

$$m_k\frac{d^2}{dt^2}X_k(t) = \sum_{j=1}^{N}\frac{\partial^2 S}{\partial x_j \partial x_k}\dot{X}_j + \frac{\partial^2 S}{\partial x_k \partial t} \, . \tag{5.B.5}$$

If we use (5.1.1.3) to replace $\dot{X}_j$ by $(1/m_j)(\partial S/\partial x_j)$ in (5.B.5), and insert (5.B.4) in the second term of (5.B.5), evaluating the functions at $\mathbf{X}$ instead of $\mathbf{x}$, the terms

$$\sum_{j=1}^{N} \frac{1}{m_j} \frac{\partial^2 S}{\partial x_j \partial x_k} \frac{\partial}{\partial x_j} S \quad \text{and} \quad -\frac{\partial}{\partial x_k} \sum_{j=1}^{N} \frac{1}{2m_k} \left(\frac{\partial S}{\partial x_j} \right)^2$$

cancel each other and we get

$$m_k \frac{d^2}{dt^2} X_k(t) = -\frac{\partial}{\partial x_k} (Q + V) ,$$

where the quantum potential Q is given by

$$Q = -\sum_{j=1}^{N} \frac{1}{2m_j} \frac{1}{R} \frac{\partial^2 R}{\partial x_j^2} . \tag{5.B.6}$$

This proves (5.1.1.5).

5.C Proof of the Equivariance of the $|\Psi(x, t)|^2$ Distribution

5.C.1 Equivariance in $\mathbb{R}^N$

We will first prove[98] that, for a system in $\mathbb{R}^N$ and $\forall A \subset \mathbb{R}^N$,

$$P\big(\mathbf{X}(t) \in A\big) = \int_A |\Psi(\mathbf{x}, t)|^2 d\mathbf{x} , \tag{5.C.1.1}$$

provided that, at $t = 0$, $\forall A \subset \mathbb{R}^N$,

$$P\big(\mathbf{X}(0) \in A\big) = \int_A |\Psi(\mathbf{x}, 0)|^2 d\mathbf{x} . \tag{5.C.1.2}$$

In (5.C.1.1), the left-hand side involves $\mathbf{X}(t)$, the solution of (5.1.1.4), while on the right-hand side, $\Psi(\mathbf{x}, t)$ is the solution of Schrödinger's equation (5.1.1.1).
    Consider first a general differential equation

$$\frac{d\mathbf{X}(t)}{dt} = \mathbf{V}\big(\mathbf{X}(t), t\big) , \tag{5.C.1.3}$$

with $\mathbf{X} : \mathbb{R} \to \mathbb{R}^N$. We assume that $\mathbf{V}$ is such that (5.C.1.3) has solutions for all times. Let $\phi^t(\mathbf{x})$ denote the solution at time t of (5.C.1.3) with initial condition $\mathbf{X}(0) = \mathbf{x}$ at time 0. Then, given any probability density $\rho(\mathbf{x}, 0)$ on the initial conditions $\mathbf{X}(0) = \mathbf{x}$,

---

[98]This property is what is called equivariance in [141].

define[99] $\rho(\mathbf{x}, t), \forall A \subset \mathbb{R}^N$, by

$$\int_A \rho(\mathbf{x}, t)d\mathbf{x} = P\big(\mathbf{X}(t) \in A\big) = \int \mathbb{1}_A\big(\phi^t(\mathbf{x})\big)\rho(\mathbf{x}, 0)d\mathbf{x} , \qquad (5.\mathrm{C}.1.4)$$

where $\mathbb{1}_A$ is the indicator function of the set A.

Then we have

$$\frac{\partial \rho(\mathbf{x}, t)}{\partial t} + \nabla \cdot \Big[\mathbf{V}(\mathbf{x}, t)\rho(\mathbf{x}, t)\Big] = 0 . \qquad (5.\mathrm{C}.1.5)$$

To prove (5.C.1.5), observe that, by approximating integrals by sums, for any smooth function with compact support, we get from (5.C.1.4):

$$\int F(\mathbf{x})\rho(\mathbf{x}, t)d\mathbf{x} = \int F\big(\phi^t(\mathbf{x})\big)\rho(\mathbf{x}, 0)d\mathbf{x} . \qquad (5.\mathrm{C}.1.6)$$

Then, differentiating both sides of (5.C.1.6) with respect to t, and using the fact that (5.C.1.3) implies $\partial\phi^t(\mathbf{x})/\partial t = \mathbf{V}\big(\phi^t(\mathbf{x}), t\big)$, we have

$$\int F(\mathbf{x})\frac{\partial \rho(\mathbf{x}, t)}{\partial t}d\mathbf{x} = \int \nabla F\big(\phi^t(\mathbf{x})\big) \cdot \mathbf{V}\big(\phi^t(\mathbf{x}), t\big)\rho(\mathbf{x}, 0)d\mathbf{x} .$$

Using (5.C.1.6) with $F(\cdot)$ replaced by $\nabla F(\cdot) \cdot \mathbf{V}(\cdot, t)$, the right-hand side of this equals

$$\int \nabla F(\mathbf{x}) \cdot \mathbf{V}(\mathbf{x}, t)\rho(\mathbf{x}, t)d\mathbf{x} .$$

Integrating by parts, this is

$$-\int F(\mathbf{x})\nabla \cdot \Big[\mathbf{V}(\mathbf{x}, t)\rho(\mathbf{x}, t)\Big]d\mathbf{x} . \qquad (5.\mathrm{C}.1.7)$$

Since (5.C.1.7) is valid for any F, it implies (5.C.1.5).

Now, if we consider the differential equation (5.C.1.3) with $\mathbf{V}(\mathbf{X}, t) = \mathbf{V}_\Psi(\mathbf{X}, t)$, with $\mathbf{V}_\Psi(\mathbf{X}, t)$ given by (5.1.1.4), and $\rho(\mathbf{x}, t) = |\Psi(\mathbf{x}, t)|^2$, we get, using (5.C.1.4), (5.C.1.5), and (5.A.7),

---

[99] We assume, as in (5.1.3.1), that the integrals (5.C.1.4) are absolutely continuous, so that they define the function $\rho(\mathbf{x}, t)$. Equation (5.C.1.4) is similar to (5.1.3.1), but here it applies to a probability distribution on $\mathbb{R}^N$, with N the number of variables of the system, whereas in (5.1.3.1) it referred to the empirical density in $\mathbb{R}^3$. These concepts are different, even if the formulas look similar. The connection between them is made at the end of this appendix.

$$\frac{d}{dt} P\big(\mathbf{X}(t) \in A\big) = \int_A \frac{\partial \rho(\mathbf{x}, t)}{\partial t} d\mathbf{x}$$

$$= -\int_A \nabla \cdot \big[\mathbf{V}_\Psi(\mathbf{x}, t)\rho(\mathbf{x}, t)\big] d\mathbf{x}$$

$$= \frac{d}{dt} \int_A |\Psi(\mathbf{x}, t)|^2 d\mathbf{x} , \qquad (5.C.1.8)$$

which proves (5.C.1.1) if we assume that (5.C.1.1) holds at $t = 0$, which is (5.C.1.2).

5.C.2 Proof of Equivariance of the Empirical Distributions

To finish the proof of (5.1.3.2), consider a large number N of copies of non-interacting identical systems,[100] whose initial wave function[101] is

$$\Psi(\mathbf{x}, 0) = \Psi(x_1, \ldots, x_N, 0) = \prod_{i=1}^{N} \psi(x_i, 0) . \qquad (5.C.2.1)$$

The density of the associated probability distribution is

$$|\Psi(x_1, \ldots, x_N, 0)|^2 = \prod_{i=1}^{N} |\psi(x_i, 0)|^2 , \qquad (5.C.2.2)$$

which means that the variables x_i are independent identically distributed random variables with distribution $|\psi(x, 0)|^2$. This implies, by the law of large numbers (see Appendix 3.A), that the empirical density distribution $\rho(x, 0)$ of the N particles will be given, as $N \to \infty$, by $\rho(x, 0) = |\psi(x, 0)|^2$.

By (5.C.1.1), the probability distribution density of $(x_1, \ldots, x_N)$ at time $t > 0$ will be

$$|\Psi(x_1, \ldots, x_N, t)|^2 = \prod_{i=1}^{N} |\psi(x_i, t)|^2 , \qquad (5.C.2.3)$$

and the empirical density distribution $\rho(x, t)$ of the N particles at time t will be given, as $N \to \infty$, by $\rho(x, t) = |\psi(x, t)|^2$. This completes the proof of (5.1.3.2).

---

[100] We consider here each $x_i \in \mathbb{R}$, but one could also consider $\mathbf{x}_i \in \mathbb{R}^3$.

[101] This wave function could be the effective wave function of those non-interacting identical systems, defined by (5.1.7.5).

5.D Examples of de Broglie–Bohm Dynamics

5.D.1 The Gaussian Wave Function

Consider the time evolution of a Gaussian wave function (2.A.2.23), with initial condition $\Psi(x, 0) = \pi^{-1/4} \exp(-x^2/2)$, which we copy here[102]:

$$\Psi(x, t) = \frac{1}{(1 + it/m)^{1/2}} \frac{1}{\pi^{1/4}} \exp\left[-\frac{x^2}{2(1 + it/m)}\right] . \qquad (5.D.1.1)$$

It is easy to see that, if we write $\Psi(x, t) = R(x, t) \exp\left[i S(x, t)\right]$, we have (up to a constant term)

$$S(x, t) = \frac{tx^2}{2m(1 + t^2/m^2)} ,$$

so that (5.1.1.3) gives rise to the equation of motion

$$\frac{d}{dt} X(t) = \frac{t X(t)}{m^2 + t^2} , \qquad (5.D.1.2)$$

which can be easily integrated by a separation of variables:

$$X(t) = \frac{X(0)}{m} \sqrt{m^2 + t^2} = X(0) \sqrt{1 + \left(\frac{t}{m}\right)^2} . \qquad (5.D.1.3)$$

This gives the explicit dependence of the position at time t on the initial condition and shows that the particle does not move if it is initially at $X(0) = 0$, but moves asymptotically, when $t \to \infty$, according to $X(t) \sim X(0)t/m$.

Equation (2.A.3.3) can be checked explicitly here. Indeed, putting $m = 1$ as in Appendix 2.A.3, we get from (2.A.2.24)

$$t|\Psi(pt, t)|^2 = t\frac{1}{\sqrt{\pi(1 + t^2)}} \exp\left(-\frac{p^2 t^2}{1 + t^2}\right) , \qquad (5.D.1.4)$$

whose limit as $t \to \infty$ is $\pi^{-1/2} \exp(-p^2) = |\widehat{\Psi}(p, 0)|^2$.

It is also easy to verify explicitly the property (5.1.3.2) of equivariance. From (5.D.1.3), we get $\phi^t(x) = x\sqrt{1 + (t/m)^2}$. If we make the change of variable $y = \phi^t(x)$ in the integral on the right-hand side of (5.1.3.1), we get, with the initial distribution $\rho(x, 0) = |\Psi(x, 0)|^2 = \pi^{-1/2} \exp(-x^2)$,

---

[102]For more details on this example and on similar ones, see Holland [268, Chap. 4] or Bohm and Hiley [70, Sect. 3.6].

$$\rho(y, t) = \frac{1}{\sqrt{\pi\left[1 + (t/m)^2\right]}} \exp\left[-\frac{y^2}{1 + (t/m)^2}\right] . \qquad (5.D.1.5)$$

Changing the y variable back to x, this coincides with $|\Psi(x, t)|^2$ given by (2.A.2.24), with the prefactor $1/\sqrt{1 + (t/m)^2}$ coming from $dx = dy/\sqrt{1 + (t/m)^2}$.

5.D.2 The Particle in a Box

Consider a particle in a box,[103] in one dimension, of size L, chosen to be the interval $[0, L]$. The particle is free in the box, but constrained to remain in it. Let us look for solutions of the Schrödinger equation (5.1.1.1), with $N = 1$, of the form $\exp(-iEt)\Psi(x)$, whose associated probability distribution $|\exp(-iEt)\Psi(x)|^2 = |\Psi(x)|^2$ is independent of time.[104]

We model the presence of the box by setting the potential in (5.1.1.2) such that $V = 0$ inside the box and $V = \infty$ outside the box, which means that $\Psi(x)$ vanishes outside V. So we have to solve

$$\frac{1}{2}\frac{d^2}{dx^2}\Psi(x) = -E\Psi(x) , \qquad (5.D.2.1)$$

with $\Psi(0) = \Psi(L) = 0$. We have set the mass $m = 1$.

The general solution of (5.D.2.1) is

$$A \exp\left(i\sqrt{2E}x\right) + B \exp\left(-i\sqrt{2E}x\right) , \qquad (5.D.2.2)$$

and the conditions $\Psi(0) = \Psi(L) = 0$ imply $A = -B$ and $E_n = n^2\pi^2/L^2$, for $n \in \mathbb{Z}$, $n \neq 0$. From (5.D.2.2), we get the eigenvectors

$$\Psi_n(x) = \sqrt{\frac{2}{L}} \sin(k_n x)\mathbb{1}_{[0,L]}(x) , \qquad (5.D.2.3)$$

where $\mathbb{1}_{[0,L]}$ is the indicator function of $[0, L]$, and $k_n = n\pi/L$. The sine function being odd, we can restrict ourselves to $n = 1, 2, \ldots, \infty$, since the solution should be nonzero and is defined up to a multiplicative factor which does not affect the motion of the particle, as can be seen from (5.1.1.4).[105] The factor $\sqrt{2/L}$ ensures that $\int_{\mathbb{R}} |\Psi_n(x)|^2 dx = 1$.

If we take a function $\Psi_n(x)$ as initial condition in (5.1.1.1), then the time-dependent solution is

---

[103]This Appendix is based on [268, Sect. 6.5].

[104]Of course, these are eigenvectors of the Hamiltonian, and the corresponding E are its eigenvalues.

[105]The theory of Fourier series implies that the set $\left(\Psi_n(x)\right)_{n=1}^{\infty}$ forms a basis of $L^2([0, L], dx)$, but we will not use this fact.

$$\Psi_n(x, t) = \sqrt{\frac{2}{L}} e^{-iE_n t} \sin(k_n x) \mathbb{1}_{[0,L]}(x) . \tag{5.D.2.4}$$

Now comes the apparent paradox: the phase S_n of $\Psi_n(x, t)$ is constant in x, so $\partial S_n / \partial x = 0$, and by (5.1.1.3), the particle is at rest in the box. Let us see how to solve that paradox. As we said, one method to measure the momentum is to remove the walls of the box and to detect the position x of the particle after a certain time t, and give x/t as a "measure" of momentum.

To see what happens when the walls of the box are removed, we have to compute the time evolution of the wave function with initial condition (5.D.2.3). Since the walls are removed, the time evolution is governed by the free Schrödinger equation (5.1.1.1). The solution is therefore obtained from (2.A.2.21) and we need to compute $\widehat{\Psi}(p, 0)$ from (5.D.2.3), leaving out the index n on Ψ. Thus,

$$\widehat{\Psi}(p, 0) = \frac{1}{\sqrt{2\pi}} \int_{\mathbb{R}} e^{-ipx} \sqrt{\frac{2}{L}} \sin(k_n x) \mathbb{1}_{[0,L]}(x) dx \tag{5.D.2.5}$$

$$= \frac{1}{\sqrt{2\pi}} \int_0^L e^{-ipx} \sqrt{\frac{2}{L}} \sin(k_n x) dx .$$

Writing $\sin(k_n x) = (e^{ik_n x} - e^{-ik_n x})/2i$, making the change of variable $x = x' + L/2$, and integrating over $x' \in [-L/2, L/2]$, we obtain

$$\widehat{\Psi}(p, 0) = \frac{-i}{\sqrt{\pi L}} \left[e^{i(k_n - p)L/2} \frac{\sin\left[(p - k_n)L/2\right]}{p - k_n} - e^{-i(k_n + p)L/2} \frac{\sin\left[(p + k_n)L/2\right]}{p + k_n} \right] . \tag{5.D.2.6}$$

This can be written as:

$$\widehat{\Psi}(p, 0) = \frac{-i}{\sqrt{\pi L}} e^{-ipL/2} e^{ik_n L/2} \left[\frac{\sin\left[(p - k_n)L/2\right]}{p - k_n} - e^{-ik_n L} \frac{\sin\left[(p + k_n)L/2\right]}{p + k_n} \right]$$

$$= \frac{-i}{\sqrt{\pi L}} e^{-ipL/2} e^{ik_n L/2} \left[\frac{\sin\left[(p - k_n)L/2\right]}{p - k_n} + (-1)^{n+1} \frac{\sin\left[(p + k_n)L/2\right]}{p + k_n} \right] , \tag{5.D.2.7}$$

since $e^{-ik_n L} = e^{-in\pi} = (-1)^n$. From (2.A.2.21) we get, using the variable $x' = x - L/2$,

$$\Psi(x', t) = \frac{-i}{\pi \sqrt{2L}} e^{ik_n L/2} \int_{\mathbb{R}} \exp\left(-it\frac{p^2}{2} + ipx'\right) \left[\frac{\sin\left[(p - k_n)L/2\right]}{p - k_n} \right. \tag{5.D.2.8}$$

$$\left. + (-1)^{n+1} \frac{\sin\left[(p + k_n)L/2\right]}{p + k_n} \right] dp .$$

From Plancherel's theorem [see also (2.A.2.26)], we know that

$$\int_{\mathbb{R}} |\Psi(x',t)|^2 dx' = \int_{\mathbb{R}} |\widehat{\Psi}(p,t)|^2 dp = \int_{\mathbb{R}} |\widehat{\Psi}(p,0)|^2 dp = \int_{\mathbb{R}} |\Psi(x,0)|^2 dx = 1 .$$

Now, for $t \neq 0$, if we write $\Psi(x',t) = R(x',t)e^{iS(x',t)}$, there is no reason why we should have $\partial S(x',t)/\partial x' = 0$ and an analysis of (5.D.2.8) shows that we do indeed have $\partial S(x',t)/\partial x' \neq 0$ for $t \neq 0$. In fact, we can see this indirectly using (2.A.3.2) and (2.A.3.3), which imply

$$\lim_{t \to \infty} \int_{At} |\Psi(x',t)|^2 dx' = \int_{A} |\widehat{\Psi}(p,0)|^2 dp . \tag{5.D.2.9}$$

This means that the support of $\Psi(x',t)$ spreads as $t \to \infty$ (away from $x' = 0$, i.e., away from the center of the box $x = L/2$), and by equivariance (5.1.3.1), this means that the particles will move towards either $+\infty$ or $-\infty$.

To illustrate this, consider what happens in the limit of large n, L (mathematically, $n, L \to \infty$) with $k_n = n\pi/L$ fixed.[106] It is well known (from the theory of Fourier transforms) that

$$\frac{\sin(Ly/2)}{y} = \frac{1}{2} \int_{-L/2}^{L/2} e^{isy} ds \longrightarrow \pi\delta(y) , \tag{5.D.2.10}$$

as $L \to \infty$, in the sense of distributions. So in that limit, we may change variables $y = p - k_n$ or $y = p + k_n$ in (5.D.2.8) and, to a good approximation, replace the integral in (5.D.2.8) by the value of the factor $\exp(itp^2/2 + ipx)$ at $p = \pm k_n$. We then obtain

$$\Psi(x',t) \approx \frac{-ie^{ik_nL/2}}{\sqrt{2L}} \left[e^{-itk_n^2/2 + ik_n x'} + (-1)^{n+1} e^{-itk_n^2/2 - ik_n x'} \right] , \tag{5.D.2.11}$$

which represents two plane waves propagating in opposite directions with absolute velocity k_n, since $\partial(\pm k_n x')/\partial x' = \pm k_n$.

Of course, one has to be a bit careful with this limit, since we have $L \to \infty$ and $1/\sqrt{L}$ in (5.D.2.11). But the function in square brackets in (5.D.2.11) is not square integrable over $\mathbb{R}$ (while we know that $\int_{\mathbb{R}} |\Psi(x',t)|^2 dx' = 1$ for all t), and moreover, its phase is constant in x'. So what the limit means is that, for large t, $\Psi(x',t)$ is the sum of two square integrable functions, one whose Fourier transform is localized near $p = k_n$, and the other whose Fourier transform is localized around $p = -k_n$. And by (5.D.2.8) and equivariance, we find that the particles that are in the support of one of those two wave functions move with a speed approximately equal to $\pm k_n$.

---

[106]This may seem artificial as a limit, but if we put back factors of $\hbar$ in our equations, it simply means that we take L large compared to the de Broglie wavelength of the particle.

The result (5.D.2.11) is exactly what Einstein asked for, when he thought of the classical limit as being made of a motion going back and forth between the walls of the box,[107] except that here this only happens once we remove the walls of the box.[108]

5.E The Effective Collapse of the Quantum State and Decoherence

Let us describe the measurement process of Appendix 2.D in more detail (here we follow [70, Sect. 6.1]). When the particle is deflected by the magnetic field, either up or down, its presence is detected by a macroscopic device. One example of such a device would be, for a charged particle, an ion chamber. A cascade of ionization of atoms will eventually produce a pulse that can be observed, at a macroscopic level, by a galvanometer.

Consider two measuring devices, one detecting the particle if it goes up, and one detecting it if it goes down. Let the initial wave function for the two devices be

$$\Phi_0(x_1, \ldots, x_N, y_1, \ldots, y_N) = \Phi_0^{up}(x_1, \ldots, x_N)\Phi_0^{down}(y_1, \ldots, y_N), \quad (5.E.1)$$

where $x_1 \ldots x_N$ are the coordinates of the particles making up the first device and $y_1 \ldots y_N$ are those for the second device, while N is of the order of the Avogadro number ($N \sim 10^{23}$). Let $\Phi_f^{up}(x_1, \ldots, x_N)$ and $\Phi_f^{down}(y_1, \ldots, y_N)$ (f for "final") denote the wave functions of the detector after the particle has been detected. Note that these wave functions have a support that is disjoint from the support of Φ_0^{up}, Φ_0^{down} for each variable x_i, y_i. Indeed, if we think of an ion chamber, then the electrons lost by the ions will move towards an electric plate in Φ_f^{up}, while they will remain bound to their atoms in Φ_0^{up} (and the same holds for Φ_0^{down}, Φ_f^{down}). But this can occur only if the supports of the wave functions are disjoint, for each electron.

If the initial quantum state of the particle is given by (5.1.4.7), then the combined initial quantum state of the particle and the ion chamber is

$$\Phi_0 = \Phi_0(x_1, \ldots, x_N, y_1, \ldots, y_N)\Psi(z)\big(c_1|1\uparrow\rangle + c_2|1\downarrow\rangle\big).$$

Then, the final quantum state [see (5.1.4.8)] will be:

$$\Phi_f^{up}\Phi_0^{down}\Psi(z-t)c_1|1\uparrow\rangle + \Phi_0^{up}\Phi_f^{down}\Psi(z+t)c_2|1\downarrow\rangle, \quad (5.E.2)$$

where we have left out the arguments of the wave functions $\Phi_f^{up}\Phi_0^{down}$ and $\Phi_0^{up}\Phi_f^{down}$. This means that, for the $\Psi(z-t)c_1|1\uparrow\rangle$ part of the quantum state, the up detector

---

[107] See Sect. 5.1.4 and [172].

[108] One can also check that, in this limiting situation, the acceleration of the particles vanishes, since the velocity is given by k_n, and the quantum potential defined by (5.B.6) is constant, whence its contribution to (5.1.1.5) vanishes.

will be activated and the down part will remain in its original state, and vice versa for the $\Psi(z+t)c_2|1\downarrow\rangle$ part.

Now, since the two parts in (5.E.2) have disjoint supports, the original particle can only be in one of them, and so will be guided only by that part of the quantum state in whose support it happens to lie. However, we cannot generally neglect the part of the wave function in the support of which the particle is not located, because the two parts could subsequently be brought together, in such a way that their supports overlap. This is exactly what causes the interference phenomena, exemplified by the Mach–Zehnder interferometer, as discussed in Sect. 2.2. In our example, the quantum state of the particle, viz.,

$$\Psi(z)\big(c_1|1\uparrow\rangle + c_2|1\downarrow\rangle\big) \,,$$

evolves into

$$\Psi(z-t)c_1|1\uparrow\rangle + \Psi(z+t)c_2|1\downarrow\rangle \,, \tag{5.E.3}$$

where $\Psi(z-t)$ and $\Psi(z+t)$ have disjoint supports (for t not too small). In a quantum state like (5.E.3), the particle is guided by $\Psi(z-t)$ *or* $\Psi(z+t)$, depending on where it is, but we cannot "reduce" the quantum state and keep only that part, because $\Psi(z-t)$ and $\Psi(z+t)$ may later evolve into wave functions with overlapping supports.

But how can one do that with a quantum state like (5.E.2) involving of the order of 10^{23} particles? It is not enough to make $\Psi(z-t)$ and $\Psi(z+t)$ overlap, we need to do that also with $\Phi_f^{up}\Phi_0^{down}$ and $\Phi_0^{up}\Phi_f^{down}$. But this is impossible in practice. Let us see why.

Remember that these wave functions represent a collection of $N\sim 10^{23}$ particles, all of which have different locations, some moving more or less freely and some being bound to their atoms. Hence, the wave functions must have different supports for each particle. Consider, for simplicity, a wave function that factorizes:

$$\Phi_0^{up}(x_1,\ldots,x_N) = \prod_{i=1}^{N} \Phi_{0i}(x_i) \,, \tag{5.E.4}$$

and similarly for Φ_f^{up}, Φ_0^{down}, and Φ_f^{down}. The support of Φ_{0i} is disjoint from the support of Φ_{fi} for *every* particle since, in the support of Φ_{0i}, the particle is bound to the atom and, in the support of Φ_{fi}, it is moving.

For the two wave functions to interfere again, we need the wave functions $\Phi_{0i}(x_i)$ of all the particles to overlap in the future. Indeed, Φ_0^{up} and Φ_f^{up} have disjoint supports if

$$\Phi_0^{up}(x_1,\ldots,x_N)\Phi_f^{up}(x_1,\ldots,x_N) = 0 \,, \tag{5.E.5}$$

$\forall(x_1,\ldots,x_N)\in\mathbb{R}^N$. Of course, the same equation also holds for $\Phi_0^{down}(x_1,\ldots,x_N)$ $\Phi_f^{down}(x_1,\ldots,x_N)$. But for (5.E.5) to be true, it is enough to have $\Phi_{0i}(x_i)\Phi_{fi}(x_i)=0$, $\forall x_i\in\mathbb{R}$, for a *single* value of i. Hence, for interference to occur, we need to get the wave functions of every particle in (say) the ion chamber to overlap again.

It is clear that this is, in practice, impossible. The same is true for the particles constituting the live cat and the dead cat, the pointer pointing up or down, the exploded and unexploded bomb, etc. Now, of course, once the position of the original particle is in the support of one of the two terms in (5.E.2), we can simply reduce the quantum state to that term, since we can be sure that the other term will never again interfere with it.

But we can know in which of the supports of the two terms the particle happens to be located simply by looking at the (by definition) macroscopic result. Hence, in some sense, we do "collapse" the quantum state after we look at the result of an experiment. However, this is an entirely practical matter. We can still consider, if we wish, that the true quantum state is and remains forever given by the time evolution of (5.E.2). Put simply, there is one of the terms that does not guide the motion of the particle, either now or ever again.

5.F Proof of the Factorization Formula

We now prove (5.1.7.5). Note that (5.1.7.4) implies that

$$\Psi(x_1, \ldots, x_M, \mathbf{Y}) = \psi(x_i)\Phi_i(\hat{x}_i, \mathbf{Y}) , \qquad (5.F.1)$$

where $\hat{x}_i = (x_1, \ldots, x_M)$ with x_i missing and we have put in $\Phi_i(\hat{x}_i, \mathbf{Y})$ the arguments that were not indicated in (5.1.7.4).

To prove (5.1.7.5), consider $M = 2$. By (5.F.1), we have

$$\Psi(x_1, x_2, \mathbf{Y}) = \psi(x_1)\Phi_1(x_2, \mathbf{Y})$$
$$= \psi(x_2)\Phi_2(x_1, \mathbf{Y}) . \qquad (5.F.2)$$

Dividing by $\psi(x_1)\psi(x_2)$, we get

$$\frac{\Phi_1(x_2, \mathbf{Y})}{\psi(x_2)} = \frac{\Phi_2(x_1, \mathbf{Y})}{\psi(x_1)} , \qquad (5.F.3)$$

which means that both sides are functions of $\mathbf{Y}$ alone, which we can write $\Phi(\mathbf{Y})$. Thus,

$$\frac{\Psi(x_1, x_2, \mathbf{Y})}{\psi(x_1)\psi(x_2)} = \Phi(\mathbf{Y}) ,$$

which is (5.1.7.5) for $M = 2$. The argument easily extends to all M.

Chapter 6
Are There Any Alternative Theories?

By "alternative" we mean alternative to de Broglie–Bohm, of course. We have presented the de Broglie–Bohm theory as being a completed and rational version of ordinary quantum mechanics, so that we do not "count" the latter as an alternative to the de Broglie–Bohm theory, but rather as a truncated version of it, which is perfectly sufficient for all practical purposes. However, the field of "foundations" of quantum mechanics seems to be full of various theories or interpretations of quantum mechanics that could count as competitors to de Broglie–Bohm.

But that is not quite right. One can, roughly speaking, divide the alternatives into four categories (three of which were advocated, along with de Broglie–Bohm theory, during a "science festival" debate[1] in New York in June 2014). We will discuss these in the historical order of their invention: the many-worlds interpretation, the spontaneous collapse (of the wave function) theories, the decoherent histories approach, and QBism.[2] However, we shall leave aside ideas such as "quantum logic", "quantum probabilities", or the "modal interpretation", partly because they are not that popular any more.[3] Our goal is to show that none of these alternatives reaches the level of clarity and explanatory power of the de Broglie–Bohm theory.

It will be useful to introduce here words like *ontology* and *beables*, the first being a traditional philosophical term, already used in Chap. 3, and the second an invention

---

[1]The debate, entitled *Measure for Measure: Quantum Physics and Reality* can be viewed at www.worldsciencefestival.com/2014/06/measure-measure-can-reconcile-waves-particles-quantum-mechanics/. The positions defended during the debate were the de Broglie–Bohm theory (by Sheldon Goldstein), the many-worlds interpretation (by Sean Carroll), spontaneous collapse theories (by David Albert), and quantum Bayesianism (by Rüdiger Schack).

[2]In a section entitled: "The proliferation of interpretations", Freire discusses more or less the same four categories, see [195, pp. 319–327]. In [301], Laloë discusses more interpretations than we do, and discusses in detail the ones that we consider. For an encyclopedic review of "interpretations", from a viewpoint very different from ours, see Auletta [22].

[3]The idea of quantum logic and quantum probabilities goes back to Birkhoff and von Neumann in 1936 [58], see [518] or [281, Chap. 8] for a review of quantum logic and [311] for the modal

© Springer International Publishing Switzerland 2016
J. Bricmont, *Making Sense of Quantum Mechanics*,
DOI 10.1007/978-3-319-25889-8_6

by John Bell. They both refer to what exists or at least to what a theory postulates as existing. The word beable was invented to contrast with "observables", because a physical theory should be concerned with what there is and not only with what is observed. Of course, physicists do have an ontology, made of electrons and protons, quarks and gluons, etc. But because of the ambiguous status of the quantum state, it is not exactly clear what all this means, except when one falls back on observations.

In the non-relativistic de Broglie–Bohm theory, the ontology comprises particles that have trajectories. In the quantum field version, it could be a mixture of fields and particles. For each of the theories below, it will be important to specify what ontology is actually posited. If one is not explicit about which beables a given theory assumes to exist, one easily falls back into predicting "results of measurements", for which ordinary quantum mechanics is just fine.

6.1 The Many-Worlds Interpretation

The many-worlds interpretation was introduced in 1957 by Hugh Everett III, then a graduate student at Princeton University working under the supervision of John Wheeler. We will first discuss what we call the "naive view of the many-worlds interpretation". Then, we will give a mathematical formulation of it, due to Allori et al. [14], and finally discuss a more sophisticated version of the many-worlds interpretation which is favored by some of its defenders.

6.1.1 The Naive Many-Worlds Interpretation

The naive view[4] is that, when the proverbial cat (or any other macroscopic device) finds itself in a superposed state, such as (5.1.6.2), then instead of having a collapse by fiat as in ordinary quantum mechanics or an effective collapse as in the de

---

(Footnote 3 continued)
interpretation. Albert and Loewer offer a criticism of the modal approach in [7] (see also the appendix to [8]). Both of these approaches try to assign properties (other than positions) to quantum systems even when they are not measured, but in our opinion, the problem is that they try to ignore the active role of measurement devices and therefore run into difficulties because of the no hidden variables theorem. As Bell remarked:

> When one forgets the role of the apparatus, as the word "measurement" makes all too likely, one despairs of ordinary logic — hence "quantum logic". When one remembers the role of the apparatus, ordinary logic is just fine.

John Bell [46, p. 34]

[4]The reader can watch the 2012 popular presentation of the many-worlds interpretation by David Wallace called *The Long Earth: Multiverse Physics* (available at www.youtube.com/watch?v=GRJT9qY21nA), to see that this "naive view" is not very different from the one that some defenders of the many-worlds interpretation present to a general audience.

Broglie–Bohm theory, both terms simply continue to exist. But how can that be possible? We always see the cat alive *or* dead but not both! The short answer is that they both exist, but in different "universes" or "worlds".

Hence, still thinking naively, whenever an experiment leads to a macroscopic superposition, the universe splits into two or more worlds, depending on the number of distinct macroscopic states in the sum in (5.1.6.2), one for each possible result. But why do I always perceive only one of the results? It is simple: I, meaning my body, my brain (and thus also my consciousness) gets entangled with the states of the cat, so there are two or more copies of me also, one seeing the dead cat in one world, another seeing the live cat in another world. And that, of course, is also true for everything else: every molecule in the entire universe gets to be copied twice (maybe not instantaneously, but that is another question), since the two different copies of me, or of the cats, could in principle interact later with those molecules and that might affect their state. Or one can think of different light rays reflected from the cat, whose details will depend on whether the cat is alive or dead, and which will propagate arbitrarily far.

The physicist and proponent of the many-worlds interpretation, Lev Vaidman, describes this multiplication of worlds rather vividly:

> "I" am an object, such as the Earth, a cat, etc. "I" is defined at a particular time by a complete (classical) description of the state of my body and of my brain. "I" and "Lev" do not refer to the same things (even though my name is Lev). At the present moment there are many different "Lev"s in different worlds (not more than one in each world), but it is meaningless to say that now there is another "I". I have a particular, well defined past: I correspond to a particular "Lev" in 2012, but not to a particular "Lev" in the future: I correspond to a multitude of "Lev"s in 2022. In the framework of the MWI it is meaningless to ask: Which Lev in 2022 will I be? I will correspond to them all. Every time I perform a quantum experiment (with several possible results) it only seems to me that I obtain a single definite result. Indeed, Lev who obtains this particular result thinks this way. However, this Lev cannot be identified as the only Lev after the experiment. Lev before the experiment corresponds to all "Lev"s obtaining all possible results.

Lev Vaidman [480]

Bryce DeWitt, another proponent of the many-worlds interpretation, also stresses that this multiplication of worlds has to be taken literally and not metaphorically:

> This universe is constantly splitting into a stupendous number of branches,[5] all resulting from the measurementlike interactions between its myriad components. Moreover, every quantum transition taking place on every star, in every galaxy, in every remote corner of the universe is splitting our local world on earth into myriads of copies of itself.

Bryce S. DeWitt [131], reprinted in [132, p. 161]

To say that this view is weird is accepted by all sides. Indeed, after the passage quoted here, Bryce DeWitt also writes:

---

[5]The word "branches" refers to the different terms in a macroscopic superposition such as (5.1.6.2). (Note by J.B.).

I still recall vividly the shock I experienced on first encountering this multiworld concept. The idea of 10^{100+} slightly imperfect copies of oneself all constantly splitting into further copies, which ultimately become unrecognizable is not so easy to reconcile with common sense.

Bryce S. DeWitt [131], reprinted in [132, p. 161]

That last phrase might qualify as being the understatement of the century.

Even in the original paper [180] (reprinted in [132, p. 146]), Everett stressed that "*all* elements of a superposition (all 'branches') are 'actual', none any more 'real' than the rest." Everett felt obliged to write this because "some correspondents" had written to him saying that, since we experience only one branch, we have only to assume the existence of that unique branch. This shows that some early readers of Everett were already baffled by the radical nature of the "many-worlds" proposal.

But why am I not conscious of this proliferation of copies of myself? To explain that, one appeals to decoherence: since, for macroscopic systems, the different branches of the wave function do not interfere with each other, in practice at least,[6] the copies of me do not "see" each other. They could pass through the same place without having any effect on each other, since each branch of the wave function evolves independently of the other branches, in the sense that no branch interferes with any other branch.

Defenders of the many-worlds interpretation argue that the same objection (about my not being conscious of those copies of myself) was raised against Copernicus and Galileo (why don't we feel the motion of the Earth around the Sun?),[7] against Darwin (how come we are so different from apes or other animals if we have common ancestors?), and against the atomic theory of matter (how come things look full when they are mostly empty?), and that, in each case, science answers those objections by explaining why things look to us in certain ways, while in reality they are not as they seem. Decoherence is supposed to be a similar answer for this multiplication of unobservable worlds and identities. But there are several differences between the many-worlds interpretation and those historical examples. The atomic theory made many novel predictions, and Galileo and Darwin's theories had great explanatory power (and led also later to many confirmed predictions).

So, one may ask: why postulate the existence of all those worlds which I am not conscious of? Doesn't Ockham's razor[8] suggest the rejection of such a proliferation of worlds? The usual answer is that, in the many-worlds interpretation, there is only the wave function and its Schrödinger evolution, nothing else. It is sometimes said that the many-worlds interpretation amounts to just taking quantum mechanics "literally": there is a quantum state and its time evolution through the Schrödinger

---

[6]See Sect. 5.1.6 and Appendix 5.E.

[7]This reply was given in a *Note added in proof* to the original paper by Hugh Everett [180], reprinted in [132, pp. 146–147].

[8]This is the principle, going back to the scholastic philosopher William of Ockham, that one should not introduce in our explanations more entities than what is necessary. Obviously, the correct application of this "principle" depends very much on the context.

equation or some relativistic/quantum field theoretical extension of it, and that is all there is. Proponents then appeal to Ockham's razor (but to reach the conclusion opposite to the one eliminating the unexperienced worlds) to justify the "paucity" of theoretical assumptions: no particle positions, no wave function collapse, no hidden variables, etc.

First of all, in the naive view, this paucity of assumptions is not true: there are also cats and brains multiplying themselves to infinity in the usual three-dimensional space and *that* is not merely a wave function. After all, the latter is simply a function defined on configuration space. We will come back to the distinction between a wave function and three-dimensional distributions of matter in Sect. 6.1.3.

But even putting aside this objection, as well as the weirdness of the multiplication of "worlds", one should ask whether the scheme of many worlds is coherent. A first problem concerns the Born rule. Suppose that the probability of having the cat alive or dead is $(1/2, 1/2)$. And suppose that we repeat a similar experiment many times (let us assume that we have a large supply of cats, or, let's say, of pointers). Then, first of all, for every possible sequence of outcomes, there will be some of my "descendants" (by that I refer to the many copies of me, or of Lev Vaidman, that exist in all the future worlds) that will see it. There will be a sequence of worlds in which the cats are always alive and one where they are always dead. There are also many sequences of worlds where the cats are alive one quarter of the time and dead three quarters of the time, and that is true for any other statistics different from $(1/2, 1/2)$. So that we can be certain that many of our descendants will *not* observe Born's rule in their worlds. Of course, this is also true for many of our "cousins", i.e., descendants, like us, of some of our ancestors, if many identical experiments have been made in the past in each of those worlds.

But one could argue, on the basis of the law of large numbers that, at least in the vast majority of worlds, the Born rule will be obeyed, since in the vast majority of worlds the frequencies will be close to $(1/2, 1/2)$. It is just as in coin tossing (discussed in Sect. 3.4.4 and in Appendix 3.A): each world splits into two worlds, one for each possible outcome, so that after N iterations of this process, there will be, in the vast majority of worlds, approximately $N/2$ splittings where the cats end up alive and $N/2$ splittings where they end up dead. So one can use a sort of typicality argument, as we did in the de Broglie–Bohm theory, to justify the Born rule: we and our descendants just happen to live in a typical sequence of worlds, so that the quantum frequencies are observed in them.

But what happens if, instead of being $(1/2, 1/2)$, the probabilities predicted by quantum mechanics are, say, $(3/4, 1/4)$, or anything else different from $(1/2, 1/2)$? We will still have two worlds coming out of each splitting, since the number of worlds corresponds to the number of terms in the macroscopic superposition and is independent of the coefficients c_n in (5.1.6.2). But then the same use of the law of large numbers leads to the conclusion that, in the vast majority of worlds, the quantum predictions will *not* be observed, since our descendants will still see $N/2$ splittings where the cats end up alive and $N/2$ splittings where they end up dead, instead of the $(3/4, 1/4)$ frequencies.

To make the point more explicit, let us introduce two different probability distributions:

(1) *The counting probability distribution*

   Consider a sequence of similar experiments made in each of the successive worlds and let us count the number of worlds in which a given sequence of results occur. In the example above, with two possible outcomes, that probability distribution will give most of its weight to sequences where one result happens half of the time and the other result also.

(2) *The "Born rule" probability distribution*

   One can define a probability distribution, in the example above, as the one associated with a sequence of independent random variables taking two values with probabilities (3/4, 1/4). Then, of course, the law of large number implies that, *relative to that probability distribution*, the frequencies will coincide with the Born rule. But what does this mean concerning what my "descendants" will actually observe?

In his original article [180], Everett introduces the "Born rule" probability, from which of course he deduces the "right" frequencies (appealing to the law of large numbers and to what we call "typicality", as in statistical mechanics), but without explaining what that probability distribution has to do with the observed frequencies in the many worlds, which for him are, as we saw, all equally real. In fact, Everett "deduces" the "Born rule" probability from a natural condition of additivity of a measure defined on the coefficients that we denote by c_n in (5.1.6.2). But the problem is that the number of worlds that are created at each splitting depends on the number of terms in (5.1.6.2), and not on the values of the coefficients c_n. One might wish to try to modify the many-worlds interpretation by making the number of worlds depend on those coefficients, but then it would be an entirely different theory, with new problems, e.g., what to do if some of the c_n have irrational values?

In [238], Neill Graham gives a numerical example of the difference between the two probability distributions: he considers three possible outcomes, with one of them having probability 0.8, and he finds that in most worlds the event "with probability 0.8" occurs only one-third of the time (which is what the counting probability predicts, since there are three possible outcomes here, instead of two). Numerically, for $N = 45$ trials, Graham computes that there are over 800 million times more worlds where the Born rule is *not satisfied* than those where it is [132, p. 233]. So the problem is not just that there exist worlds in which the Born rule is not satisfied (which would be the case even if all outcomes were equiprobable), but that this rule is not satisfied in the overwhelming majority of worlds, or, in the language of statistical mechanics, in a "typical" world. From this, Graham concludes:

   Everett gives no connection between his measure and the actual operations involved in determining a relative frequency, no way in which the value of his measure can actually influence the reading of, say, a particle counter. Furthermore, it is extremely difficult to see what significance such a measure can have when its implications are completely contradicted by a simple count of the worlds involved, worlds that Everett's own work assures us must all be on the same footing.

Neill Graham [132, p. 236]

Another way to visualize the problem is to imagine a biased coin, with probability 3/4 of falling heads and 1/4 of falling tails.[9] What sense could we make of those probabilities if we were told that both results, heads and tails, happen, but in "different worlds", each of which includes of course observers that "see" the coin being heads in one world and being tails in another world? Upon repetition of the coin tossing in each successor world, most of our descendants would see the coin falling heads approximately half of the time and tails the other half.

This is *prima facie* a serious problem for the many-worlds interpretation. There have been many proposals to solve this problem but it would be too long and too technical to discuss all of them here.[10] For example, after giving the numerical example that we mentioned, Graham introduces a "relative frequency" operator (or "observable"), which supposedly counts the frequency with which certain worlds, characterized by given results of measurement, occur. He then computes the average and the variance of that operator with respect to what is in reality the Born rule probability distribution.[11] Unsurprisingly, he recovers the quantum predictions. But that is just a way to introduce the Born rule probability distribution by *fiat*, as Everett did, without explaining what relevance it has to the many worlds in which the future copies of myself actually live.

In [501], David Wallace objects that what we call the counting probability distribution is not time invariant. Indeed, suppose that if I do an experiment at time t_1 leading to two worlds, and in one of them my descendant gets, say, a hat. Next, repeat the experiment at time t_2 in that world but not in the other one, leading to another splitting of the world with the hat into two worlds. So we have a total of three worlds at time t_2, two with a hat and one without. Then the (counting) probability of me having a hat at time t_1 is $1/2$, while it will be $2/3$ at time t_2. But that misses the point of the objection to the many-worlds interpretation based on counting the worlds. All our evidence for the truth of quantum mechanics is based on certain observed statistics. If I can invent scenarios, like the one with probabilities (3/4, 1/4) above, that will guarantee that most of my descendants will *not* observe the correct statistics, then why should I expect to observe them in the future? And since the present is just the future of the past and since such scenarios may have happened in the past, why

[9]One can, if one wants, produce such a coin by mechanically coupling it to the result of a quantum experiment with two possibles outcomes, having those probabilities (3/4, 1/4), so that the coin is heads if one outcome occurs and tails otherwise.

[10]For a review of such proposals and critiques of them, see e.g., [26, 289, 328, 435, 501]. For a critique of older proposals, see [288].

[11]To be precise, Graham does this as if what one measures are these frequency operators. Measuring those operators will then orient the splittings of the worlds in a way that is compatible with the Born rule probability distribution. But the actual measurements made in a laboratory are related to ordinary operators, such as the spin, and not to those frequency operators.

should I expect to observe the correct statistics now?[12] Of course, since all results do occur (in different worlds), the correct statistics will also occur, but will be very rare. This objection has nothing whatsoever to do with the counting probability distribution being a "nice" probability from some a priori viewpoint, such as requiring it to be time invariant.[13]

Another way to "solve" the problem is to relate the numbers $|c_n|^2$ to a certain "measure of existence" or "reality" of the world corresponding to the coefficient c_n in (5.1.6.2). Lev Vaidman introduces this notion as follows:

> A believer in MWI can define a *measure of existence of a world*, the concept which yields his subjective notion of probability. The measure of existence of a world is the square of the magnitude of the coefficient of this world in the decomposition of the state of the Universe into the sum of orthogonal states (worlds). *The probability postulate* of MWI is: If a world with a measure[14] μ splits into several worlds then the probability (in the sense above) for a sentient being to find itself in a world with measure μ_i (one of these several worlds) is equal to μ_i/μ.
>
> Lev Vaidman [479]

Another supporter of the many-worlds interpretation, the French physicist Thibault Damour, explains that:

> Each configuration q will have more or less "reality" according to the value of the amplitude $A(q)$. In other words, we interpret A as an *existence amplitude*, and not (like in the Born–Heisenberg–Bohr interpretation) as a *probability amplitude*.[15] Indeed, the notion of probability amplitude for a certain configuration q suggests, from the very beginning, a random process by which only one configuration among an ensemble of possible configurations, is realized, passing from the possible to the actual. By contrast, the notion of existence amplitude suggests the simultaneous existence (within a multiply exposed frame) of all possible configurations, each actually "existing", but with more or less intensity.
>
> Thibault Damour [104, p. 155], quoted in [325]

But although one can imagine oneself living in "another world" than the one we live in (for example, if some event had or had not occurred in our past), it is difficult to see what it could mean to "exist" in another world, but with a different "intensity" than the one we live in, and those suggestions flatly contradict Everett's dictum that

---

[12]Of course, we don't know if this scenario is realized in the existing worlds, and we have no idea what the worlds look like, because the set of branchings is extremely complicated and impossible to compute, which is also an argument put up by Wallace. But then we have no idea either whether Born's rule is observed in these worlds. If one can convince oneself that, in simple cases, the Born rule won't be satisfied in most worlds, it does not help to reply that the actual splittings into multiple worlds are very complicated and that we have no idea how many worlds actually exist.

[13]In [180], Everett imposes an additivity requirement that yields the Born rule probability distribution, but does not explain what this has to do with the empirical statistics observed in most worlds.

[14]Here μ and μ_i refer to what we denote by $|c_n|^2$, with c_n coming from (5.1.6.2). (Note by J.B.).

[15]Here $A(q)$ denotes what we call $\psi(x)$, q being the position variable. If one deals with a sum like (5.1.6.2), then $A(q)$ corresponds to c_n (Note by J.B.).

"*all* elements of a superposition (all 'branches') are 'actual' none any more 'real' than the rest" [132, p. 146]. Are all the worlds equally real or are some worlds more equally real than others?

David Deutsch and David Wallace have tried to justify the "Born rule probability distribution" by appealing to betting strategies by a rational agent (see, e.g., [128, 498, 500]). But putting aside the fact that rational (or irrational) decisions made by agents has nothing to do with accounting for empirical frequencies, which is what physics is supposed to do,[16] if all outcomes occur and are all real, then it is not clear what sense it makes to bet on one of my descendant, rather than another (this point is developed in [328]).

A final problem with the naive many-worlds interpretation is that what we just described so far is vague. Can one define this multiplication of worlds in precise mathematical terms? There is a way to do that, due to Allori et al. [14] (but not one favored by most supporters of the many-worlds interpretation) and we will discuss it now.

6.1.2 A Precise Many-Worlds Interpretation

Allori et al. [14] associate a continuous matter density with the wave function of a system of N particles. For each $\mathbf{x} \in \mathbb{R}^3$, and $t \in \mathbb{R}$, one defines

$$m(\mathbf{x}, t) = \sum_{i=1}^{N} m_i \int \delta(\mathbf{x} - \mathbf{x}_i) |\Psi(\mathbf{x}_1, \ldots, \mathbf{x}_N, t)|^2 d\mathbf{x}_1 \ldots d\mathbf{x}_N, \qquad (6.1.2.1)$$

where $\Psi(\mathbf{x}_1, \ldots, \mathbf{x}_N, t)$ is the usual wave function of the system at time t. This equation makes a connection between the wave function defined on the high-dimensional configuration space and an object, the matter density, existing in our familiar space $\mathbb{R}^3$. Here, the ontology does not consist only of the wave function, but also of the matter density (one can have several matter densities in the relativistic version, but let us leave that aside). In our three-dimensional world, there is just a continuous density of mass: there is no structure, no atoms, no molecules, no brain cells, etc., just an amount of "stuff", with high density in some places and low density elsewhere.

The authors of [14] analyze this model and claim that it is empirically equivalent to ordinary quantum mechanics as far as the probabilistic predictions are concerned (see below for a discussion of that claim). But they also show that the theory is nonlocal, as it must be because of the EPR–Bell argument.

Note that each term in a sum like (5.1.6.2) will give a different contribution to the matter density, provided that the terms in (5.1.6.2) do not overlap: if we

---

[16]Note also that the many-worlds interpretation is often advertised as being relevant for cosmology, because there were no observers at the time of the Big Bang. But then, there was nobody betting either.

write $\Psi(\mathbf{x}_1, \ldots, \mathbf{x}_N, t) = \sum_n c_n \varphi_0(z - \lambda_n t)|e_n\rangle$ (remember that z is a function of $\mathbf{x}_1, \ldots, \mathbf{x}_N$, for example, their center of mass), and if the different terms in the sum have disjoint support in configuration space,[17] we get that $|\Psi(\mathbf{x}_1, \ldots, \mathbf{x}_N, t)|^2$ is approximately equal to $\sum_n |c_n|^2 |\varphi_0(z - \lambda_n t)|e_n\rangle|^2$, since the cross terms vanish if the supports of the different terms do not overlap.

So this version of many worlds makes precise the naive picture sketched in the previous subsection. Lev Vaidman (and all of us) have many copies of ourselves in these different worlds. And because the coefficient $|c_n|^2$ will enter into (6.1.2.1), one can say that the different copies of ourselves living in the different worlds have different "intensities". But that does not tell us what it means to live in a low intensity world.

Indeed, it is not clear how to think of this model. If we assume that the cat which is alive does not move and stays at the same place as the dead cat, then the matter density will be somewhat mushy, since all there is is the function $m(\mathbf{x}, t)$, which is the sum of the contributions from the two cats [which are like different terms in (5.1.6.2)], and it is not clear how to "see" each cat in the resulting sum. Certain details of the cats will be different, one has her blood circulating, the other not, for example. And, of course, if the live cat was moving, its contribution to the matter density would be distinct from that of the dead cat. Besides, the wave functions of our bodies and our brains also get entangled with the wave functions of the cats, so that they also have two contributions to their matter density. The authors of [14] offer an analogy: the world, in that theory, is like a TV set that is not correctly tuned so that a mixture of several channels is seen at any given same time.

We leave it to the reader to decide whether this analogy makes the theory more understandable, but it is still not clear how one is supposed to think about our perceptions: is a "mind" attached to each matter density or only to the total density? If it is attached to each density, then there are many minds, one for each individual, each being conscious of the cat being alive or dead, but not both. But then we would not "see" the incorrectly tuned TV set; each of our minds would see a clear picture, but a different one for each mind. Only "god", looking at the world from the outside, sees an incorrectly tuned TV set.[18]

Let us also note that, if the god of the physicists was trying not to be malicious[19] and if the many-worlds version of [14] is true, then he failed badly: indeed, it means that we were wrong all along when we "discovered" atoms, nuclei, electrons, etc., and that we are lying to schoolchildren when we tell them that matter is mostly void with a few pieces of matter (the atoms) here and there. Indeed, in that theory, matter is continuous after all, with higher and lower density in some places, and we have simply been fooled by this version of the many-worlds interpretation of quantum mechanics into thinking that it is not.

---

[17]The reason why this is so is given in Appendix 5.E.

[18]According to Roderich Tumulka, this is how one should think of the version of many worlds in [14] (private communication).

[19]The famous quote "God is subtle but he is not malicious" is attributed to Einstein.

Finally, it is not clear how this many-worlds theory solves the problem of the probabilities discussed above. The coefficients c_n in (2.D.6) will determine different weights $|c_n|^2$ associated with the decoherent branches corresponding to the different worlds. So coming back to our example with two outcomes, one having probability 3/4 and the other 1/4, the density of matter will be different in the world where ones "sees" the outcome having probability 3/4 and in the one having probability 1/4. But what difference does it make? In what way does having a smaller or larger matter density affect my states of mind? And if it does not, we are back to the problem that, if one repeats many times the experiment whose outcomes have probabilities (3/4, 1/4), most of my descendants (some of course having a small matter density) will see massive violations of the Born rule. One can, as is done in [14] (and was done by Everett in [180]), *define* a "probability" on those sequences of worlds that is built from the coefficients $|c_n|^2$, and which is what we call the Born rule probability distribution. Relative to that probability distribution, the ordinary quantum predictions will be typical. But what that probability distribution means for the sequences of worlds generated by repeating the same experiment with probabilities (3/4, 1/4) in each one of them remains unclear.

The authors of [14] argue that our counting probability distribution defined above "is not an option", because it is difficult to determine, in general, the number of different (macroscopic) worlds (as we saw, the same objection is made by Wallace in [501]). But if one reads the "founding fathers" of the many-worlds interpretation, Everett et al. (in [132]), one has the definite impression that it is exactly this multiplication of macroscopic worlds that their theory was trying to account for, and that they discuss it in idealized situations, with the cat being alive in one world and dead in another. Of course, the example we gave above with two macroscopic worlds emerging after each experiment, which, by assumption, is repeated in each successive world, with probabilities equal to (3/4, 1/4), is an idealization, but why is it less legitimate than other idealizations in physics? And that example does produce the wrong statistics in the vast majority of worlds. Unless one can understand why a world with a small matter density is less real than one with a larger matter density, the problem will remain.

In any case, the version of the many-worlds interpretation due to [14], with its extravagant ontology consisting of a constant multiplication of every material entity, does not appeal to many partisans of the many-worlds interpretation. Defenders of the many-worlds interpretation tend to dismiss both this approach and the naive one and prefer to speak of a pure wave function ontology, something we will discuss now.

6.1.3   The Pure Wave Function Ontology

In this ontology, the world *is* made of a universal wave function and nothing else.[20] There are, in actual fact, no cats, pointers, brains, etc., situated in ordinary

---

[20]For a more detailed discussion, see the section on what is called GRW0 in [13], and also [322, 323, 325, 361].

three-dimensional space, but only a mathematical object, a complex function defined
on an abstract space, containing all the possible configurations of particles and fields
in the universe.[21] In the language of Bell [42], there are no "local beables" in that
theory, i.e., things that exist (beables) and that are localized in $\mathbb{R}^3$.

It is also important to realize that the ontology of the "pure wave function" many-
worlds interpretation is a purely mathematical object: a function defined on a space
of "configurations" (of particles and fields), but without any particles or fields being
part of the ontology. Indeed, if we put particles and fields, existing in ordinary three-
dimensional space, into our ontology, then we get back to the "naive" picture of
worlds constantly splitting and multiplying themselves.

But it is very difficult to see how to make sense of this pure wave function ontology,
in particular how to relate it to our familiar experience of everyday objects situated
in three-dimensional space. As Bell has emphasized, there is no sense in saying that
the wave function has a certain value at a given point $\mathbf{x} \in \mathbb{R}^3$, unless one specifies a
lot of other points $\mathbf{x}_2, \ldots \mathbf{x}_N \in \mathbb{R}^3$ and one considers the value of $\Psi(\mathbf{x}_1, \ldots, \mathbf{x}_N, t)$,
with $\mathbf{x} = \mathbf{x}_1$.[22] If we think of the live and dead cat, in this ontology, it just means
that the wave function has a high value on configurations corresponding to the dead
cat and also on configurations corresponding to the live cat, but without there *being*
particles constituting the cat (either alive or dead). Just as in the well-known Magritte
painting "this is not a pipe", a wave function with a high density on a set of points that
would constitute a cat, *if those points corresponded to particle positions*, is simply
not a cat if there are no particles.

This critique is of course rejected by supporters of the many-worlds interpretation.
For example, Lev Vaidman writes[23]:

> If a component of the quantum state of the Universe, which is a wave function in the shape
> of a man, continues to move (to live?!) exactly as a man does, in what sense is it not a man?
> How do I know that I am not this "empty" wave?

Lev Vaidman [479]

Similarly, David Wallace summarizes his views in what he calls a "slogan":

> A tiger is any pattern which behaves as a tiger.

David Wallace [498]

The counterargument is that Vaidman, like tigers and everything else, is located
somewhere in our three-dimensional space and is composed of various parts, also

---

[21]In fact, because of the spin variables, we have a family of such functions, but we will leave that
aside.

[22]N being potentially the number of particles in the universe, and setting aside here the existence
of fields.

[23]This is his main objection to the de Broglie–Bohm theory, where of course only one world actually
exists, characterized by the term in the sum (5.1.6.2) in whose support the particles are actually
located. The other terms in the sum are sometimes called "empty waves".

located in three-dimensional space, and that "a wave function", even one "in the shape of a man" or a "pattern which behaves as a tiger", is not.

This basic disagreement about the actual structure of the world leads some supporters of the many-worlds interpretation to consider that the de Broglie–Bohm theory is just "many worlds in denial".[24] Indeed, for them, since the empty branches of the quantum state continue to exist after they decohere from the branch in whose support the particles are located, all these branches are equally real, and "adding particles" to one branch does not change anything.

But as stressed by Valentini in [489], the de Broglie version of the theory, before 1927, did not "add particles" to anything. At that time, quantum mechanics did not exist, and de Broglie was simply proposing a theory about the motion of particles.[25] He was not solving the measurement problem, because that "problem" arose only within the post-1927 "orthodox" version of quantum mechanics.

From the de Broglie–Bohm point of view, it is the motion of the particles themselves, *in three-dimensional space*, that is fundamental, with the wave function "merely" guiding this motion. The existence of empty branches of the wave function can be considered as somewhat similar to the fact that a Hamiltonian or electromagnetic fields are defined even where there are no particles on which they "act".

Finally, we have to leave it to the reader to decide whether the "pure wave function ontology" makes sense or not. But to make that judgment, he or she should eliminate all familiar three-dimensional pictures of the world, and think of the world as being only an abstract mathematical object that takes certain "large" values on sets of points that would correspond to particles constituting cats, tables, or brains, if such things existed. In particular, one should not be misled by the expression "configuration space", since there are no configurations in that theory.

The many-worlds interpretation with the pure wave function ontology is sometimes presented as being a local theory, hence compatible with relativity. The argument is that, since all there is in the world is the wave function and since the potentials entering the Schrödinger equation are local (meaning that for an interaction, say, between two bodies, the strength of the interaction decays with the distance between those bodies, unlike the EPR–Bell nonlocal effects), then everything evolves locally. The problem with that argument is that, since there are no "local beables" in that theory, no objects localized in space, and not even a spacetime, the issue of nonlocality cannot even be formulated. Indeed, nonlocality has to do with the fact that, acting on an object localized somewhere (like a detector) can affect what happpens to another object localized far away from the first object (like another detector). But if the world is only an abstract mathematical object, the wave function of the universe, which is intrinsically nonlocal since it is defined on the configuration space of the

---

[24]See Valentini [489], for references to that criticism made by Deutsch [127], Zeh [525], and Brown and Wallace [86].

[25]De Broglie even said [24, p. 346]: "It seems a little paradoxical to construct a configuration space with the coordinates of points that do not exist." See Sect. 7.6.1 for a further discussion of what de Broglie thought.

entire universe, what does it mean for a theory whose ontology is reduced to that object to be local?

Let us mention also another way to (sort of) make sense of the many-worlds interpretation: the many minds version of David Albert and Barry Loewer [5, 6, 8], where the "material world" is just the pure wave function, but each individual is equipped with an infinity of minds and the splitting occurs inside the set of minds. Again, we leave the reader make up his or her mind(s) about the plausibility of such a scenario.

The multiplicity of ways to try to make sense of the many-worlds interpretation suggests that it is not so easy to do, i.e., to give a precise meaning to the naive picture described at the beginning of this section.[26] This is probably why it is the naive view that remains in the mind of most people who speak about the many-worlds interpretation or who claim to "like" it.

Regarding the many-worlds interpretation with the pure wave function ontology, Travis Norsen speaks of solipsism "for all practical purposes" [361], although the defenders of the many-worlds interpretation usually present themselves as strong supporters of realism. This is because the world of the pure wave function ontology has a status similar to that of the brain in a vat story mentioned in Chap. 3, and this is what Norsen calls "solipsism FAPP". Indeed, it may be that we are radically deluded about living in a three-dimensional world and that the only thing that exists is this universal wave function. Perhaps we can tell some story that explains why our conscious experiences are so radically mistaken about the world (although it is not at all clear how such a story would go), just as it may be that we are brains in a vat, with our brains manipulated from the outside in just such a way as to make our conscious experiences what they are, even though we are radically mistaken about the way the world really is.

Norsen makes this point about the pure wave function ontology, but it also applies to the many-minds version of the many-worlds interpretation and also, to some extent, to the continuous matter density version, because in all these theories we are deeply mistaken about everything, not only the very small or the very large or the distant past, but even about the familiar macroscopic reality. But why should we believe those theories any more than believing the brain in a vat story?

As we argued in Chap. 3, common sense realism comes first, i.e., before scientific realism: all our reasons to "believe" our scientific theories are based on observations of the macroscopic world (including of course, positions of pointers and meter readings). If a scientific theory tells us that we are radically deluded about the existence and the basic properties of the macroscopic world (like saying that the moon is not there when nobody looks [332] or that reality is only an abstract wave function), then why believe it?

---

[26]In [289], Adrian Kent gives a list of 22 papers trying, in various ways, to make sense of the many-worlds interpretation.

6.2   The Spontaneous Collapse Theories

In [209], Ghirardi, Rimini, and Weber (GRW) introduced a modified version of quantum mechanics.[27] We saw in Sect. 2.5 that, thanks to the linearity of the Schrödinger equation, there is never a collapse of the wave function and this leads to the existence of macroscopic superpositions.

In order to solve that problem, GRW introduce a mechanism by which wave functions spontaneously collapse. To be precise, in their model, the wave function evolves according to the Schrödinger equation most of the time, but there is a set of spacetime points $(\mathbf{y}_i, t_i)$ chosen at random, such that the wave function $\Psi(\mathbf{x}_1, \ldots, \mathbf{x}_N, t)$ for a system of N particles is multiplied at the chosen times t_i by a Gaussian function in the variable $\mathbf{x}_i$ (chosen uniformly among $1, \ldots, N$), centered in space at the chosen space points $\mathbf{y}_i$. The probability distribution of these random points is determined by the wave function of the system under consideration (which, theoretically, is the entire universe) at the times when they occur, and is given by the familiar $|\Psi|^2$ distribution. This ensures that the predictions of the GRW theory will (almost) coincide with the usual ones.

The above-mentioned multiplication factors localize the wave function in space, and for a system of many particles in a superposed state, effectively collapse the quantum state onto one of the terms.[28] Now, the trick is to choose the parameters of the theory so that spontaneous collapses are rare enough for a single or for a few particles to ensure that they do not lead to detectable deviations from the quantum predictions, but frequent enough for a large number of particles, say $N = 10^{23}$, to ensure that they will not stay in a superposed quantum state such as (5.1.6.2) for more than a split second.

There are two parameters in this theory, to be thought of as new constants of Nature: the rate at which collapses occur and the width of the Gaussian multiplication factor. For instance, in [209], the rate of collapse is chosen equal to $10^{-16} \, \text{s}^{-1}$ and the standard deviation of the Gaussian is chosen equal to 10^{-5} cm. Of course, other choices are possible (provided that collapses occur rarely enough for small systems and often enough for large ones), and there is no evidence that those parameters are the right ones.[29] The fact that the multiplying function is Gaussian is also an arbitrary "choice".

The GRW theory is not the same as the de Broglie–Bohm theory or ordinary quantum mechanics, since the spontaneous collapse theory leads to predictions that differ from the usual ones, even for systems made of a small number of particles.[30]

---

[27] For reviews and further discussions of those theories, see [13, 15, 32, 210–212, 230, 383–385].

[28] See Appendix 6.A for an explanation of why these macroscopic superpositions quickly disappear.

[29] Indeed, if there was such evidence, it would mean that quantum mechanics is wrong, since quantum mechanics assumes that the time evolution of the quantum state is always linear and that it never spontaneously collapses.

[30] For systems with many particles, the predictions are also different of course, unless one incorporates the collapse rule by *fiat* into the rules of quantum mechanics.

But the parameters of the theory are simply adjusted so as to avoid being refuted by experiments, which is not exactly an appealing move.

Moreover, as for the many-worlds interpretation, there is also the problem of making sense of a pure quantum state ontology (even when the latter collapses, since the collapsed quantum state is still just a function defined on a high-dimensional space), which is called the "bare" GRW theory in [13]. The proponents of spontaneous collapse theories should be credited for having recognized this problem, in contrast to those who favor the many-worlds interpretation. Two solutions have been proposed to give a meaning to the GRW theory beyond the pure wave function ontology: the matter density ontology [210] often denoted GRWm, and the flash ontology [44] denoted GRWf.

In the first solution, one associates the continuous matter density defined in (6.1.2.1) with the wave function of a system of N particles. So the ontology here is exactly the same as in the version of many worlds due to [14], but the time evolution of the quantum state is of course different: whereas it was the pure Schrödinger evolution in the many-worlds case, it is that evolution interrupted at some random times by random collapses in the GRWm theory. This has the advantage that there is, in practice, only one macroscopic world.

In the flash ontology, one has a world made only of spacetime points at the center of the Gaussian multipliers of the wave function that "collapse" it. As Bell puts it, all there is in the world, according to that theory, is a "galaxy" of spacetime points:

> [...] the GRW jumps [...] are well localized in ordinary space. Indeed each is centered on a particular spacetime point (x, t).[31] So we can propose these events as the basis of the 'local beables' of the theory. These are the mathematical counterparts in the theory to real events at definite places and times in the real world (as distinct from the many purely mathematical constructions that occur in the working out of physical theories, as distinct from things which may be real but not localized, and distinct from the 'observables' of other formulations of quantum mechanics, for which we have no use here). A piece of matter then is a galaxy of such events.
>
> John Bell [49, p. 205]

But again, if either the GRWm or the GRWf theories are true, then the god of the physicists is quite malicious. In the GRWm case, this is true for the same reason, explained above, as for the version of many worlds described in [14]. On the other hand, if the flash ontology is true, then we have been fooled into thinking that there exists something most of the time (like atoms): if we take the visible universe since the Big Bang, then it has contained only finitely many flashes. Since the flashes are all there is in that ontology, this means that, most of the time, the universe is just empty.

To get a feeling for how odd those theories are,[32] consider what happens, from their point of view, in the thought experiment of Einstein's boxes discussed in Sect. 4.2.

---

[31]This is one of the points that we denoted $(\mathbf{y}_i, t_i)$ above. (Note by J.B.).

[32]See Maudlin [319, Chap. 10] for a similar discussion of the GRWm and GRWf ontologies, as opposed to the de Broglie–Bohm ontology.

First note that, in the de Broglie–Bohm theory, nothing surprising happens: the particle is in one of the half-boxes all along and is found where it is. But in the GRWm theory, there is one-half of the matter density of a single particle in each half-box, at least if we assume that no collapse of the wave function of that particle occurs, which is likely since collapses are very rare for a single particle. But when one opens one of the half-boxes, the evolution of the wave function of the particle becomes coupled with a macroscopic measuring device and many collapses occur quite rapidly, so that the matter density suddenly jumps from being one-half of the matter density of a single particle in each half-box to being the total matter density of a particle in one half-box and nothing in the other. Of course, this happens randomly, so that the matter density may jump from the half-box that one opens to the other half-box. In any case, there is a nonlocal transfer of matter in the GRWm theory, while there is no such thing in the de Broglie–Bohm theory, and not even anything nonlocal when one deals with one particle.[33]

In the GRWf theory, there is simply nothing in either half-box, just a wave function traveling so to speak with the half-boxes. When one opens one of the half-boxes and the wave function of the particle becomes coupled with a macroscopic device, there is suddenly a "galaxy of flashes" appearing (randomly) in one of the two half-boxes, which we interpret as meaning that the particle is in that half-box. Again, the theory is more nonlocal, in a sense, than de Broglie–Bohm, since, by opening one half-box, one may trigger a galaxy of flashes in the other half-box.

These two alternatives, GRWm and GRWf, do not quite qualify as being "Duhem–Quine monstrosities", but they come dangerously close to that unenviable status.

One of the arguments in favor of these theories is that they have relativistic extensions, without introducing a privileged reference frame (or foliation) as one does in the relativistic versions of the de Broglie–Bohm theory discussed in Sect. 5.2.2 [33, 475–477]. But the problem is to understand why one has to be so realistic about relativity if one is willing to admit that all our scientific view of the micro-world is radically wrong.[34] After all, in the de Broglie–Bohm versions of quantum field theories, the predictions are "relativistic" in the sense that they agree with the usual ones, and one can debate to what extent the introduction of a sort of true simultaneity (which can be done in a Lorentz covariant way) is or is not compatible with the spirit of relativity. But why insist on that spirit if one is willing to jettison so much of our understanding of the world, and in particular the existence of atoms as localized objects (as opposed to a continuous distribution of matter) or the existence of a world that is not most of the time empty?

The GRW proposal, even putting aside the issue of its two bizarre ontologies, is terribly ad hoc, since the parameters of the theory are simply chosen so as to avoid the possibility of experimental refutations. So it is difficult to see why one would adopt a theory that does make different predictions than the de Broglie–Bohm theory, and

---

[33]Bell-type arguments establishing the existence of nonlocal causal effects apply only to systems composed of at least two particles.

[34]See Maudlin [319, 3rd edn, Chap. 10] for a detailed discussion of this issue.

yet manages to avoid being refuted, at least given current technologies (this may of course change in the future), by suitably adjusting its parameters.

The GRW approach stems from the desire to "solve the measurement problem", but it is not clear that this proposal would ever have been made if the fact that the de Broglie–Bohm theory solves this problem in a natural way had been better known. Actually, one may wonder, why not simply add a particle ontology to the GRW models, as in de Broglie–Bohm? One could have a dynamics like (5.1.1.3) even if the wave function satisfied a modified Schrödinger equation (with random collapses). Wouldn't that ontology be more natural than either the continuous matter density or the flashes? The obvious answer is that, if one adopts a particle ontology, then there is no need to modify Schrödinger's equation through the collapses, since the latter are accounted for as "effective collapses" within the de Broglie–Bohm theory, as we saw in Sect. 5.1.6. This strongly suggests that the spontaneous collapse models are the result of a desire to solve an artificial problem (at least, it is artificial from the point of view of the de Broglie–Bohm theory), namely, the problem of "collapses" or "measurements".

Of course, one should consider the possibility of a more natural modification of Schrödinger's equation, say by a nonlinear term, which would be a new theory. Lajos Diósi [134] and Roger Penrose [390] have proposed such modified equations. Penrose draws an analogy between Newton's gravitation, where the gravitational potential is a solution of a linear equation (Poisson's equation) and general relativity, whose equations are nonlinear, to suggest a nonlinear modification of quantum mechanics. For another nonlinear modification of Schrödinger's equation, see Weinberg [502, 503], but such deterministic nonlinear modifications would allow superluminal transmission of messages, as shown by Gisin [213] (as we saw in Sect. 5.2.1, the same phenomenon of superluminal transmission of messages is possible in the de Broglie–Bohm theory if one assumes the existence of an ensemble of particles that are not in quantum equilibrium). There are also spontaneous collapse models that modify Schrödinger's equation (apart from the collapses) and try to incorporate the effect of gravitation [135, 136, 391].

All these models may be thought of as being toy models for an as yet unknown totally new theory that would some day supersede quantum mechanics, perhaps also incorporating gravity. However, that new theory would not be an "interpretation" of quantum mechanics but a replacement of it. Of course, it should have testable consequences and one would still have to see which ontology (besides the wave function) would naturally supplement such a nonlinear evolution.

6.3   The Decoherent Histories Approach

In a series of papers and books, Robert Griffiths [241–248], Roland Omnès [365–368], Murray Gell-Mann, and Jim Hartle [201, 202, 204, 253] have introduced and advocated a *decoherent histories* or *consistent histories* approach to quantum mechanics. The idea is to assign probabilities in a coherent way to histories of real events, meaning events that happen in the world whether we measure them or not, like the particle trajectories in de Broglie–Bohm, but for more general sorts of events, for example the spin being up or down in a given direction, or any other "observable" having a given value. Naturally, those assigned probabilities agree with the quantum mechanical ones, whenever observations are made.

Robert Griffiths explains quite clearly the goal of the decoherent or consistent histories approach. He writes, about "the role of measurements in standard quantum mechanics and in consistent histories":

> For example, one can show that a properly constructed measuring apparatus will reveal a property that the measured system had before the measurement, and might well have lost during the measurement process. The probabilities calculated for measurement outcomes (pointer positions) are identical to those obtained by the usual rules found in textbooks. What is different is that by employing suitable families of histories one can show that measurements actually measure something that is there, rather than producing a mysterious collapse of a wave function.
>
> Robert B. Griffiths [249]

From this, it should be clear that one is talking about probabilities of real events, not just of measurements. From a consistent history point of view, there is no problem with the collapse postulate:

> From the consistent histories perspective, wave function collapse is a mathematical procedure for calculating certain kinds of conditional probabilities that can be calculated by alternative methods, and thus has nothing to do with any physical process. That is, "collapse" is something which takes place in the theorist's notebook, not in the experimentalist's laboratory. Consequently, there is no conflict between quantum mechanics and relativity theory.
>
> Robert B. Griffiths [249]

Not surprisingly, Griffiths also claims that there is no nonlocality in quantum mechanics[35]:

> Einstein, Podolsky, and Rosen (EPR) in a celebrated paper [164] showed that by measuring the property of some system A located far away from another system B one can, under suitable conditions, infer something about the system B. By itself the possibility of such an indirect measurement is not at all surprising, as one can see from the following example. Colored slips of paper, one red and one green, are placed in two opaque envelopes, which are then mailed to scientists in Atlanta and Boston. The scientist who opens the envelope in Atlanta and finds a red slip of paper can immediately infer, given the experimental protocol, the color of the slip of paper contained in the envelope in Boston, whether or not it has already been opened. There is nothing peculiar going on, and in particular there is no mysterious influence of one "measurement" on the other slip of paper.
>
> Robert B. Griffiths [249]

---

[35] See Sect. 7.5 for similar claims made by Gell-Mann.

Stated in the language of our Chap. 4, the idea expressed here by Griffiths is that answers (being here the colors of the slips of paper) pre-exist the questions being asked at X and Y (i.e., Atlanta and Boston), and the answers are carried by the particles (the opaque envelopes) coming from the source. But that is precisely what Bell has shown to be impossible, by considering the statistics of answers to different questions at X and Y.

Something, somewhere, must be wrong!

To understand where the problem comes from, let us start by discussing how probabilities are assigned in the decoherent histories approach. This cannot be done in a naive way, because of interference. Suppose, for example, that in the double-slit experiment[36] we want to associate a probability to the following pair of "histories": "the particle goes through the upper slit and is detected at a certain place x on the screen", and "the particle goes through the lower slit and is detected at x on the screen". The probability of being detected at x should be the sum of the probabilities of those two sequences of events, but the probability of being detected at x when both slits are open is different from the sum of the probabilities of being detected at x and going through each slit, when only one slit is open. This means that a priori there is a difficulty in assigning probabilities consistently to those histories.[37] Of course, if one considers two histories made of different sequences of events: "the particle is detected at the upper slit and is detected again at x on the screen", and "the particle is detected at the lower slit and is detected again at x on the screen", then one can assign probabilities consistently to such a pair of histories, because the detection at the slits in effect collapses the wave function and thus destroys the interference phenomenon. Then the situation is the same as when only one slit is open (see Fig. 2.8).

So the above-mentioned authors introduce conditions of decoherence between members of a family of histories in such a way as to allow the application of the usual rules of probability to that family. Then, the probabilities assigned to events agree with the quantum mechanical predictions when measurements are performed. We will not discuss in detail those conditions (see Appendix 6.B for a sketch of the decoherence condition and the probabilistic formulas), because what they try to avoid are inconsistencies that might arise when one attributes probabilities to events occurring at different times (as in the example of the particle going through the upper or lower slit and being detected at a certain place x on the screen) and the problem that we want to focus on, in the decoherent histories approach, occurs already when one considers events happening at a single time.

Indeed, and this cannot be stressed enough, the originality and the merits of the decoherent histories approach are that it tries to make sense of probabilities of real events happening in the world, whether we observe them or not. If it did not try to do that, then there would be nothing new, since the formulas for sequences of events occurring at different times in that approach are just the usual quantum mechanical

---

[36]Discussed in Appendix 2.E. We refer the reader to Fig. 2.8 to illustrate what follows. This example is discussed from a decoherent histories point of view by Hartle in [253, Sect. 2].

[37]However, we will see below how this can be done in a more subtle way in the de Broglie–Bohm theory.

ones, obtained from the combination of Schrödinger evolution and the collapse rule. Moreover, those real events are defined by the values taken by arbitrary "observables". In principle, this is once again to the credit of the decoherent histories approach, since if it could be done, it would be close to the spirit of quantum mechanics, or at least to the way it is usually presented, i.e., putting all "observables" on an equal footing.

However, and this is the source of the difficulties, by considering as real events the values taken by arbitrary observables, this approach treads on the dangerous terrain of the no hidden variables theorems. And this leads to a problem, raised in papers by Sheldon Goldstein [222] and by Angelo Bassi and Giancarlo Ghirardi [29] (see [30, 31, 246] for further discussions), a problem which exists even if one considers "histories" occurring *at a single time* and for which the decoherence condition is trivially satisfied. The arguments presented by these authors are based on the no hidden variables theorems discussed at the end of Chap. 2.

Goldstein gives an example of four decoherent histories, based on an example due to Lucien Hardy [252]. Consider four quantities A, B, C, D, associated with a pair of spin $1/2$ particles, which have the following properties[38]:

(1) A and C can be measured simultaneously,[39] and if one gets $A = 1$, then $C = 1$.
(2) B and D can be measured simultaneously, and if one gets $B = 1$, then $D = 1$.
(3) C and D can be measured simultaneously, but it never happens that both $C = 1$ and $D = 1$.
(4) A and B can be measured simultaneously, and it sometimes happens[40] that both $A = 1$ and $B = 1$.

Each of these four statements corresponds to a decoherent history in a trivial way, since both quantities, in each of the pairs involved here, can be measured simultaneously, and we are considering everything at a single time. So one can assign probabilities to those events in a consistent way. For example, in (1), the probability that $C = 1$, given that $A = 1$, is equal to 1.

But these four statements cannot all be true, because when it happens that $A = 1$ and $B = 1$, as it sometimes does, by (4), one must have, by (1) and (2), that $C = 1$ and $D = 1$, which is impossible because of (3). Of course, each of the statements above is true if they all refer to results of measurements. But then no contradiction arises, because the measurements to which they refer are different and cannot be performed simultaneously since they do perturb the system (as reflected by the collapse rule in ordinary quantum mechanics and understood through the analysis in Sect. 5.1.4 in the context of the de Broglie–Bohm theory).

As stressed by Goldstein, the inconsistency is a problem, but only from the decoherent history (DH) point of view:

---

[38]See Appendix 6.B for a definition of those quantities, and the proof of those properties. Here we are only interested in the logic of the argument; for more detail, see [219, 222, 252, 336].

[39]This means that the matrices representing those quantities commute (see Appendices 2.B, 2.C, and 6.C).

[40]"Sometimes" here means that it occurs in about 9 % of the cases, but that precise value does not matter for the logic of the argument.

It is important to appreciate that, for orthodox quantum theory (and, in fact, even for Bohmian mechanics), the four statements above, if used properly, are not inconsistent, because they then would refer merely to the outcomes of four different experiments, so that the probabilities would refer, in effect, to four different ensembles.

However, the whole point of DH is that such statements refer directly, not to what would happen were certain experimental procedures to be performed, but to the probabilities of occurrence of the histories themselves, regardless of whether any such experiments are performed.

Sheldon Goldstein [222]

Following a similar logic, Bassi and Ghirardi [29] consider six decoherent histories[41] and deduce a contradiction, but without having to rely on a probabilistic statement like (4) above.

The response from Gell-Mann and Hartle and from Griffiths [223] consists basically in saying that there is no decoherent history including the four operators A, B, C, and D or the six histories of Bassi and Ghirardi. This is true, because A and D cannot be measured simultaneously, nor can B and C. But that answer misses the point, which is that each decoherent history is supposed to be a statement about real events happening in the world, to which truth values can be assigned. So each of the above statements is meant to be true, but they cannot all be true, since, taken together, they lead to a contradiction.

It is worth quoting in full Goldstein's reply to Gell-Mann and Hartle and to Griffiths:

Concerning this issue, Murray Gell-Mann and James Hartle complain that I have "misunderstood one important point" — namely, that "it is essential to restrict statements relating the probabilities of occurrence of histories to a given family containing them" because "inconsistencies can arise if statements relating the probabilities of occurrence of histories are made while referring to different families in the course of a given argument." I am puzzled by their response. Each of my four individual statements concerns only probabilities for a single family (with, of course, a different one for each statement). And the fact that "inconsistencies can arise ..." is precisely the point of the example I used in the article and am using here.

Robert Griffiths is more explicit about the cause of my having made "the erroneous assertion that the consistent histories formalism is ...inconsistent" — namely, my not "paying serious attention to [Omnès's] Rule 4." Here is the rule, as given on page 163 of the reference Griffiths mentions[42]: "Any description of the properties of an isolated physical system must consist of propositions belonging together to a common consistent logic. Any reasoning to be drawn from the consideration of these properties should be the result of a valid implication or of a chain of implications in this common logic." What Omnès calls a "consistent logic" amounts more or less to a (decoherent) family.

I have always had great difficulty with this rule. I don't understand what it actually means, in terms of both detail and basic meaning. Does the description provided by the four statements in my example, which requires reference to four families, violate this rule because the four statements are on adjacent lines? What if they were on different pages, or were made by different people? It can hardly be expected that, when thinking about the same system, all

----

[41]Based on operators related to those used in the proof of the no hidden variables theorems in Chap. 2.
[42]The reference is [367]. (Note by J.B.).

people at all times will — by some peculiar harmony — formulate statements concerning only the same common family.

Besides, why are my four statements not a counterexample? They are a description of precisely the sort that Rule 4 informs us "must" not be. This raises the question as to exactly what is meant in the rule by "must," and, in its next sentence, by "should".

The real problem, I believe, is this: If we "must" or "should" restrict our descriptions and reasoning in the manner described by Rule 4, it must be because of the meanings of the statements under consideration and the way the language expressing them is intended to function. For example, if (as would be appropriate in orthodox quantum theory) we were to use the four statements above as an elliptical way of talking about results of possible experiments, then it is apparent that we could get into trouble by considering, at one time, several of these statements, should we slide into the mistake of thinking that the several statements refer to a common experiment.

However, if, for DH, descriptions such as those provided by the statements above are to be understood with their usual meanings, then Rule 4 is simply false, to the extent that it has any meaning at all. And if the proponents of DH have some other meaning in mind for such statements, they should so inform us and supply this meaning — something that, as far as I am able to tell, they nowhere do.

It may be argued that Rule 4 should be regarded as merely a rule — that is, as merely defining a certain game. But then why must I play this game when analyzing the implications of DH?

It is true that, to deduce or recognize that the four statements above are inconsistent, we must consider a collection of statements involving more than a single family. If we obey Rule 4 in our analyses, we will encounter, as Gell-Mann and Hartle say, "no inconsistencies along the way." But the statements will remain inconsistent even if we invoke and adhere to rules that demand, in effect, that we ignore the inconsistency.

                                                                    Sheldon Goldstein [222]

One possible way to avoid this inconsistency would be to restrict the set of histories considered real, by choosing three of the four statements (1)–(4) above and dropping the fourth one. But it is not obvious how to do that, because the four statements are equally plausible (if one adopts the decoherent histories approach). As we said in Chap. 2, the natural interpretation of the no hidden variables theorems is that none of these statements referring to pre-existing spin values (or pre-existing values for the similar quantities A, B, C, D, discussed here) that would be revealed by observations is actually true.

The fundamental problem of the decoherent history approach is once again naive realism about operators, that is to say, thinking that the physical quantities represented by operators do have values whether one observes them or not and thus, forgetting Bohr's insistence on the "active role" of the measuring devices.

Finally, one may contrast the decoherent history approach and the de Broglie–Bohm theory. After all, by assigning probabilities, at each time, to particle positions, the de Broglie–Bohm theory also assigns probabilities to certain "histories", namely the particle trajectories. One can do this as follows: assign to a trajectory $\mathbf{X}(t)$, satisfying Eqs. (5.1.1.1) and (5.1.1.4), with initial condition $\mathbf{X}(0) = \mathbf{x}$, a probability density given by $|\Psi(\mathbf{x}, 0)|^2$. By equivariance, this assigns the usual quantum mechanical probability density for $\mathbf{X}(t) = \mathbf{x}$, given by $|\Psi(\mathbf{x}, t)|^2$, at all times, and hence also predicts the quantum mechanical probabilities for all observables. But note that

these histories (the particle trajectories) are not decoherent, in the sense of the decoherent history approach! Indeed, they do assign probability densities to the history
of a particle going through the upper slit and ending at a given point x on the screen
and to the history of a particle going through the lower slit and ending at the same
point x on the screen, but in a consistent way (unlike what happens if one assigns
these probabilities in what we called a naive way at the beginning of this section).

It is easy to see why, by looking at Fig. 5.1. If the point x is in the upper half of the
screen, then the probability of going through the lower slit and ending at x is zero,
while the conditional probability of going through the upper slit, given that one ends
at x, is one, so there is no inconsistency in these assignments of probabilities: the
probability density of ending at x *is* (for this assignment of probabilities) the sum of
the probability densities of going through the lower slit and ending at x (which is zero)
and of of going through the upper slit and ending at x (which equals the probability
density of ending at x). This may sound counterintuitive, but this assignment of
probabilities is consistent and its counterintuitive nature arises solely from a quantum
mechanical "intuition" that is deeply linked with results of measurements.

If, on the other hand, one observes the slit the particle goes through, then one does
get decoherent histories and the de Broglie–Bohm probabilities become the familiar
quantum mechanical ones; but the de Broglie–Bohm theory accounts for that by
taking into account the effective collapse of the wave function that such observations
induce, so that the probability distribution after the slits is no longer given by the
$|\Psi(\mathbf{x}, t)|^2$ density without measurement, but by the corresponding density of the
collapsed wave function behind the slit the particle went through (t being the time
of passage through the slits), which is the same as what happens with only one slit
opened (see Fig. 2.8). Note that in this situation, there is a nonzero probability for
the particle that goes through the upper slit to end up on the lower half of the screen,
and vice versa, as opposed to what happens when both slits are open.

This example illustrates one of the merits of the de Broglie–Bohm theory, namely
that it renders explicit the "active role" of measuring devices, because the probabilities when measurements are performed are different from those when there are
no measurements, while in the decoherent histories approach, no such distinction is
made.

6.4   QBism

This is the most recent "interpretation" of quantum mechanics, and it is presented
as being linked to what Alain Aspect calls the "second quantum revolution", i.e.,
the "quantum information" revolution[43]: this includes quantum cryptography, which
already exists and allows more secure encryptions than anything that can be done

---

[43] See Aspect's introduction to the 2004 edition of [49]. For the theory of quantum information, see,
e.g., [352, 404].

classically, as well as quantum computation, which will be able in principle (but not yet in practice) to perform some calculations much faster than classical computers.

The field of quantum information is intimately related to the phenomenon of entanglement and is, in a sense, a product of the work by EPR, Schrödinger, and Bell. But since everything in the field of quantum information is an application of ordinary quantum mechanics (possibly a revolutionary application, but nevertheless based on the usual principles), one cannot use those remarkable applications to justify a particular way to understand quantum mechanics.

In QBism, Q stands for quantum and B for Bayesianism. As explained in Chap. 3, the subjective Bayesian view of probabilities considers them as judgements: if a coin is thrown and we do not look at the result, it is rational, because of the symmetry of the situation, to assign probabilities (1/2, 1/2) to each side. But if one looks at the result, those probabilities jump to (1, 0) or (0, 1). If Alice has seen the result, but Bob has not, then her probabilities have changed but Bob's probabilities have not. If Alice tells Bob the result (and if he thinks she is reliable), then Bob updates his probabilities. Thus, probabilities simply reflect our degree of knowledge of the world and nothing else, so different individuals assign different probabilities to the same event. This holds also for more complicated situations: a doctor may assign different probabilities than a patient to the latter's health and a meteorologist will assign different probabilities than the man in the street to tomorrow's weather.

So far so good: there is nothing surprising or paradoxical or counterintuitive here. QBism wants to extend this "subjective" view of probabilities to the quantum probabilities. Since they are defined in terms of the quantum state, why not also give the latter a purely subjective meaning? For a QBist, the quantum state is simply something that *I assign* to the world. Fuchs and Peres write:

> [...] the time dependence of the wavefunction does not represent the evolution of a physical system. It only gives the evolution of our probabilities for the outcomes of potential experiments on that system. This is the only meaning of the wavefunction.
>
> Christopher A. Fuchs, Asher Peres [196]

Therefore, there is nothing surprising for a QBist in the collapse rule: just as I update my probabilities when I learn something about the world, I update my quantum state when I learn something, for example after a "measurement", as Fuchs, Mermin, and Schack explain:

> Quantum states determine probabilities through the Born rule. Since probabilities are the personal judgments of an agent, it follows that a quantum state assignment is also a personal judgment of the agent assigning that state. The notorious 'collapse of the wave-function' is nothing but the updating of an agent's state assignment on the basis of her experience.
>
> Christopher A. Fuchs, N. David Mermin, and Rüdiger Schack [199]

QBism is a sort of return to Copenhagen, but with a vengeance.[44] It is more radical than the usual Copenhagen interpretation, because according to the latter (at least as it is generally understood), a physical system *has* a quantum state; this state is an objective property of the world, even though it is only a tool allowing us to predict results of measurements.

One should not confuse QBism with the idea, suggested by some people, that it is our consciousness that produces the collapse of the quantum states, viewed as an objective fact.[45] Indeed, for QBists, there is nothing objective about the quantum state to start with. It represents only a certain information that we have.

But there is then an obvious question: information about what? If the world is fundamentally quantum mechanical, and if the most basic quantum concept is the quantum state and if even that is not objective, then what is? In ordinary quantum mechanics, one can be ignorant about the true quantum state, in which case one assigns a probability distribution over those states, which is then similar to a classical probability. But for Qbists there is no such thing as a true, but unknown, quantum state, just as there is no true probability of the coin being heads or tails (before we look). There are just different degrees of information.[46]

But for the coin, there is a fact of the matter as to whether it is heads or tails (after it has fallen). Our information, or lack of it, refers to something "out there". If one assigns a probability 1 to heads or tails, depending on the result, it is because one has learned a fact about the world. Even when one assigns probabilities $(1/2, 1/2)$, it is because one knows that the coin is symmetric, which is again a fact about the coin.

It is sometimes suggested that the information is only about the macroscopic world, or even only about results of measurements:

> The thread common to all the nonstandard "interpretations" is the desire to create a new theory with features that correspond to some reality independent of our potential experiments. But, trying to fulfill a classical worldview by encumbering quantum mechanics with hidden variables, multiple worlds, consistency rules, or spontaneous collapse, without any improvement in its predictive power, only gives the illusion of a better understanding. Contrary to those desires, quantum theory does not describe physical reality. What it does is provide an algorithm for computing probabilities for the macroscopic events ("detector clicks") that

---

[44]One of David Mermin's papers is entitled *QBism puts the scientist back into science* [340], which is exactly what we have been trying to avoid in this book, for the obvious reason that one would like science to say something about places and times where and when there are no scientists around.

[45]See, e.g., the quotes by Wigner in Sect. 1.3.

[46]See, for example, [94] for an explicit rejection of the idea that there can be, from a Qbist point of view, an unknown quantum state: "If a quantum state is a state of knowledge, then it must be known by somebody". In [474], Tumulka gives a nice counterargument to the idea that the wave function is a state of knowledge: a computer could prepare a quantum system in a given state of the form (2.B.1), corresponding to a quantity A, with some of the coefficients c_n equal to zero and others not. It then prints on a piece of paper the set of coefficients c_n equal to zero, puts the paper in a sealed envelope, and erases the quantum state from its memory. Hence, nobody knows what the state is. Yet we know with certainty that a measurement of A will yield one of the eigenvalues corresponding to the terms in which $c_n \neq 0$ (and that could be checked by opening the envelope after the measurement). So in a sense "Nature" still knows the state, after it has been erased from the computer's memory, even if no human being nor any machine knows it.

are the consequences of our experimental interventions. This strict definition of the scope of quantum theory is the only interpretation ever needed, whether by experimenters or theorists.

Christopher A. Fuchs, Asher Peres [196]

But then we run into the usual problem of how to construct a real macroscopic world out of a non-real microscopic one. Dennis and Norsen underline this in their reply to Fuchs and Peres:

Instead, FP's answer is that the probabilities calculated in quantum theory refer only to macroscopic events. But this simply places a different name on the same ambiguity that had previously been shuffled under the idea of "measurement". Where exactly is the cut between micro- and macroscopic, and why should such a cut enter into the fundamental laws of physics? If one electron is not objectively real, and two electrons are not objectively real, why should a collection of 10^{23} electrons be real? Things like temperature and elasticity may be emergent properties, but surely existence as such is not. Nothing real can emerge from that which doesn't exist.

Eric Dennis and Travis Norsen [121]

But in many instances, QBists claim that, in their view, quantum mechanics is only about subjective experiences, and not about an outside world, even a macroscopic one. For example, concerning the question of what information is about, Mermin writes:

Partly from my associations with quantum computer scientists and partly from endless debates with constructivist sociologists of science, I have come to feel that "Information about what?" is a fundamentally metaphysical question that ought not to distract tough-minded physicists. There is no way to settle a dispute over whether the information is about something objective, or is merely information about other information. Ultimately it is a matter of taste, and, like many matters of taste, capable of arousing strong emotions, but in the end not really very interesting.

David Mermin [338]

And Fuchs similarly declares:

'Whose information?' 'Mine!' Information about what? 'The consequences (for me) of my actions upon the physical system!' Its all 'I-I-me-me mine', as the Beatles sang.

Christopher Fuchs [197], quoted in [361]

Fuchs, Mermin, and Schack even seem to think that "reality" depends on which "agent" one considers:

This means that reality differs from one agent to another. This is not as strange as it may sound. What is real for an agent rests entirely on what that agent experiences, and different agents have different experiences. An agent-dependent reality is constrained by the fact that different agents can communicate their experience to each other, limited only by the extent that personal experience can be expressed in ordinary language. Bob's verbal representation of his own experience can enter Alice's, and vice-versa. In this way a common body of reality can be constructed, limited only by the inability of language to represent the full flavor — the "qualia" — of personal experience.

A QBist takes quantum mechanics to be a personal mode of thought — a very powerful tool that any agent can use to organize her own experience. That each of us can use such a tool to organize our own experience with spectacular success is an extremely important objective fact about the world we live in. But quantum mechanics itself does not deal directly with the objective world; it deals with the experiences of that objective world that belong to whatever particular agent is making use of the quantum theory.

Christopher A. Fuchs, N. David Mermin, and Rüdiger Schack [199]

Unavoidably, this radical idealism leads to a form of relativism:

A child awakens in the middle of the night frightened that there is a monster under her bed, one soon to reach up and steal her arm—that we-would-call-imaginary experience has no less a hold on onticity than a Higgs-boson detection event would if it were to occur at the fully operational LHC. They are of equal status from this point of view — they are equal elements in the filling out and making of reality.

Christopher Fuchs [197], quoted in [361]

Of course, QBists claim not to be solipsists (people who write articles and books to convince others of their views rarely claim to be solipsists), but if physics is all about updating my subjective experiences, then in what sense does it differ from solipsism? Travis Norsen calls it solipsism FAPP in [361], but it is rather solipsism in denial, since that approach does not deal with anything except one's subjective experiences; so the difference with solipsism is only that one denies being solipsist, without saying anything about what could exist outside one's conscious states of mind.

Indeed, if one is not a solipsist, then one will try to describe the world "outside of us" as having certain objective properties (a coin has two faces and a dice six) and "information" would then be about those properties.

If the proverbial Alice and Bob decide to become QBists, then they should not ask themselves which objective properties the world possesses that make the updating of their quantum states work so well (at least for statistical predictions).

QBists also claim that their approach is perfectly local, which in a sense is true, but so is solipsism. If all Alice does is update her subjective states of mind, then in the EPR–Bell situation, if she measures the spin in one direction, she will learn later that, if Bob has measured the spin in the same direction, his result is perfectly correlated with hers. But if all there is to science is to adjust one's state of mind in order to predict one's future sensations, Alice is not supposed to be puzzled by these correlations nor to try to explain them through local causes, in which case she would realize that Bell's theorem prevents her from doing so. Such an enterprise is just not part of science, according to QBists.

It is not true to say that QBism has given a local explanation for those EPR–Bell correlations; it rather defends a view of science that refuses to address this sort of question, which is very different.

There is something odd about insistence on relativistic invariance, whether by QBists or by defenders of the many-worlds interpretation, while in their theories, either there is no reality or reality is reduced to an abstract function defined on a high-dimensional space. Maudlin has drawn attention to this paradox:

> Physicists have been tremendously resistant to any claims of nonlocality, mostly on the assumption (which is not a theorem) that nonlocality is inconsistent with Relativity. The calculus seems to be that one ought to be willing to pay any price — even the renunciation of pretensions to accurately describe the world — to preserve the theory of Relativity. But the only possible view that would make sense of this obsessive attachment to Relativity is a thoroughly realistic one! These physicists seem to be so certain that Relativity is the last word in spacetime structure that they are willing even to forego any coherent account of the entities that inhabit spacetime.

Tim Maudlin [321, p. 305]

Moreover, QBists tend to present their viewpoint as non-philosophical, just bare quantum mechanics, without any "interpretation". Indeed, one paper by Fuchs and Peres is entitled: *Quantum theory needs no 'interpretation'* [196]. Eric Dennis and Travis Norsen provide harsh but apposite criticism of this attitude:

> Alan Sokal once described the motivation for his famous prank[47] as follows [452]: "What concerns me is the proliferation not just of nonsense and sloppy thinking per se, but of a particular kind of nonsense and sloppy thinking: one that denies the existence of objective realities, or (when challenged) admits their existence but downplays their practical relevance." Sokal was, of course, referring to the depths of irrationality to which some fields in the humanities and social sciences have descended.
>
> It is frightening that the same anti-realism Sokal ridiculed there can be put forward as a supposedly natural and uncontroversial interpretation of quantum theory. It is even more frightening that this anti-realism is glibly passed off as non-philosophical, for this suggests not only that many physicists have accepted a fundamentally anti-scientific set of philosophical ideas but that, in addition, they have done so unwittingly.
>
> Does the community of *physicists* — those individuals whose lives are dedicated to observing, understanding, and learning to control the physical world — really accept as solid, hard-nosed science the idea (hedged or not) that there is no physical world — that, instead, its all in our minds?

Eric Dennis and Travis Norsen [121] (italics in the original)

6.5 Conclusions

Although there is a general perception that there are many theories competing with the de Broglie–Bohm theory or that this theory is just one "interpretation" among others, we claim to have shown here that it is the only theory[48] that is free from any of the following problems:

---

[47]In 1996, the physicist Alan Sokal managed to have an article, full of "postmodernist" nonsense, published in the journal *Social Text*. See [51, 451, 453, 455] and Chap. 8 (Note by J.B.).

[48]Putting aside the question of underdetermination discussed in Sect. 5.4.1.

- Inconsistency, like the decoherent histories approach.
- Making different predictions than quantum mechanics, hence also than the de Broglie–Bohm theory, as the spontaneous collapse theories do, while adjusting its parameters so that no empirical contradiction can be detected.
- Putting the observer back at the center of our physical theories, as QBism does, or as the defenders of the decoherent histories approach also do (sometimes).
- Being ill-defined or hard to make sense of, like the many-worlds interpretation, or the "bare" GRW theory (the one without a matter density ontology or a flash ontology).

In Chap. 1, we criticized the "Copenhagen" approach for making a rhetorical use of TINA, "there is no alternative"; aren't we doing the same here, but with de Broglie–Bohm, instead of Copenhagen? Not really, because even though we try to show that there is no existing alternative to de Broglie–Bohm that reaches the level of clarity and explanatory power of the latter, we do not claim that the de Broglie–Bohm theory is the "end of history", mainly because a complete understanding of how to combine nonlocality and relativity is still missing, and this may be the source of a future upheaval of our world view.

Finally, one may ask: isn't the bride too beautiful? If the de Broglie–Bohm theory has so many qualities, why is it so generally ignored? To answer this, we must turn to the history of quantum mechanics and take into account some psychological and sociological factors. We will do this in the remaining two chapters.

Appendices

6.A Spontaneous Collapse

We will explain here why macroscopic superpositions quickly disappear in GRW-type models, due to spontaneous localizations: consider a superposition of the form (5.E.2), with Φ_0^{up}, Φ_f^{up}, Φ_0^{down}, Φ_f^{down} of the form (5.E.4). As already observed, the support of Φ_{0i} is disjoint from that of Φ_{fi} for *every* particle. So suppose that the wave function of a single particle is "hit" by a spontaneous collapse. This means that its wave function becomes effectively either Φ_{0i} or Φ_{fi}: indeed, the Gaussian that multiplies the wave function will be centered around either the support of Φ_{0i} or the support of Φ_{fi} (because the statistical distribution of the hits is determined by the square of the absolute value of the wave function, which has two parts $|\Phi_{0i}|^2$ and $|\Phi_{fi}|^2$ here, with almost disjoint supports) and, since the latter are far apart, one of the wave functions will be multiplied by a tail of the Gaussian and will become almost equal to zero.

But then, because the whole wave function in (5.E.2) is a product of the form (5.E.4), the fact that one factor in (5.E.4) (almost) vanishes makes the whole product (almost) zero. We are then left effectively with one of the two terms in (5.E.2), and the macroscopic superposition has been eliminated. Now, when the number N in

(5.E.4) is of the order of the Avogadro number ($\sim 10^{23}$), we can arrange the rate of collapse so that the probability of a collapse per unit of time is very small for a single particle (thus, there will be no observable violation of the quantum predictions for small systems), yet a collapse will occur with probability very close to 1 for at least one of the N particles in a very small interval of time.[49] And as we just saw, that is enough to "collapse" the macroscopic superposition.

6.B Decoherent Histories: The Formalism

A *history* $\alpha = \big((\alpha_1, t_1), (\alpha_2, t_2), \ldots, (\alpha_n, t_n)\big)$ is a sequence of times and projection operators at the prescribed times[50]:

$$\{P^1_{\alpha_1}(t_1), P^2_{\alpha_2}(t_2), \ldots, P^n_{\alpha_n}(t_n)\},$$

satisfying

$$\sum_{\alpha_k} P^k_{\alpha_k}(t_k) = \mathbf{1}, \tag{6.B.1}$$

where $\mathbf{1}$ is the unit operator, and

$$P^k_{\alpha_k}(t_k) P^k_{\alpha'_k}(t_k) = \delta_{\alpha_k, \alpha'_k} P^k_{\alpha_k}(t_k), \tag{6.B.2}$$

which means that they form a complete set of orthogonal projection operators. The operator $P^k_{\alpha_k}(t_k)$ is the projection operator $P^k_{\alpha_k}$ evolved up to time t_k. This is called the "Heisenberg picture", but it can be explained by saying that we have the pure Schrödinger evolution between the different times $t_1, \ldots, t_n$, without collapses.

These projection operators project onto certain subspaces associated with eigenvalues of "observables". We can simply think of them, in the language of Appendix 2.B, as "collapsing" the quantum state, given certain results of measurement. For example, one could take (with $n = 2$) a history α of a spin variable, where at time $t_1, \sigma_x = 1$, and at time $t_2, \sigma_z = -1$ [222].[51] We can associate a product of projection operators with such a history:

$$C_\alpha = P^n_{\alpha_n}(t_n) \cdots P^2_{\alpha_2}(t_2) P^1_{\alpha_1}(t_1).$$

Given an initial state Ψ, one can associate a probability with a history by the formula

$$\mathcal{P}(\alpha) = \langle C_\alpha \Psi | C_\alpha \Psi \rangle, \tag{6.B.3}$$

---

[49] If an event has probability p to occur in each of N independent random events, then the probability that it never occurs is equal to $(1 - p)^N \sim 0$ if $pN \gg 1$.

[50] We follow here the pedagogical article by Hartle [253].

[51] We use the expression $\sigma_x = 1$ as a shorthand for "the value of the spin σ_x equals 1".

which is the standard quantum mechanical formula for the probability of a sequence of *results of measurement*, corresponding to the outcomes $\alpha_1, \alpha_2, \ldots, \alpha_n$ at times $t_1, \ldots, t_n$. But remember that the goal of Griffiths, Omnès, Gell-Mann, and Hartle is to define probabilities of *events happening in the world*, whether we observe them or not.

Using our example with spin, with the Hamiltonian equal to zero, i.e., a time evolution which is constant, and taking the state $|\Psi\rangle$ to be the eigenstate of $\sigma_z = +1$, $\mathcal{P}(\alpha) = 1/4$. (This is the product of two factors of 1/2: the probability that $\sigma_x = 1$ at time t_1, given that the initial quantum state has spin up in direction z, and the probability that $\sigma_z = -1$ at time t_2, given that at time t_1 the quantum state is the eigenstate of σ_x with eigenvalue 1.) The problem is that one could consider another history, given by, at time t_1, $\sigma_x = -1$ and, at time t_2, $\sigma_z = -1$, which also has probability $\frac{1}{4}$. But this could lead to inconsistencies. Indeed, if we ignore what happens at $t = t_1$ (or if we sum over the two possibilities $\sigma_x = 1$ and $\sigma_x = -1$, both terms being positive), the probability would be zero: if the time evolution is constant and the initial quantum state has spin up in direction z, then the probability of having $\sigma_z = -1$ at a later time is zero. As observed in [222], there is no such inconsistency in ordinary quantum mechanics, because a measurement of the spin in direction x at time t_1 will change the state, by collapsing it onto the eigenvector corresponding to the observed eigenvalue.

As we saw in Sect. 6.3, a similar problem occurs in the double-slit experiment. This is why one introduces a decoherence condition: a family of histories $\mathcal{H}$ is *decoherent* if

$$\langle C_{\alpha'}\Psi | C_\alpha\Psi\rangle \sim 0, \tag{6.B.4}$$

for all $\alpha \neq \alpha'$ in $\mathcal{H}$.

To illustrate this condition, consider two histories with a single spin, first α, where at time t_1, $\sigma_x = 1$ and at time t_2, $\sigma_z = -1$, and then α', where at time t_1, $\sigma_x = -1$ and at time t_2, $\sigma_z = -1$. They are not decoherent in the quantum state $|\Psi\rangle$ which is the eigenstate of $\sigma_z = +1$, since $\langle C_{\alpha'}\Psi | C_\alpha\Psi\rangle = 1/4 \neq 0$.

Then one can check that the formula (6.B.3), applied to the histories of a consistent family defines a consistent probability satisfying the usual rules. For example, if one considers a history containing three events, one wants to show that

$$\sum_{\alpha_2} \mathcal{P}(\alpha_3, \alpha_2, \alpha_1) \sim \mathcal{P}(\alpha_3, \alpha_1), \tag{6.B.5}$$

which, by (6.B.3), means

$$\sum_{\alpha_2} \langle \Psi | P^1_{\alpha_1} P^2_{\alpha_2} P^3_{\alpha_3} P^3_{\alpha_3} P^2_{\alpha_2} P^1_{\alpha_1} \Psi\rangle \sim \langle \Psi | P^1_{\alpha_1} P^3_{\alpha_3} P^3_{\alpha_3} P^1_{\alpha_1} \Psi\rangle = \mathcal{P}(\alpha_3, \alpha_1). \tag{6.B.6}$$

This holds, because, by (6.B.4), we have

$$\sum_{\alpha_2} \langle \Psi | P_{\alpha_1}^1 P_{\alpha_2}^2 P_{\alpha_3}^3 P_{\alpha_3}^3 P_{\alpha_2}^2 P_{\alpha_1}^1 \Psi \rangle \sim \sum_{\alpha_2', \alpha_2} \langle \Psi | P_{\alpha_1}^1 P_{\alpha_2'}^2 P_{\alpha_3}^3 P_{\alpha_3}^3 P_{\alpha_2}^2 P_{\alpha_1}^1 \Psi \rangle, \qquad (6.B.7)$$

since the sum over the terms with $\alpha_2' \neq \alpha_2$ is negligible. But then, by (6.B.1), we get (6.B.6).

6.C Decoherent Histories: Proof of Inconsistency

As we said in Sect. 6.3, if we consider "histories" defined by eigenvalues of commuting observables at a single time, then the decoherence condition is trivially satisfied [if C_α, $C_{\alpha'}$ in (6.B.4) contain only one factor, then, because of (6.B.2), (6.B.4) holds automatically], and the probability formula (6.B.3) is supposed to be, from the decoherent histories point of view, the probability of real events. We will give here explicit formulas for matrices A, B, C, D used by Goldstein [222] to prove that the different histories are inconsistent, but our presentation will follow Mermin [336].

Consider a basis $(|e_1\rangle, |e_2\rangle)$ of vectors for a two-dimensional spin space associated with a first particle and a basis $(|f_1\rangle, |f_2\rangle)$ associated with a second particle. Consider then the quantum state (the products between states are tensor products)

$$|\Psi\rangle = a|e_1\rangle|f_2\rangle + a|e_2\rangle|f_1\rangle - b|e_1\rangle|f_1\rangle, \qquad (6.C.1)$$

with the normalization

$$2a^2 + b^2 = 1, \qquad (6.C.2)$$

where a, b will be chosen below. Introduce the vectors

$$|g\rangle = c|e_1\rangle + d|e_2\rangle, \qquad (6.C.3)$$

$$|h\rangle = c|f_1\rangle + d|f_2\rangle, \qquad (6.C.4)$$

where $c^2 + d^2 = 1$ and c, d are chosen so that $\langle g f_1 | \Psi \rangle = \langle e_1 h | \Psi \rangle = 0$, which means that

$$ad - bc = 0. \qquad (6.C.5)$$

Let P_{e_1} denote the projection operator on the vector $|e_1\rangle$, and similarly for the other vectors. We define A, B, C, D as follows:

(1) $A = P_h$,
(2) $B = P_g$,
(3) $C = P_{e_2} = \mathbf{1} - P_{e_1}$,
(4) $D = P_{f_2} = \mathbf{1} - P_{f_1}$,

where in the last two identities we use the fact that $(|e_1\rangle, |e_2\rangle)$, $(|f_1\rangle, |f_2\rangle)$ are basis vectors of a two-dimensional space.

All these operators are projection operators, so their only eigenvalues are 0 and 1. We will write $A = 1$, $B = 0$ to mean that the result of the measurement of A gives 1, of B gives 0, etc. It is easy to check that all the pairs (A, C), (B, D), (C, D), and (A, B) commute, since they operate on different particles, while A does not commute with D, and B does not commute with C, since they project onto non-orthogonal vectors.[52]

If we measure these operators when the quantum state is (6.C.1), we get:

(1) If $A = P_h = 1$, then since $\langle e_1 h | \Psi \rangle = 0$, we must have $P_{e_1} = 0$, and this means that $C = P_{e_2} = \mathbf{1} - P_{e_1} = 1$.

(2) If $B = P_g = 1$, then since $\langle g f_1 | \Psi \rangle = 0$, we must have $P_{f_1} = 0$, which means that $D = P_{f_2} = \mathbf{1} - P_{f_1} = 1$.

(3) $CD = P_{e_2} P_{f_2} = 0$, because the state $|e_2\rangle | f_2 \rangle$ is absent from the sum (6.C.1). Thus, if one projects $|\Psi\rangle$ onto $|f_2\rangle$, one gets $|e_1\rangle | f_2\rangle$, and P_{e_2} acting on that latter state is zero. This means that one cannot have both the results $C = P_{e_2} = 1$ and $D = P_{f_2} = 1$.

(4) $P_h = 1$ and $P_g = 1$, meaning $A = 1$, $B = 1$, has a nonzero probability of occurring. The probability is

$$|\langle gh | \Psi \rangle|^2 = \left| \left(c\langle e_1| + d\langle e_2| \right) \left(c\langle f_1| + d\langle f_2| \right) \Psi \right|^2 = (2acd - bc^2)^2 = b^2 c^4, \tag{6.C.6}$$

where in the last equality we have used (6.C.5). By (6.C.5) and (6.C.2) and the fact that $c^2 + d^2 = 1$, one gets $b^2 = (1 - c^2)/(1 + c^2)$, and the maximum of $(1 - c^2)c^4/(1 + c^2)$ is reached for $c^2 = (\sqrt{5} - 1)/2$ (the reciprocal of the golden mean), which yields a probability of about 9 %.

Another way to state this result is that, if we want to attribute "hidden variables" taking the values 0 or 1 for each "observable" A, B, C, D and if we assume that they would be revealed by the measurement of those observables, then, if the quantum state is (6.C.1), we run into a contradiction. As shown by Mermin [336], this argument can be used to give another proof of EPR–Bell: since the two particles can be arbitrarily far from each other, if we assume no action at a distance, those hidden variables associated with the observables A, B, C, D must exist (this is the EPR part of the argument) and what we have shown here is that this assumption leads to a contradiction (the Bell part of the argument).

The decoherent histories approach would not directly claim that the four observables A, B, C, D have pre-existing values (because they do not all commute with each other), but they would attribute such values to each decoherent history corresponding to each commuting pair (A, C), (B, D), (C, D), and (A, B). But then, we get four statements that are mutually contradictory.

---

[52] As we said, the decoherence condition (6.B.4) is trivially and exactly satisfied (no need for $\sim$). To see this explicitly, consider the pair (A, C). We can introduce four "histories" (at a single time) defined by the projectors $P_h P_{e_2}$, $P_h P_{e_1}$, $(\mathbf{1} - P_h) P_{e_2}$, $(\mathbf{1} - P_h) P_{e_1}$. The sum of those projectors satisfies (6.B.1) and two different projectors (or histories) satisfy (6.B.4), since all those projectors are mutually orthogonal by (6.B.2).

Chapter 7
Revisiting the History of Quantum Mechanics

We will not attempt to give here a comprehensive history of quantum mechanics. That would take several volumes. But we will try to explain why the standard view of that history, the view that is taught to students (including me when I was one) and is probably shared by the majority of physicists, is widely off the mark.

That view can be summarized as follows (and this will be illustrated by many quotes from very respectable physicists below). Einstein did not like the indeterminism of ordinary quantum mechanics. He was old-fashioned and attached to a classical view of the world.[1] Therefore, he invented several clever arguments in order to prove that quantum mechanics was wrong, in particular that the Heisenberg uncertainty relations could be violated, but Niels Bohr successfully answered all his objections.

Moreover, in 1932, von Neumann [496] actually proved mathematically that "hidden variables", that might give a more complete description of quantum systems than the one provided by quantum mechanics, could not be introduced without leading to a contradiction with the experimental predictions of quantum mechanics. His result was strengthened in the 1960s by Kochen and Specker [291] and by Bell [35].

In 1952, Bohm [62], elaborating on earlier ideas of de Broglie, tried to introduce a "hidden variables" theory (thus running against von Neumann's result). However, this theory was nonlocal, therefore non-relativistic, and could only deal (maybe) with the special case of non-relativistic quantum mechanics.

Finally, in 1964, John Bell proved that any theory reproducing certain quantum mechanical predictions must be either nonlocal or non-realistic [35]. Given that nonlocality obviously conflicts with relativity, we must reject realism, since the predictions in question have been verified experimentally.

We claim that all of the above is historically wrong. Moreover, in order to understand why, one does not need to have a deep understanding of physics, but simply

---

[1] We already saw in Chap. 1 that Einstein was reproached for this.

© Springer International Publishing Switzerland 2016
J. Bricmont, *Making Sense of Quantum Mechanics*,
DOI 10.1007/978-3-319-25889-8_7

to grasp the English language (since most texts originally in German or French have been translated.[2]) Someone who agrees with the mainstream view of quantum mechanics, dislikes the de Broglie–Bohm theory, and thinks that Bell has not proven the existence of nonlocal effects, could nevertheless see how wrong the standard view of history is, simply by carefully reading de Broglie, Einstein, Schrödinger, Bohm, and Bell, and trying to understand what they actually said.

We want to show in this chapter that the views of Einstein, Schrödinger, and Bell were widely misunderstood, while those of de Broglie and Bohm were simply ignored most of the time.[3]

7.1 The Bohr–Einstein Debate

That debate is considered to be one of the most famous intellectual debates of the 20th century, and probably the most famous one between physicists. We will examine three episodes in that debate. The first, during the fifth Solvay Conference in 1927, which is the founding event of the "Copenhagen" interpretation of quantum mechanics; another, less well known one, during the sixth Solvay Conference in 1930; and finally the discussion about the EPR paper of 1935. There was a final rehearsal of these debates in the collective volume devoted in 1949 to Einstein "philosopher–scientist" [436]. But we must first outline what were the main positions of Einstein and Bohr and what their disagreement was about.

7.1.1 What Was the Debate Really About?

The most often quoted phrase of Einstein is probably "God does not play dice".[4] But this is misleading. It is true that Einstein did express concerns about determinism, for example, in 1924, in a letter to Born's wife Hedwig:

---

[2]However, as we will see below, these translations must sometimes be double-checked.

[3]For books giving other heterodox accounts of the history of quantum mechanics, see Bacciagaluppi and Valentini [24], Beller [52], Cushing [102], Freire [195], and Wick [512]. The book by Jammer [281] is a comprehensive overview of the history of quantum mechanics. The book by Belinfante [34] offers a review of attempts to build hidden variables theories, including de Broglie–Bohm, but also many others (that run into problems because of the no hidden variables theorems). The "bibliographic guide" by Cabello [88], containing a mere 10 340 references, might be useful.

[4]The complete quote comes from a letter to Max Born in 1926 [79, p. 91]: "Quantum mechanics is very worthy of regard. But an inner voice tells me that this is not yet the right track. The theory yields much, but it hardly brings us closer to the Old One's secrets. I, in any case, am convinced that He does not play dice." Of course, as Einstein emphasized several times, his "God" had nothing to do with the personal gods of the "revealed" religions.

I find the idea quite intolerable that an electron exposed to radiation should choose of its own free will, not only its moment to jump off, but also its direction. In that case, I would rather be a cobbler, or even an employee in a gaming-house, than a physicist.

Albert Einstein, letter to Hedwig Born, 29 April 1924 [79, p. 82]

But that was in 1924, before quantum mechanics was developed, and it is not true that Einstein was preoccupied above all with determinism, especially in his debates with Bohr. That determinism was Einstein's main concern was even strongly denied by Pauli, who was, in general, on the Copenhagen side of the arguments (if one wants to use this somewhat reductionist terminology) and who wrote in 1954 to Max Born:

Einstein does not consider the concept of "determinism" to be as fundamental as it is frequently held to be (as he told me emphatically many times) [...] he disputes that he uses as a criterion for the admissibility of a theory the question: "Is it rigorously deterministic?"

Wolfgang Pauli [79, p. 221]

But if determinism was not Einstein's main concern, what was? A clue to the answer is given in another letter, written in 1942, which also speaks of God not playing dice:

It seems hard to sneak a look at God's cards. But that he plays dice and uses "telepathic" methods (as the present quantum theory requires of him) is something that I cannot believe for a single moment.

Albert Einstein [168] quoted in [140, p. 68]

The telepathy here is of course the action at a distance analyzed in Chap. 4. Indeed, Einstein saw, unlike most of his critics, that, if the world is local, then quantum mechanics is incomplete and a certain form of "determinism" must hold, namely that, in the Bohm version of the EPR situation, the spin values must pre-exist their measurement, or, in the original EPR version, the positions and velocities of the correlated particles must pre-exist.[5] If a measurement creates a "random" jump, then a measurement at A would have to create the correlated result at B, and that would be nonlocal. And this, Einstein could not accept: but it was locality that was for him the "sacred principle" as Bell puts it [49, p. 143], not determinism per se.

The dilemma for Einstein was: either quantum mechanics is complete, in which case it is nonlocal, or it is local but incomplete, and hidden variables have to be added to it. In his 1949 *Reply to criticisms*, Einstein did pose the question in the form of a dilemma, implied by the EPR paradox:

[...] the paradox forces us to relinquish one of the following two assertions:

(1) the description by means of the ψ-function is complete,

(2) the real states of spatially separated objects are independent of each other.

Albert Einstein [436, p. 682]

---

[5] See Sect. 7.1.4 for more details on that version.

Historian of science Don Howard [274], and also Bacciagaluppi and Valentini [24], claim that Einstein's main concern, right from the beginning of the study of quantum phenomena (even before quantum mechanics was formulated),[6] was the fact that quantum mechanics seemed to require a sort of non-independence of distant events.

Let us now try to formulate Bohr's viewpoint. This is not something everybody agrees on, but, from what I understand, he insisted that quantum mechanics, and in particular the uncertainty relations, impose limitations to how much we can know about the microscopic world. In order to talk about that world, we need to use "classical" concepts, meaning concepts that deal with macroscopic objects or experimental devices, and since one cannot measure mutually incompatible properties in a single experiment, one has to resort to "complementary" views of the microscopic world: if we set up one type of experiment, we can talk about, for example, positions, and through another experiment, about momenta; or we could talk about spins in different directions, one for each experimental setup. In each setup, we can have well-defined answers, but is is pointless to try to combine these different "complementary" aspects of reality into a coherent mental image.

The root of the difference between Einstein and Bohr is that Einstein was arguing at the level of ontology, of *what there is*, insisting that a complete description of physical systems *must* go beyond the description given by ordinary quantum mechanics, if the world is local. Bohr, on the other hand, was answering systematically at the level of epistemology, of *what we can know*. Bohr was not answering Einstein, he was simply not listening to his objections. Let us show that this was already the case at the 1927 Solvay Conference.

7.1.2 The 1927 Solvay Conference

At that Solvay Conference, Einstein considered a particle going through a hole, as shown in Fig. 7.1. In the situation described in the picture, the wave function spreads itself over the half circle, but one always detects the particle at a given point, on the hemispherical photographic film denoted by P in Fig. 7.1. If the particle is *not* localized anywhere before its detection (think of it as a sort of "cloud", as spread out as the wave function), then it must condense itself at a point, in a nonlocal fashion, since the part of the particle that is far away from the detection point must "jump" there instantaneously. If, on the other hand, the particle *is* localized somewhere before its detection, then quantum mechanics is incomplete, since it does not include the position of the particle in its description.

Einstein contrasted two conceptions of what the wave function could mean:

Conception I. The de Broglie–Schrödinger waves do not correspond to a single electron, but to a cloud of electrons extended in space. The theory gives no information about individual processes, but only about the ensemble of an infinity of elementary processes.

---

[6]When he introduced photons as particles in 1905, and also in his work on Bose–Einstein statistics in 1925.

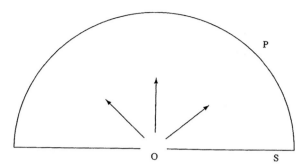

Fig. 7.1 Einstein's objection at the 1927 Solvay Conference. Reproduced with the permission of the Solvay Institutes and of Cambridge University Press. See [24, p. 440], or [456, p. 254] for the "original" (published in French translation at the time of the Solvay Conference). In fact Einstein raised a similar issue as early as 1909, at a meeting in Salzburg [24, p. 198]

> Conception II. The theory claims to be a complete theory of individual processes. Each particle directed towards the screen, as far as can be determined by its position and speed, is described by a packet of de Broglie–Schrödinger waves of short wavelength and small angular width. This wave packet is diffracted and, after diffraction, partly reaches the screen P in a state of resolution.[7]

> Albert Einstein [24, p. 441]

The distinction made here was repeated by Einstein all his life. Either one adopts a statistical view (conception I) and a more complete theory can be envisioned, or one declares the quantum description "complete", but this means that the particle is spread out in space before being detected on the screen.

After explaining the merits of conception II, Einstein raises the following objection:

> But the interpretation, according to which $|\psi|^2$ expresses the probability that *this* particle is found at a given point, assumes an entirely peculiar mechanism of action at a distance, which prevents the wave continuously distributed in space from producing an action in *two* places on the screen.

> Albert Einstein [24, p. 441]

Indeed, if the particle is spread out in space before being detected (which is what the expression "complete description" *means*), then the fact that it is always detected at a given point implies that it condenses itself on that point and that its presence vanishes elsewhere. Thus something nonlocal must be taking place. Einstein adds:

> In my opinion, one can remove this objection [action at a distance] only in the following way, that one does not describe the process solely by the Schrödinger wave, but that at the same time one localises the particle during the propagation. I think that Mr de Broglie is right to

---

[7]That last expression is not very clear, but the original German text has been lost and the French translation [456], from which this is translated into English, is not clear either. (Note by J.B.).

search in this direction.[8] If one works solely with the Schrödinger waves, interpretation II of $|\psi|^2$ implies to my mind a contradiction with the postulate of relativity.

                                                        Albert Einstein [24, p. 441]

Thus, the essence of Einstein's objection to the orthodox view of quantum mechanics (including the conflict between nonlocality and relativity) was already expressed in 1927!

It is interesting to read Bohr's response:

I feel myself in a very difficult position because I don't understand what precisely is the point which Einstein wants to [make]. No doubt it is my fault.

                                                        Niels Bohr [24, p. 442]

Einstein' objections became sharper as time went on, specially with the EPR argument. But there is an intermediate episode between the Solvay Congress of 1927 and that 1935 argument.

7.1.3   The Photon and the Box Experiment

At the sixth Solvay meeting, held in Brussels in October 1930, there were several discussions between Bohr and Einstein that were later (in 1949) recounted by Bohr. We do not have traces of these discussions, except in Bohr's recollections almost twenty years later; his version of the story was, however, corroborated by Heisenberg [510, p. 8]. But we have no record of Einstein's point of view, except for a letter from Ehrenfest to Bohr that we will discuss below, and that contradicts the Bohr–Heisenberg version.

The historian of science Don Howard nicely summarizes Bohr's version of the discussion:

A box containing a photon has an opening covered by a shutter that is activated by a timer attached to a clock inside the box by means of which we could accurately time the emission of the photon from the box. The whole box is suspended by a spring by means of which arrangement we could weigh the box both before and after the photon's emission with whatever accuracy we desire, thus determining the photon's energy via the mass–energy equivalence relation.[9] As Bohr tells the story, Einstein introduced the photon-box thought experiment for the purpose, yet again, of exhibiting violations of Heisenberg indeterminacy. Simply perform both measurements: weigh the box to fix the emitted photon's energy and open the box to check the clock and fix the time of emission. Bohr tells us that, at first, Einstein had him completely stumped. He could find no flaw in Einstein's reasoning. Only in the wee hours of the morning did it come to him. Ironically, general relativity would

---

[8]See Sect. 7.6.1 below for a discussion of de Broglie's ideas and of the reactions to them. (Note by J.B.).

[9]What is described here is a *Gedanken* or thought experiment. One is not claiming that such measurements could actually be made accurately enough in practice. (Note by J.B.).

save quantum mechanics, specifically the general relativistic effect of a gravitational field on clock rates.[10] A quick calculation showed Bohr that the change in the box's mass when the photon is emitted changes, in turn, its vertical location in the Earth's gravitational field, and that the effect of the latter change on the rate of the clock in the box induces precisely the uncertainty in the clock's rate needed to insure satisfaction of the Heisenberg indeterminacy principle [73, pp. 227–228]. Bohr uses general relativity against Einstein to save quantum mechanics! A wonderful story. But is it true?

Don Howard [275]

The Heisenberg indeterminacy relation mentioned here is the time–energy relation.[11] If Var(E) is the variance of the distribution of the photon energy and Var(t) the variance of the time at which the photon is emitted, their product is bounded from below by

$$\mathrm{Var}(E)\mathrm{Var}(t) \geq \frac{1}{4} , \qquad (7.1.3.1)$$

putting $\hbar = 1$. This bound is of course similar to (2.C.1.4) in Appendix 2.C.

Bohr's argument is based on the fact that, in the general theory of relativity, the way clocks run is affected by gravitational fields, and he uses that property to "derive" (7.1.3.1).[12] This sounds like a knock-down argument against Einstein, since he himself discovered the general theory of relativity! But there is much less here than meets the eye. First of all, the fact that Bohr's argument uses a theory discovered by Einstein may be psychologically impressive, but has no significance whatsoever for the correctness of his argument. Second, Bohr did not really derive (7.1.3.1); he reduced the validity of the time–energy uncertainty relation to that of the position–momentum relation (2.C.1.4). If the latter is correct, shows Bohr, then using a general relativistic argument, the time–energy uncertainty relation will also be correct.

But this raises a natural question: how could the internal consistency of quantum mechanics depend on the truth of general relativity? Quantum mechanics was developed in a perfectly "classical" context, that is, it accepted the spacetime structure of classical mechanics, with its Galilean invariance. It did not assume the truth of general relativity or of any theory of gravitation whatsoever. Moreover, even now, it remains an open problem to combine quantum mechanics with general relativity. So it is difficult to see why one would need to use an argument based on general relativity to "save" the consistency of quantum mechanics.[13]

---

[10] In general relativity, a gravitational field will affect the way a clock runs; this is somewhat related to the relativity of simultaneity in special relativity. (Note by J.B.).

[11] There is some debate in the literature as to the exact meaning of this relation, and in particular as to whether time can be put on the same footing as energy, position, or momentum. But this will not matter for our discussion here. For a proof of the relation in the context of Bohr's argument, we refer to [353].

[12] For a picture of the experiment and more details, see [73, p. 227].

[13] It is true that Einstein's argument, as recalled by Bohr, did use the link between mass and energy ($E = mc^2$) of special relativity, but there is no reference in that argument to general relativity.

However, all of this is somewhat academic from the point of view of physics, since there exists a correct derivation of the time–energy uncertainty relation in the situation discussed by Bohr, due to Nikolić [353]. That derivation is based on a proper quantum mechanical reasoning and appeals to neither special relativity nor general relativity. But from a historical point of view, this story is interesting, since it shows that, although Bohr's argument was beside the point, it is often regarded by physicists as one of his "victories" over Einstein, for the purely psychological reason that Bohr used Einstein's own theory! But things are even worse. There is evidence that Einstein's goal in discussing the photon in the box experiment was totally different from the one that Bohr attributed to him in 1949.

The evidence comes from a letter sent by the Austrian physicist Paul Ehrenfest (who was close to both Bohr and Einstein) to Bohr in July 1931.[14] As explained by Don Howard:

> Ehrenfest and Einstein seem to have had a long and thorough chat about the debate with Bohr at the previous fall's Solvay meeting. Ehrenfest reports to Bohr a most surprising comment from Einstein:
>
> He [Einstein] said to me that, for a very long time already, he absolutely no longer doubted the uncertainty relations, and that he thus, e.g., had BY NO MEANS invented the "weighable light-flash box" (let us call it simply L-F-box)[15] "contra uncertainty relation", but for a totally different purpose. (Ehrenfest to Bohr, 9 July 1931 [159] […]).
>
> What was that totally different purpose? It was nothing other than an anticipation of Einstein's later argument for the incompleteness of quantum mechanics.
>
> As Ehrenfest explains to Bohr, Einstein's idea was this. Let the photon leave the box and be reflected back from a great time and distance, say one-half light year. At about the time when the photon is reflected, we can either weigh the box or check the clock, making possible our predicting either the exact time of the photon's return or its energy (literally, its color), which is to say that, depending upon which measurement we choose, we ascribe a different theoretical state to the photon, one with definite energy, one entailing a definite time of arrival. Crucial is the fact that the event of performing the measurement on the box—weighing it the second time or checking the clock—is space-like separated[16] from the event of the photon's distant reflection, because then our choice of a measurement to perform can have no effect on the real state of affairs of the photon, meaning that the photon's real state of affairs when it returns will be one and the same, regardless of the measurement we performed on the box. This is all just quantum mechanics, in Einstein's view. But then quantum mechanics

---

[14]Jammer [281, p. 171] discusses this letter at length, but according to Howard [275], mistranslates part of it: Einstein said that "for a very long time already, he absolutely no longer doubted the uncertainty relations", while Jammer says that, according to the letter, Einstein "no longer intends to use the box experiment 'against the indeterminacy relations'", which suggests that he did originally use it for that purpose. On the other hand, Heisenberg corroborates the story told by Bohr in his "commentary" on the Bohr–Einstein dialogue in [510, p. 8]. So we cannot be sure of what was really said in 1930, but we are mostly interested here in what Einstein thought well before the EPR article.

[15]This is what we call the photon and the box experiment. (Note by J.B.).

[16]This means that the events are sufficiently far apart so that the time between their occurrence is shorter than the time that would be taken by a ray of light to travel between them. (Note by J.B.).

has associated two different theoretical states with one real state of affairs, which is possible only if the quantum theory's state descriptions are incomplete.

Don Howard [276]

This looks awfully similar to the later EPR argument. By choosing to make one measurement or the other (and *not* both as Bohr thought), when the photon is far away, we can affect its state (having a definite energy or a definite time of return). But then, barring actions at a distance, the photon must have possessed both properties, irrespective of our measurements made on the box. However, there is no indication here that Einstein wanted to disprove the uncertainty relation (7.1.3.1). Saying that both properties exist is not the same thing as saying that one can measure them simultaneously with an arbitrary precision.

In actual fact, as explained by Nikolić [353], the correct quantum mechanical discussion of the photon and the box experiment does imply a form of nonlocality similar to what Bell proved in the EPR situation. The states of the photon and the box are "entangled", in a way similar to the states (4.4.2.1) used in Chap. 4, and acting on the box (by a measurement) affects the state of the photon.

Bohr does actually discuss the Ehrenfest's letter in his contribution to the volume on Einstein "philosopher–scientist", but his answer misses the point:

In fact, we must realize that in the problem in question we are not dealing with a *single* specified experimental arrangement, but are referring to *two* different, mutually exclusive arrangements. [...]

In both theses cases, as also assumed by Einstein, the observable effects are expected to be in complete conformity with the predictions of the theory.

Niels Bohr [73, pp. 229–230]

But Einstein's point was not about the "observable effects", and he quite agreed that there were "*two* different, mutually exclusive arrangements". But it was precisely because he could *choose* one of the two experimental arrangements and learn *one property or the other* of the photon while it was "one-half light year" away that, without action at a distance, the photon must have had both properties to start with.

7.1.4 The Einstein–Podolsky–Rosen Argument

We explained the EPR argument in Chap. 4. We used there spin variables and this version of the argument is due to Bohm [61]. The original EPR article did not use spin, but position and momentum.[17] The authors considered two particles starting from the same place and moving in opposite directions, in such a way that their total momentum was conserved and equal to zero. Thus, by measuring the momentum

---

[17]There exists an unpublished manuscript by Einstein, written in 1954 or 1955, that uses the spin variables to reach the usual incompleteness conclusion, assuming locality [432].

of one particle, one could know the momentum of the other particle. But if one measured instead the position of that particle, one would know the position of the other particle (since they move at the same speed in opposite directions). However, if the two particles are far apart, and if locality holds, the choice that we make of the quantity to measure on one particle cannot affect the state of the other particle. Thus, that second particle must possess a well-defined momentum *and* a well-defined position, and this shows that quantum mechanics is incomplete, since the quantum state does not include those variables.

As already emphasized,[18] Einstein's concern with "determinism" was not due to determinism itself, but to his attachment to locality. Indeed, he understood that locality required that variables such as the spin values or positions and momenta *must* exist if measures on one side (on A) do not affect the physical situation on the other side (on B).

However, the EPR article was maybe not as transparent as one might have liked it to be. In particular, there was no need to invoke *both* position and momentum. The fact that one of them, say position, could be determined for one particle by a measurement on the other particle, arbitrarily far away, is enough to prove incompleteness of quantum mechanics (since the quantum state does not in general assign a determinate position to either particle), if one takes locality for granted.[19] The article, according to Einstein, was written by Podolsky "for reasons of language", because Einstein's English was far from perfect. But Einstein complained in a letter to Schrödinger on 19 June 1935 [165, 276] that "it has not come out as well as I really wanted" and that "the main point was, so to speak, buried by the erudition".

One rather common misconception is, again, to think that the goal of EPR was to beat the uncertainty principle and to show (using the notation of Chap. 4) that one could measure, say, the spin in direction 1 at X and in direction 2 at Y, and therefore know the values of the spin of both particles in directions associated with operators that do not commute (and cannot therefore be simultaneously measured, according to standard quantum mechanics).[20]

Actually, Einstein explicitly denied that his goal was to refute the uncertainty principle in his letter to Schrödinger of 19 June 1935.[21] His argument, expressed in the language of Chap. 4, was that, in the state (4.2.2.1), if one measures the spin in direction 1 at X, one may get, for example, the state $|A\ 1\ \uparrow\rangle|B\ 1\ \downarrow\rangle$ and if one measures the spin in direction 2, one may get, say, the state $|A\ 2\ \downarrow\rangle|B\ 2\ \uparrow\rangle$. But then

---

[18]See also Maudlin [327].

[19]This is discussed in detail by Maudlin [319, 3rd edn, pp. 128–132]. However, even taking these criticisms into account, if one takes locality for granted, the EPR argument against completeness of quantum mechanics was perfectly valid.

[20]Of course, since EPR were speaking of position and momentum instead of spin, it is those quantities that would have been simultaneously measured. But the misconception is in part due to the fact that EPR consider these two quantities, while one would have been enough, as shown by the "boxes" argument. See [355] for further discussion of this point.

[21]It was in this same letter that Einstein gave the "boxes" argument discussed in Chap. 4, and that argument again proves the incompleteness of quantum mechanics, if Nature is local, without getting into the more complex EPR argument.

at Y, one gets two different states $|B\ 1\ \downarrow\rangle$ and $|B\ 2\ \uparrow\rangle$, and this leads to different predictions for the future behavior of the system. Hence, by choosing which quantity to measure at X, one changes the state at Y in different ways: action at a distance! And that is what Einstein objected to. However, Einstein emphasized that "he couldn't care less" whether the collapsed states at X and Y were eigenstates of observables that could not be measured simultaneously, like the spin in directions 1 and 2 [186, p. 38].

In fact, the reason why one cannot measure simultaneously the spin in direction 1 at X and in direction 2 at Y is that either of those measurements will affect the quantum state (by "collapsing" it) and it will affect it in both places, i.e., nonlocally, so that a measurement of the spin in direction 1 at X will generally change the results of a spin measurement in direction 2 at Y. In the EPR situation, it is because of the nonlocal character of the collapse that one cannot perform these simultaneous measurements. But if things were perfectly local, then one could measure the spin in direction 1 at X (or the position of the particle at X) and the spin in direction 2 at Y (or the momentum of the particle at Y), since neither of these measurements would affect the state of the other particle and thus one would know, because of the perfect correlations, the spin in both directions (or the position and the momentum) at X and at Y.

The "onslaught" of EPR "came down upon us as a bolt from the blue", wrote Léon Rosenfeld, who was in Copenhagen at that time and who reported EPR's argument to Bohr. The latter then worked intensely "week after week" on a reply.[22] There is a widespread belief among physicists that Bohr came up with an adequate answer to the EPR paper [72]. But, unlike Born, whose misunderstanding of Einstein (see Sect. 7.2) is clearly stated, Bohr's answer is hard to understand,[23] despite the fact that, according to Rosenfeld, one of his favorite aphorisms was a line from the poet Friedrich von Schiller [510, p. 143]: "Only fullness leads to clarity, and truth dwells in the depths."

So what was Bohr's reply? EPR had written that:

If, without in any way disturbing a system, we can predict with certainty (i.e., with probability equal to unity) the value of a physical quantity, then there exists an element of physical reality corresponding to this physical quantity,

Albert Einstein, Boris Podolsky, Nathan Rosen [164], reprinted in [510, pp. 138–141]

This again means that if, by doing something at A, we can learn something about the situation at B, then what we learn must exist before we learn it, since A and B can be far apart. Here of course, EPR assumed locality.

Bohr replied:

[...] the wording of the above-mentioned criterion [...] contains an ambiguity as regards the meaning of the expression "without in any way disturbing a system". Of course there

---

[22] See Rosenfeld's commentary in [510, p. 142].

[23] As an amusing side remark, we note that, in the reprint of Bohr's paper in [510], which was the most accessible source of that paper before Internet, pp. 148 and 149 are the wrong way round.

is in a case like that just considered no question of a mechanical disturbance of the system under investigation during the last critical stage of the measuring procedure. But even at this stage there is essentially the question of *an influence on the very conditions which define the possible types of predictions regarding the future behavior of the system* [...] their argumentation does not justify their conclusion that quantum mechanical description is essentially incomplete [...] This description may be characterized as a rational utilization of all possibilities of unambiguous interpretation of measurements, compatible with the finite and uncontrollable interaction between the objects and the measuring instruments in the field of quantum theory.

Niels Bohr [72], quoted in [49, p. 155] (italics in the original)

Bell dissects this passage as follows:

Indeed I have very little idea what this means. I do not understand in what sense the word 'mechanical' is used, in characterizing the disturbances which Bohr does not contemplate, as distinct from those which he does. I do not know what the italicized passage means — 'an influence on the very conditions [...]'. Could it mean just that different experiments on the first system give different kinds of information about the second? But this was just one of the main points of EPR, who observed that one could learn *either* the position *or* the momentum of the second system. And then I do not understand the final reference to 'uncontrollable interactions between measuring instruments and objects', it seems just to ignore the essential point of EPR that in the absence of action at a distance, only the first system could be supposed disturbed by the first measurement and yet definite predictions become possible for the second system. Is Bohr just rejecting the premise — 'no action at a distance' — rather than refuting the argument ?

John Bell [49, pp. 155–156] (italics in the original)

Actually, Einstein also thought that Bohr rejected EPR's premise that "the real situation of B could not be influenced (directly) by any measurement taken on A" (see [436, pp. 681–682]). But this is by no means clearly stated and, because of Bohr's reaction at the 1927 Solvay Conference (admitting that he did not understand Einstein) and because of his comment in 1949 on Ehrenfest's 1935 letter, where he manifestly missed Einstein's point, one may doubt that Bohr really understood Einstein. Indeed, the latter's main point was that certain quantities must exist, because of locality, and not that we can know or control them.

7.1.5 Who Won the Bohr–Einstein Debate?

Léon Rosenfeld, who was a staunch supporter of Bohr, wrote about the latter's response to the EPR paper:

Einstein's problem was reshaped and its solution reformulated with such precision that the weakness in the critics' reasoning became evident, and their whole argument, for all its false brilliance, fell to pieces.

Léon Rosenfeld [423]

Although most physicists would probably not go that far, the idea is widespread that Bohr carried the day in his debate with Einstein.

Regarding the EPR paper, the physicist Abraham Pais, who produced a well-known biography of Einstein [370], and another of Bohr [371], writes:

> 'No reasonable definition of reality could be expected to permit this.'[24] The only part of this article which will ultimately survive, I believe, is this last phrase which so poignantly summarizes Einstein's views on quantum mechanics in his later years. The content of this paper has been referred to on occasion as the Einstein–Podolsky–Rosen paradox. It should be stressed that this paper contains neither a paradox nor any flaw of logic. It simply concludes that objective reality is incompatible with the assumption that quantum mechanics is complete. This conclusion has not affected subsequent developments in physics and it is doubtful that it ever will.
>
> Abraham Pais [369, p. 904]

It is odd that Pais, who had the privilege of being a colleague of Einstein at the Institute for Advanced Study in Princeton,[25] so seriously misunderstood Einstein's point, and also that he seems to consider "objective reality" as a sort of superfluous concept.

There is a sense in which Einstein was wrong: Nature is not local. But that was proven long afterwards, in 1964, by Bell. As Bell puts it [49, p. 150]: "It would be wrong to say that 'Bohr wins again'; the argument [of Bell] was not known to the opponents of Einstein, Podolsky and Rosen." But there is slim evidence that Bohr really understood the idea of locality in the EPR argument. As we will see in Sect. 7.3, Schrödinger was far more explicit about the consequences of EPR.

Heisenberg, on the other hand, did understand the problem, even in 1929, long before the argument of Einstein's boxes[26] was given, in 1935. He considered an example similar to Einstein's boxes (also due to Einstein), but with the wave function of a photon split into two parts (transmitted and reflected) by a beam splitter[27]:

> We imagine a photon which is represented by a wave packet built up out of Maxwell waves. It will thus have a certain spatial extension and also a certain range of frequency. By reflection at a semitransparent mirror, it is possible to decompose it into two parts, a reflected and a transmitted packet. There is then a definite probability for finding the photon either in one part or in the other part of the divided wave packet. After a sufficient time the two parts will be separated by any distance desired; now if an experiment yields the result that the photon is, say, in the reflected part of the packet, then the probability of finding the photon in the other part of the packet immediately becomes zero. The experiment at the position of the reflected packet thus exerts a kind of action (reduction of the wave packet) at the distant

---

[24]Pais is referring to the sentence in the EPR paper which stresses that the properties of one system cannot depend on the type of measurements made on a distant system, which was EPR's main point. (Note by J.B.).

[25]As mentioned in Sect. 1.3, it was to Pais that Einstein addressed his famous rhetorical question [369, p. 907]: "Whether I [Pais] really believed that the moon exists only when I look at it?".

[26]Discussed in Sect. 4.2.

[27]This experiment was actually performed by the Hungarian physicists A. Ádám, L. Jánossy, and P. Varga in 1955 [1]. The experiment was suggested to them by Schrödinger. See [97] for a discussion of their results and [355] for the relation between that situation and Einstein's boxes.

point occupied by the transmitted packet, and one sees that this action is propagated with a velocity greater than that of light. However, it is also obvious that this kind of action can never be utilized for the transmission of signals so that it is not in conflict with the postulates of the theory of relativity.

Werner Heisenberg [257, p. 39], quoted in [521]

As discussed in Chaps. 4 and 5, it is not true that there is no conflict with relativity, unless that theory is understood as the mere impossibility of sending messages; the point is that a physical theory ought to be causal and a nonlocal causality is not easy to reconcile with relativity. But at least, Heisenberg understood that the reduction of the wave function, in this example or in Einstein's boxes, involves some kind of action and "that this action is propagated with a velocity greater than that of light". Nowhere does Bohr say this as explicitly as Heisenberg did. It is true that elsewhere, Heisenberg has a more epistemic understanding of the wave function. For example, in 1958, he wrote that "the discontinuous change in our knowledge in the instant of registration [...] has its image in the discontinuous change of the probability function" ([259], quoted in [521]), but at least in 1929, he seemed to recognize the problem posed by the reduction of non-localized wave functions.

Bell's judgment on the "debate" was as tranchant as Rosenfeld's, but in the opposite direction:

I felt that Einstein's intellectual superiority over Bohr, in this instance, was enormous; a vast gulf between the man who saw clearly what was needed, and the obscurantist.

John Bell in [56, p. 84]

It is interesting to observe, as the physicist Howard Wiseman does:

When reviewing the Einstein–Bohr debates in 1949 [73], Bohr concluded his summary of his reply to EPR by quoting his defence based upon complementarity. Astonishingly, he immediately followed this by an apology for his own 'inefficiency of expression which must have made it very difficult to appreciate the trend of the argumentation [...]'! But rather than taking the opportunity to explain himself more lucidly, he instead directed the reader to concentrate on the earlier debates with Einstein regarding the consistency of quantum mechanics.

Howard Wiseman [521]

We may note also that, in his 1949 "Reply to criticisms" [170], Einstein does not address Bohr's criticisms in any detail. Maybe this was due to the fact that he did not have a very high opinion of Bohr's "philosophy"; as we saw, in his private letters to Schrödinger, Einstein referred to Bohr as the "Talmudic philosopher" for whom "reality is a frightening creature of the naive mind" [165].

Coming back to Einstein's argument, his reasoning was always of the form: if p (locality) is true, then q (incompleteness of quantum mechanics) is true. The reasoning was correct. It turns out that we know now that p is false, but q is nevertheless true, at least in the sense that the de Broglie–Bohm theory proves that a more complete theory than ordinary quantum mechanics is possible. And that is certainly something that Bohr always denied. So for Bohr also, history turned out in a way that he "would have liked least", to use Bell's phrase about how Einstein would have reacted to the existence of nonlocal effects [36, p. 11]. Maybe some day physicists will admit that Einstein lost a battle (about locality) but won the war (about incompleteness).

7.2 Born and Einstein

Born and Einstein were lifelong friends; they both had to leave Germany, to avoid persecution, Born settling in Edinburgh, Einstein in Princeton. They exchanged a long correspondence, about physics, personal matters, and politics, which was edited by Born in 1971, long after Einstein's death in 1955. In that correspondence, they also discussed quantum mechanics. It is rather fascinating to read their exchange of letters and to observe the systematic misunderstanding of Einstein's point of view by Born.

For example, Einstein wrote an article in 1948 and sent it to Born (it is reproduced in the correspondence), begging him to read it as if he had just arrived "as a visitor from Mars". Einstein hoped that the article would help Born to "understand my principal motives far better than anything else of mine you know" [79, p. 168]. In that article, Einstein restated his unchanging criticisms:

> If one asks what, irrespective of quantum mechanics, is characteristic of the world of ideas of physics, one is first of all struck by the following: the concepts of physics relate to a real outside world, that is, ideas are established relating to things such as bodies, fields, etc., which claim a 'real existence' that is independent of the perceiving subject — ideas which, on the other hand, have been brought into as secure a relationship as possible with the sense-data. It is further characteristic of these physical objects that they are thought of as arranged in a spacetime continuum. An essential aspect of this arrangement of things in physics is that they lay claim, at a certain time, to an existence independent of one another, provided these objects 'are situated in different parts of space'. […]

> The following idea characterizes the relative independence of objects far apart in space (A and B): external influence on A has no direct influence on B.

> Albert Einstein [169], reproduced in [79, pp. 170–171]

Then Einstein repeats his argument[28] that different choices of measurement at A create (through the collapse rule) a different state at B.

Here is Born's reaction to that very same article:

> The root of the difference between Einstein and me was the axiom that events which happens in different places A and B are independent of one another, in the sense that an observation on the states of affairs at B cannot teach us anything about the state of affairs at A.

> Max Born [79, p. 176]

---

[28]Outlined in Chap. 4, with X instead of A and Y instead of B.

As Bell says:

> Misunderstanding could hardly be more complete. Einstein had no difficulty accepting that affairs in different places could be correlated. What he could not accept was that an intervention at one place could *influence*, immediately, affairs at the other.

> These references to Born are not meant to diminish one of the towering figures of modern physics. They are meant to illustrate the difficulty of putting aside preconceptions and listening to what is actually being said. They are meant to encourage *you*, dear listener, to listen a little harder.

John Bell [49, p. 144] (italics in the original)

What Born said was that making an experiment in one place *teaches* us something about what is happening at another place, which is unsurprising. If, in the anthropomorphic example given in Chap. 4, both people had agreed on a common strategy, one would learn what B would answer to question 1, 2, or 3, by asking A that same question. But, and that was Einstein's point, it would mean that the answers were predetermined.

In fact, in his comments about Einstein, Born gives the following example:

> When a beam of light is split in two by reflection, double-refraction, etc., and these two beams take different paths, one can deduce the state of one of the beams at a remote point B from an observation at point A.

Max Born [79, p. 176]

But here Born is giving a classic example, where the properties of the *beam* (which is not a single particle) do pre-exist their measurement.[29] This is exactly contrary to the idea that the quantum state is a complete description, since the latter does not, in general, specify any value of the position, momentum, spin, angular momentum, energy, etc., i.e., when it is a superposition of different eigenstates of the operator corresponding to the given physical quantity [as, for example, in the state (4.4.2.1)].

This suggests that Born thought that, in our language, there are pre-existing answers, namely that the spins are up or down before the measurements, and that particles have correlated positions and velocities. In other words, it seems that Born did in fact agree with Einstein that quantum mechanics is incomplete, but simply did not understand what Einstein meant by that.

---

[29]From an ordinary quantum mechanical point of view, this is simply because the beam may have definite properties, since they are statistical properties of a large number of particles. Of course, if individual particles do not have any properties whatsoever, it is not clear how an ensemble of particles can acquire them.

7.3   What Did Schrödinger Really Worry About?

Schrödinger is, together with Einstein, one of the two best-known critics of the Copenhagen interpretation of quantum mechanics. With the example of his cat, which is
supposed to be "both alive and dead", Schrödinger hoped to give the ultimate argument against the orthodox view of quantum mechanics. Unfortunately, Schrödinger's
cat turned out to be what is probably the most misunderstood piece of reasoning about
the "foundations" of quantum mechanics.

Schrödinger is often quoted for having said to Bohr: "If we are going to stick
to this damned quantum-jumping, then I regret that I ever had anything to do with
quantum theory", to which Bohr replied that "the rest of us are thankful that you
did".[30]

But what was Schrödinger really criticizing? When he introduced his equation,
Schrödinger's initial idea was a sort of continuous matter theory, like the one discussed in Sect. 6.1.2, but with the mass replaced by the charge.[31] If the Schrödinger
waves remained localized in a small region of space, that would make sense—in
fact it would be easier to conceive of the particles as somewhat spread out in space
rather than having them localized exactly at a point (as they are idealized in classical
physics or in the de Broglie–Bohm theory).

Unfortunately, the solutions of Schrödinger's equation spread in space as time
goes by.[32] We see that in Fig. 7.1, in the double-slit experiment (Appendix 2.E),
or in the calculation (2.A.2.24) in Appendix 2.A. Of course, in standard quantum
mechanics, the spreading of the quantum state is dealt with through the collapse
postulate (i.e., what Schrödinger called "jumps"). But Schrödinger understood the
eternal problem with this postulate: why should we have this extra rule, and at which
level (microscopic, macroscopic, in-between) does collapse occur?

That was one of the things that bothered Schrödinger in his famous "cat" paper,
where the story of the cat occupies only a few lines. He first recalls the problem in
a similar way Einstein's remarks at the 1927 Solvay meeting (see Fig. 7.1). What
Schrödinger calls blurring is more or less what is called spreading here:

> But serious misgivings arise if one notices that the uncertainty affects macroscopically
> tangible and visible things, for which the term "blurring" seems simply wrong. The state
> of a radioactive nucleus is presumably blurred in such a degree and fashion that neither the
> instant of decay nor the direction, in which the emitted alpha-particle leaves the nucleus,
> is well established. Inside the nucleus, blurring doesn't bother us. The emerging particle is
> described, if one wants to explain intuitively, as a spherical wave that continuously emanates
> in all directions and that impinges continuously on a surrounding luminescent screen over

---

[30]This story is told by Heisenberg in [258, p. 14]. The exchange took place in September 1926,
in Copenhagen, where Schrödinger was visiting Bohr. The "jumping" that Schrödinger refers to
is probably the discontinuous evolution not covered by his equation, including the collapse of the
quantum state.

[31]See [14] for a discussion of that theory from a modern point of view.

[32]We neglected this spreading in our treatment of the Mach–Zehnder interferometer and in our
discussion of measurements in Appendix 2.D.

its full expanse. The screen, however, does not show a more or less constant uniform glow, but rather lights up at *one* instant at *one* spot [...]

Erwin Schrödinger [441], reprinted in [510, pp. 156–157]

In other words, the problem is that what we see macroscopically is localized, while the wave function is not. Schrödinger illustrates this with the cat example[33]:

One can even set up quite ridiculous[34] cases. A cat is penned up in a steel chamber, along with the following device (which must be secured against direct interference by the cat): in a Geiger counter there is a tiny bit of radioactive substance, so small, that perhaps in the course of the hour one of the atoms decays, but also, with equal probability, perhaps none; if it happens, the counter tube discharges and through a relay releases a hammer which shatters a small flask of hydrocyanic acid. If one has left this entire system to itself for an hour, one would say that the cat still lives if meanwhile no atom has decayed. The psi-function of the entire system would express this by having in it the living and dead cat (pardon the expression) mixed or smeared out in equal parts.

It is typical of these cases that an indeterminacy originally restricted to the atomic domain becomes transformed into macroscopic indeterminacy, which can then be *resolved* by direct observation. That prevents us from so naively accepting as valid a "blurred model" for representing reality.

Erwin Schrödinger [441], reprinted in [510, p. 157]

Of course, Schrödinger saw that as a *reductio ad absurdum* of the idea that quantum mechanics is complete, since an indeterminacy that could be acceptable in the microscopic domain to which we have no direct access (all sides agree on that, even the "realists") "becomes transformed into macroscopic indeterminacy" with the cat example.

In other words, for the cat, the description "both alive and dead" is certainly not complete. Now, certain people seem to think that quantum mechanics predicts that the cat is both alive and dead and that, since quantum mechanics is complete, the cat must be in such a state. This, to put it mildly, is a huge misunderstanding of Schrödinger's views. And of course, a mistreatment of logic as well: we have no evidence whatsoever that the cat is both alive and dead; moreover, the reason why superposed states were introduced for microscopic systems in the first place was to take into account interference phenomena and everybody agrees that such phenomena are impossible for the "psi-functions" of the living and dead cat (because of decoherence). It is only the dogmatic insistence that quantum mechanics is complete that leads to the idea of the cat being both alive and dead.

Moreover, since we never see such a cat, we have to assume that looking at the cat "collapses" its state into one of the two possibilities. But again, this is a pure *deux ex machina*, since "looking" is not associated with any precise physical interaction. It is far more sensible to consider, with Schrödinger, that the cat example shows that there is a problem with ordinary quantum mechanics, which moreover, can be solved

---

[33]Discussed in Sect. 2.5.1.

[34]In German, the word is "burlesque". (Note by J.B.).

by the de Broglie–Bohm theory (something Schrödinger does not say, although he had heard of de Broglie's theory).

Actually, for the sake of historical accuracy, it should be mentioned that the cat argument was suggested to Schrödinger by Einstein, in a slightly different form. In a letter to Schrödinger, Einstein considered the following situation:

> The system is a substance in chemically unstable equilibrium, perhaps a charge of gunpowder that, by means of intrinsic forces, can spontaneously combust and where the average lifespan of the whole setup is a year. In principle this can quite easily be represented quantum mechanically. In the beginning the ψ-function characterizes a reasonably well-defined macroscopic state. But, according to your equation, after the course of a year this is no longer the case at all. Rather, the ψ-function then describes a sort of blend of not-yet and already-exploded systems. Through no art of interpretation can this ψ-function be turned into an adequate description of a real state of affairs; [for] in reality there is just no intermediary between exploded and not exploded.

Einstein, letter to Schrödinger, 8 August 1935 [186, p. 78]

In fact, Schrödinger acknowledged that his "cat" paper, which he called a "lecture or general confession" [441, p. 163, Note 7], was motivated by the EPR paper. In the "cat" paper and in two others [442, 443] published in 1935–36, Schrödinger also discussed and extended the EPR idea, coining the word "entangled state" to denote states such as (4.4.2.1) that are not products of states related to the situation at X and at Y, but rather sums of such states.

In the "cat" paper, Schrödinger also considers two systems, one of which he calls small, whose properties are labelled by lower case letters, and another, called auxiliary, whose properties are labelled by capital letters, which is in a state entangled with the small system.[35] If the two systems are far apart (at X and Y), reasons Schrödinger, acting on one system cannot possibly affect the other (like EPR, he takes locality for granted). But because of entanglement, we have a perfect correlation between the results of measurements at X and at Y. Therefore, the small system, located at X, must have the properties that are correlated with those measured at Y, before measurement, unless there is some "magic" (Schrödinger's word for action at a distance).

Schrödinger compares the situation with one where some schoolchildren can be asked two questions, but which of the two questions is asked is chosen at random. It happens that the schoolchildren always answer the first question correctly; how can that happen (systematically) if they do not know the answer to *both* questions? This is like asking for either the position q or the momentum p of the particle at X. By doing the corresponding measurement at Y, far away from X, we know what the answer will be. As Schrödinger explains:

> I can direct *one* of two questions to the small system, either that about q or that about p. Before doing so I can, if I choose, procure the answer to *one* of these questions by a measurement

---

[35]An example of such a state is given by (4.4.2.1), where particle A would be called the small system and particle B the auxiliary system (of course, in that example, there is a symmetry between A and B, so the division between small and auxiliary is arbitrary).

on the fully separated other system (which we shall regard as auxiliary apparatus), or I may intend to take care of this afterwards, My small system, like a schoolboy under examination, *cannot possibly know* whether I have done this or for which questions, or whether and for which I intend to do it later. From arbitrarily many pretrials I know that the pupil will correctly answer the first question that I put to him. From that it follows that in every case he *knows* the answer to *both* questions.

Erwin Schrödinger [441], reprinted in [510, p. 164] (italics in the original)

Next, Schrödinger answers the following objection (that Bohr would probably have made): after measuring q, one cannot measure what the value of p would have been if one had not measured q, because each measurement perturbs the state, through the collapse rule:

That answering the first question I chose to ask tires or confuses the pupil to such an extent that his following answers are worthless changes nothing at all of this conclusion. No school principal would judge otherwise, if this situation repeated itself with thousands of pupils of similar provenance, however much he might wonder *what* makes all the scholars so dim-witted or obstinate after the answering of the first question. It would not occur to him that his own consultation of a textbook first gave away the right answer to the student, or even that, in cases where the teacher chooses to look up the answer after the student answers correctly, that the student's answer has changed the textbook in the pupil's favor.[36]

Thus my small system holds a quite definite answer to the q-question and to the p-question in readiness for the case that one or the other is the first to be put directly to it. Of this preparedness not an iota can be changed if I should perhaps measure the Q on the auxiliary system (in the analogy: if the teacher looks up one of the questions in his notebook and thereby indeed ruins with an inkblot *the* page where the other answer stands). The quantum mechanician maintains that after a Q-measurement on the auxiliary system my small system has a psi-function in which "q is fully sharp, but p fully indeterminate". And yet, as already mentioned, not an iota is changed of the fact that my small system also has ready an answer to the p-question, and indeed the same one as before.

Erwin Schrödinger [441], reprinted in [510, p. 164]

But as Schrödinger says, "the situation is even worse yet". Indeed he proceeds to show in [442] that the small system must have answered all kinds of other questions, corresponding to different functions of q and p, and thus, in principle, to different measurable quantities but whose values can be known by measurements made on the distant auxiliary system.

He even shows that, for some functions $f(q, p)$, the results of the corresponding measurements cannot coincide with what one would obtain by substituting the possible values of q and p as arguments of the function f. For example, some functions are such that the results are always multiples of integer numbers, while q and p

---

[36]This last sentence has been retranslated in order to try to make it more understandable. The original translation in [510] is: He would not come to think that his, the teacher's, consulting a textbook first suggests to the pupil the correct answer, or even, in the cases when the teacher chooses to consult it only after ensuing answers by the pupil, that the pupil's answer has changed the text of the notebook in the pupil's favor. (Note by J.B.).

vary over the real numbers.[37] Schrödinger concludes that [441, p. 165]: "how the numerical values of all these variables of *one* system relate to each other we know nothing at all, even though for each the system must have a quite specific one in readiness, for if we wish we can learn it from the auxiliary system and then find it always confirmed by direct measurement."

In fact, Schrödinger's argument was similar to one given by von Neumann, to be discussed in the next section.[38] But his conclusion was radically different: instead of thinking that this proves the "completeness" of quantum mechanics and the impossibility of "hidden variables", Schrödinger thought that, combining this argument with a locality requirement leads to serious puzzle for the orthodox point of view.

Schrödinger saw a deep mystery behind the findings of EPR and did not think that ordinary quantum mechanics offered an answer to them. We leave it to the reader to appreciate the enormous difference between Schrödinger's [441] and Bohr's [72] understanding of the problem raised in the EPR article, both papers being published in 1935.

7.4 The von Neumann No Hidden Variables Theorem

In 1932, von Neumann published what has become one of the most basic mathematical expositions of quantum mechanics [496]. In that book, he developed the measurement formalism[39] and also gave an argument trying to show that quantum mechanics is complete, namely that no hidden variables could be introduced that would reproduce the results of quantum mechanics.

Von Neumann considers, as we did in Chap. 2, the possibility of introducing variables $v(A)$ that would give, for any individual system, the value of the quantity A prior to its measurement.[40] Thus $v(A)$ would have to coincide with one of the eigenvalues of the matrix (or operator) associated with the quantity A. Von Neumann assumes, among other properties, that the hidden variables satisfy

$$v(A + B) = v(A) + v(B) , \qquad (7.4.1)$$

---

[37]Schrödinger considers the functions $H = p^2 + a^2 q^2$, where a is a real number. This is the Hamiltonian of the harmonic oscillator, but defines different functions for different values of a. It is well known that the eigenvalues of H are always of the form $(2n + 1)a$, for n a positive integer. This obviously cannot hold for all values of a, and any given values of p and q.

[38]Schrödinger showed that, assuming an equation like (7.4.1) below with $A = p^2$ and $B = aq^2$, leads to a contradiction. According to Bacciagaluppi and Crull, Pauli gave a similar argument in a letter to Schrödinger on 9 July 1935 [25].

[39]Described in Appendix 2.D.

[40]Von Neumann called these hidden variables "dispersion-free ensembles", meaning that the values of every physical quantity would be sharply defined (without dispersion).

for arbitrary values of A and B, even if A, B, and $A + B$ cannot be simultaneously measured.[41]

Here is a simple proof[42] that assumption (7.4.1) leads to a contradiction. Take $A = \sigma_x$, $B = \sigma_y$, two of the Pauli matrices defined in Appendix 2.F and corresponding by convention to a measurement of the spin along the x and y axes, respectively. Then $(\sigma_x + \sigma_y)/\sqrt{2}$ corresponds to measuring the spin at an angle of $45°$ between the x and y axes. All those matrices have eigenvalues equal to ± 1. Thus $v(A) = v(B) = \pm 1$, but also $v\big((\sigma_x + \sigma_y)/\sqrt{2}\big) = \pm 1$. So we have[43] $v(A + B) = \pm\sqrt{2}$, but $v(A) + v(B) = \pm 2$ or 0. Thus Eq. (7.4.1) cannot hold for all values of A and B. End of proof!

The reason why von Neumann postulated (7.4.1) is probably because it holds when we average over the hidden variables: assuming hidden variables, for a given quantum state, means that the values of those variables vary between different experiments and that the quantum state determines the probability distribution of those variables. If we average over those variables and if the average agrees with the quantum mechanical prediction, we have

$$\mathbb{E}\big(v(A)\big) + \mathbb{E}\big(v(B)\big) = \mathbb{E}\big(v(A) + v(B)\big) ,$$

where $\mathbb{E}$ denotes the average.[44] But saying that if an identity holds on average, then we may assume that it holds for every value over which the average is taken, is like saying that, if a function f satisfies

$$\int f(x)dx = 0 , \tag{7.4.2}$$

then we may assume that $f(x) = 0$ for all x. Hardly a natural assumption for a mathematician to make! Moreover, Bell constructed a simple example of hidden "spin variables" that reproduces the quantum mechanical results *for a single spin*[45] [36].

But we know from the theorem in Sect. 2.5 that there is a valid no hidden variables theorem and that one cannot attribute a value $v(A)$ to every physical quantity and every individual system prior to measurement. That theorem assumed only relations between values $v(A)$ for different quantities A that were true for quantities that can be measured simultaneously [unlike the assumption (7.4.1)]. Hence, that theorem was

---

[41]Meaning that the corresponding matrices or operators do not commute.

[42]Simpler than the original one in [496, pp. 305–324]. See [36] or [49, p. 164] for the original version of this proof.

[43]Strictly speaking, one also has to assume that the map v is continuous, because (7.4.1) only implies $v\big(a(\sigma_x + \sigma_y)\big) = a\big[v(\sigma_x) + v(\sigma_y)\big]$ for a rational, i.e., for rational approximations of $1/\sqrt{2}$.

[44]This is because the average value of measurements of any physical quantity represented by a matrix or operator A when the quantum state is ψ, is given by $\langle\psi|A|\psi\rangle$, and that quantity satisfies $\langle\psi|A + B|\psi\rangle = \langle\psi|A|\psi\rangle + \langle\psi|B|\psi\rangle$.

[45]This does not contradict the theorem of Chap. 2, because the proof of the latter used *two* spin matrices.

a logical consequence of empirical assumptions only. So in a sense von Neumann's direct conclusion about the impossibility of introducing a value $v(A)$ for every physical quantity and every individual system prior to measurement was correct, although his reasoning was based on arbitrary assumptions.

But we also saw in Chap. 5 how the de Broglie–Bohm theory manages to introduce certain "hidden variables" (the particle positions) and not those that are impossible to introduce because of the no hidden variables theorem. Yet that theory reproduces the usual quantum mechanical predictions. So von Neumann's more general conclusion, about the impossibility of hidden variables theories, was not justified.

However, von Neumann did not draw a modest conclusion from his theorem:

It is therefore not, as is often assumed, a question of a reinterpretation of quantum mechanics — the present system of quantum mechanics would have to be objectively false, in order that another description of the elementary processes than the statistical one be possible.

John von Neumann [496, p. 325]

Max Born was equally categorical (in lectures aimed at a rather general audience):

He [von Neumann] puts the theory on an axiomatic basis by deriving it from a few postulates of a very plausible and general character, about the properties of 'expectation values' indexExpectation value (averages) and their representation by mathematical symbols. The result is that the formalism of quantum mechanics is uniquely determined by these axioms; in particular, no concealed parameters can be introduced with the help of which the indeterministic description could be transformed into a deterministic one. Hence if a future theory should be deterministic, it cannot be a modification of the present one but must be essentially different. How this could be possible without sacrificing a whole treasure of well established results I leave to the determinists to worry about.

Max Born [76, p. 109]

Born, in his reply to Einstein discussed in Sect. 7.2, also referred to von Neumann's theorem: he stressed that the hopes of Einstein for a more complete theory "have not been realized", and that:

[…] physicists have good reasons for believing this to be impossible, based mainly on studies carried out by J. von Neumann.

Max Born [79, p. 178]

Pauli also referred to:

von Neumann's well known proof that the consequences of quantum mechanics cannot be amended by additional statements on the distribution of values of observables, based on the fixing of values of some hidden parameters, without changing some consequences of the present quantum mechanics.

Wolfgang Pauli [376, p. 309], quoted in [102, p. 132]

On the other hand, the flaw in von Neumann's assumption was pointed out as early as 1935 by the German philosopher Grete Hermann, but she published her result in an obscure philosophy journal [263], and it was generally ignored.

According to Abner Shimony, Einstein discussed the von Neumann theorem around 1938 with his assistant Bargmann (who told the story to Shimony), pointing to the assumption (7.4.1) and saying that "there is no reason why this premise should hold in a state not acknowledged by quantum mechanics", if the quantities A and B are not simultaneously measurable [448, p. 89].[46]

In his 1952 papers, Bohm also explained why von Neumann's result depends on assumptions that do not apply to his theory[47]:

> However, in our suggested new interpretation of the theory, the so-called "observables" are [...] not properties belonging to the observed system alone, but instead potentialities whose precise development depends just as much on the observing apparatus as on the observed system.
>
> David Bohm [62, p. 187]

De Broglie also pointed out, in 1956, that von Neumann had made assumptions that were not fulfilled in the de Broglie–Bohm theory [117, English version, p. 70].

However, as the physics Nobel prizewinner and philosopher of science P.W. Bridgman observed in 1960:

> Now the mere mention of concealed variables is sufficient to automatically elicit from the elect the remark that John von Neumann gave absolute proof that this way out is not possible. To me it is a curious spectacle to see the unanimity with which the members of a certain circle accept the rigor of von Neumann's proof.
>
> Bridgman [85, p. 204, Note 20], quoted in [398, p. 188]

The best known discussion and refutation of von Neumann's argument is in Bell's first paper on foundations of quantum mechanics [36],[48] where an explicit counterexample to the applicability of the theorem in physics is given (different from the de Broglie–Bohm theory, which is of course another counterexample).

In his 1988 interview with the magazine *Omni*, Bell was not particularly generous with von Neumann:

> Then in 1932 [mathematician] John von Neumann gave a "rigorous" mathematical proof stating that you couldn't find a non-statistical theory that would give the same predictions as quantum mechanics. That von Neumann proof in itself is one that must someday be the subject of a Ph.D. thesis for a history student. Its reception was quite remarkable. The literature is full of respectful references to "the brilliant proof of von Neumann"; but I do not believe it could have been read at that time by more than two or three people.

---

[46]Einstein and von Neumann were both at the Institute for Advanced Study in Princeton at that time, but we have no trace of a discussion between them on that issue.

[47]As we saw in Sect. 5.3.4, Bohm and Hiley also explained in a concrete example why von Neumann's theorem does not refute their approach [70, p. 121].

[48]This paper was written before [35] but published later, owing to an editorial mistake.

Omni: Why is that?

Bell: The physicists didn't want to be bothered with the idea that maybe quantum theory is only provisional. A horn of plenty had been spilled before them, and every physicist could find something to apply quantum mechanics to. They were pleased to think that this great mathematician had shown it was so. Yet the von Neumann proof, if you actually come to grips with it, falls apart in your hands! There is *nothing* to it. It's not just flawed, it's *silly*. If you look at the assumptions made, it does not hold up for a moment. It's the work of a mathematician, and he makes assumptions that have a mathematical symmetry to them. When you translate them into terms of physical disposition, they're nonsense. You may quote me on that: The proof of von Neumann is not merely false but *foolish*.

John Bell [45]

Every theorem depends on hypotheses and, when it comes to physics, the reasonableness of the latter matters greatly for the significance of the result. The sorry history of von Neumann's theorem[49] illustrates a general point made later by a self-critical pure mathematician:

The intellectual attractiveness of a mathematical argument, as well as the considerable mental labor involved in following it, makes mathematics a powerful tool of intellectual prestidigitation — a glittering deception in which some are entrapped, and some, alas, entrappers.

Jacob Schwartz [446]

As a final note on this theorem, one should mention that according to Wigner [516, p. 1009, Note 1] and also [517, p. 291], writing "as an old friend of von Neumann and in order to preserve historical accuracy", his theorem was not "primarily responsible for von Neumann's conviction of the inadequacy of hidden variables theories". The real reason for this conviction, according to Wigner, was that, if hidden variables existed, one could narrow down the range of values that they could take by making successive spin measurements along different directions: say first along the x axis, then the y axis, then again the x axis, etc. Eventually, this would lead to states that have a definite spin value in different directions, contrary to what quantum mechanics predicts.

To that, according to Wigner, Schrödinger responded that there might be "hidden variables" also in the apparatus used for the measurement that would somehow restore the randomness of the results, a response that von Neumann did not agree with [516, p. 1009, Note 1]. Schrödinger was closer to the truth, but in any case, in the de Broglie–Bohm theory, the results of these successive spin measurements will always be one-half up, one-half down, with equal probabilities, no matter how many measurements are made.[50]

---

[49]The sociologist of science Trevor Pinch has devoted an article to that history, in relation to Bohm's work, whose title is at least quite to the point [398]: What does a proof do if it does not prove?.

[50]We discussed this idea from the point of view of the de Broglie–Bohm theory at the end of Sect. 5.1.8. See also [152, p. 165]. Actually, Clauser [96] has shown that one can even use the toy model of hidden variables for a single spin devised by Bell in [36] (i.e., without making any use of the de Broglie–Bohm theory) to show that von Neumann's idea is not valid.

7.5   Misunderstandings of Bell

Taken by itself and forgetting about the EPR argument, Bell's result, discussed in
Chap. 4, can be stated as a "no hidden variables theorem": Bell showed that the
mere supposition that the measured values of the spin pre-exist their "measurement",
combined with the perfect anti-correlation when the axes along which measurements
are made are the same and the 1/4 result for measurements along different axes, lead
to a contradiction. Since the perfect anti-correlation and the 1/4 result are quantum
mechanical predictions, this means that these hidden variables or pre-existing values
cannot exist.

But Bell always presented his result *in combination with* the EPR argument, which
shows that the mere assumption of locality, combined with the perfect correlation
when the directions of measurement (or questions) are the same, implies the existence
of the supposedly "impossible" hidden variables. So for Bell, his result, combined
with the EPR result was not a "no hidden variables theorem", but a nonlocality
theorem, the result on the impossibility of hidden variables being only one step in a
two-step argument.

Viewing Bell's argument as a refutation of hidden variables theories (an almost
universal reaction, as we will see) is doubly mistaken: first because, combined with
EPR, Bell proves nonlocality; and second because the de Broglie–Bohm theory,
which Bell explained and defended all his life, proves that a hidden variables theory
is actually possible. Bell was perfectly clear about this:

> Let me summarize once again the logic that leads to the impasse. The EPRB correlations[51]
> are such that the result of the experiment on one side immediately foretells that on the other,
> whenever the analyzers happen to be parallel. If we do not accept the intervention on one
> side as a causal influence on the other, we seem obliged to admit that the results on both sides
> are determined in advance anyway, independently of the intervention on the other side, by
> signals from the source and by the local magnet setting. But this has implications for non-
> parallel settings which conflict with those of quantum mechanics. So we *cannot* dismiss
> intervention on one side as a causal influence on the other.

> John Bell [49, pp. 149–150] (italics in the original)

He was also conscious of the misunderstandings of his results:

> It is important to note that to the limited degree to which *determinism*[52] plays a role in the
> EPR argument, it is not assumed but *inferred*. What is held sacred is the principle of "local
> causality"— or "no action at a distance". [...]

> It is remarkably difficult to get this point across, that determinism is not a *presupposition* of
> the analysis.

> John Bell [49, p. 143] (italics in the original)

---

[51]Here EPRB means EPR and Bohm, who reformulated the EPR argument in term of spins [61].
(Note by J.B.).

[52]Here, "determinism" refers to the idea of pre-existing values. (Note by J.B.).

And he added, unfortunately only in a footnote:

> My own first paper on this subject (*Physics* **1**, 195 (1965))[53] starts with a summary of the EPR argument *from locality to* deterministic hidden variables. But the commentators have almost universally reported that it begins with deterministic hidden variables.
>
> John Bell [49, p. 157, footnote 10] (italics in the original)

An example of such a commentator is the famous physicist Murray Gell-Mann, who wrote:

> Some theoretical work of John Bell revealed that the EPRB experimental setup could be used to distinguish quantum mechanics from hypothetical hidden variable theories [...] After the publication of Bell's work, various teams of experimental physicists carried out the EPRB experiment. The result was eagerly awaited, although virtually all physicists were betting on the correctness of quantum mechanics, which was, in fact, vindicated by the outcome.
>
> Murray Gell-Mann [205, p. 172]

So Gell-Mann opposes hidden variable theories to quantum mechanics, but the only hidden variables that Bell considered were precisely those that were needed, because of the EPR argument, in order to "save" locality. So if there is a contradiction between the existence of those hidden variables and experiments, it is not just quantum mechanics that is vindicated, but locality that is refuted.

In one of his most famous papers, "Bertlmann's socks and the nature of reality" [40], Bell tried to give a pedagogical argument to explain his main idea. He gave the example of a person (Reinhold Bertlmann) who always wears socks of different colors.[54] If we see that one sock is pink, we know automatically that the other sock is not pink (let's say it is green). That would be true even if the socks were arbitrarily far away. So by looking at one sock, we learn something about the other sock and there is nothing surprising about that, because socks *do have* a color whether we look at them or not.[55] But what would we say if we were told that the socks have no color before we look at them? That would be surprising, of course, but the idea that quantum mechanics is complete means exactly that (if we replace the color of the socks by the values of the spin before measurement). But then, looking at one sock would "create" not only the color of that sock but also the colour of the other sock. And that implies the existence of actions at a distance, if the socks are far apart. Bell emphasized that in quantum mechanics, we are in a situation similar to this nonlocal creation of colors for the socks.

However, Murray Gell-Mann made the following comment on the Bertlmann's socks paper:

---

[53]Reference [35], reprinted as Chap. 2 in [49]. (Note by J.B.).

[54]See [49, p. 139] for a picture of Mr Bertlmann wearing one sock indicated "pink" and the other "not pink".

[55]In case some people might worry that the notion of "color" refers to our sensations, let us be more precise and say that the socks have the physico-chemical properties that will reflect a light of a certain wavelength (and the fact that they have those properties is independent of our perception), which will ultimately produce our perception of color.

The situation is like that of Bertlmann's socks, described by John Bell in one of his papers. Bertlmann is a mathematician who always wears one pink and one green sock. If you see just one of his feet and spot a green sock, you know immediately that his other foot sports a pink sock. Yet no signal is propagated from one foot to the other. Likewise no signal passes from one photon to the other in the experiment that confirms quantum mechanics. No action at a distance takes place. (The label "nonlocal" applied by some physicists to quantum-mechanical phenomena like the EPRB effect is thus an abuse of language. What they mean is that *if interpreted classically in terms of hidden variables*, the result would indicate nonlocality, but of course such a classical interpretation is wrong.)

Murray Gell-Mann [205, pp. 172–173]

This is not correct: it is true that, because of the random nature of the results, the experimental setup of EPRB cannot be used to send messages (or signals), as we saw in Sect. 4.3. But nevertheless, some action at a distance does take place.[56] And the remark about "classical interpretation" completely misses EPR's argument showing that hidden variables are necessary if locality holds.

Of course, Gell-Mann's goal in the passage quoted here is probably to dismiss pseudo-scientific exploitations of Bell's result (such as invoking it to justify the existence of telepathy), but his defense of science is misdirected: the behavior of quantum particles is *not* like that of Bertlmann's socks (which is indeed totally unsurprising), and that was the whole point of Bell's paper.

Eugene Wigner also saw Bell's result solely as a no hidden variables result:

The proof he [von Neumann] published [...] though it was made much more convincing later on by Kochen and Specker,[57] still uses assumptions which, in my opinion, can quite reasonably be questioned [...] In my opinion, the most convincing argument against the theory of hidden variables was presented by J.S. Bell.

Eugene Wigner [517, p. 291]

This is misleading, because Wigner considers only Bell's argument, which indeed shows that pre-existing spin values (or "hidden variables") cannot exist, but forgets the EPR part of the argument, which was Bell's starting point.[58]

To give yet another example of misunderstanding of Bell, coming from a well-known defender of the Copenhagen interpretation, consider what Rudolf Peierls declared in an interview, referring to Aspect's experiments [21]:

---

[56] As we saw in Sect. 6.3, Robert Griffiths, who defends a decoherent histories approach to quantum mechanics, has a similar misunderstanding of the EPR–Bell situation, with the color of the socks being replaced by colors of slips of paper. Likewise, in the "science festival" debate in New York in June 2014, *Measure for Measure: Quantum Physics and Reality* (www.worldsciencefestival.com/2014/06/measure-measure-can-reconcile-waves-particles-quantum-mechanics/), Rüdiger Schack, who defends the QBist approach, "explains" the perfect correlations in the EPR–Bell experiments, when the "questions" are the same on both sides, by appealing to the colors of Bertlmann's socks, without realizing that the whole point of Bell's article was to show that the quantum correlations are *not* like the correlations between the socks' colors and cannot be explained by pre-existing properties.

[57] Who proved another "no hidden variables" result, close to the theorem in Chap. 2 [291]; see [335] for a discussion. (Note by J.B.).

[58] See Goldstein [221] for a further discussion of Wigner's views.

If people are obstinate in opposing the accepted view they can think of many new possibilities, but there is no sensible view of hidden variables which doesn't conflict with these experimental results. That was proved by John Bell, who has great merit in establishing this. Prior to that there was a proof due to the mathematician von Neumann, but he made an assumption which is not really necessary.

Rudolf Peierls in [107, p. 77]

Basically the same mistakes were made in 1999 by one of the most famous physicists of our time, Stephen Hawking:

Einstein's view was what would now be called a hidden variables theory. Hidden variables theories might seem to be the most obvious way to incorporate the Uncertainty Principle into physics.[59] They form the basis of the mental picture of the universe, held by many scientists, and almost all philosophers of science. But these hidden variable theories are wrong. The British physicist, John Bell, who died recently, devised an experimental test that would distinguish hidden variable theories. When the experiment was carried out carefully, the results were inconsistent with hidden variables.

Stephen Hawking [254]

The misconceptions about Bell continue to the present day: in an article denouncing as a "scandal" the fact that in a Symposium held in 2000 in Vienna, in honor of the 10th anniversary of John Bell's death, not a single exponent of the de Broglie–Bohm theory was invited to speak (despite Bell's constant defense of that theory), the physicist Gerhard Grössing wrote:

In the 11 August 2000 issue of "Science", D. Kleppner and R. Jackiw of MIT published a review article on "One Hundred Years of Quantum Physics" [290]. They discuss Bell's inequalities and also briefly mention the possibility of "hidden variables". However, according to the authors, the corresponding experiments had shown the following: "Their collective data came down decisively against the possibility of hidden variables. For most scientists this resolved any doubt about the validity of quantum mechanics."

Gerhard Grössing [250]

In the February 2001 issue of *Scientific American*, Max Tegmark and John Archibald Wheeler published an article entitled *100 Years of the Quantum* (quoted by Grössing in [250]), where one reads:

Could the apparent quantum randomness be replaced by some kind of unknown quantity carried out inside particles, so-called 'hidden variables'? CERN theorist John Bell showed that in this case, quantities that could be measured in certain difficult experiments would inevitably disagree with standard quantum predictions. After many years, technology allowed researchers to conduct these experiments and eliminate hidden variables as a possibility.

Max Tegmark and John Archibald Wheeler [470, p. 76]

---

[59]By this, Hawking, means that hidden variables would be a way to maintain determinism, despite the fact that precise measurements of both position and velocities are prohibited by the uncertainty principle. (Note by J.B.).

These two last quotes clearly commit a double mistake: they ignore the fact that Bell's result concerns nonlocality and the fact that he does not at all refute the de Broglie–Bohm hidden variables theory, but rather supports it.

In the preface to the proceedings of the Symposium held in honor of the 10th anniversary of John Bell's death, Reinhold Bertlmann (mentioned in the "socks" paper [40]) and the distinguished Austrian experimentalist Anton Zeilinger expressed succinctly what is probably the view of the majority of physicists about Bell's result:

> [...] while John Bell had flung open the door widely for hidden variables theories [by showing the flaws in von Neumann's proof], he immediately dealt them a major blow. In 1964, in his celebrated paper "On the Einstein–Podolsky–Rosen Paradox", he showed that any hidden variables theory, which obeys Einstein's requirement of locality, i.e., no influence travelling faster than the speed of light, would automatically be in conflict with quantum mechanics. [...] While a very tiny [experimental] loophole in principle remains for local realism, it is a very safe position to assume that quantum mechanics has definitely been shown to be the right theory. Thus, a very deep philosophical question, namely, whether or not events observed in the quantum world can be described by an underlying deterministic theory, has been answered by experiment, thanks to the momentous achievement of John Bell.
>
> Reinhold Bertlmann and Anton Zeilinger, preface to [57]

This is all the more remarkable in that these authors do mention Bohm's theory, just before this statement in which the mistakes made in the two previous quotes are essentially repeated.

In *Nature*, in 2014, Howard Wiseman wrote:

> As Bell proved in 1964, this leaves two options for the nature of reality. The first is that reality is irreducibly random, meaning that there are no hidden variables that "determine the results of individual measurements". The second option is that reality is 'nonlocal', meaning that "the setting of one measuring device can influence the reading of another instrument, however remote".
>
> Howard Wiseman [523]

Here, Wiseman "forgets" the EPR part of the argument, that locality implies the existence of those hidden variables that Bell's argument rules out. So there are not two options, but only one: "the setting of one measuring device can influence the reading of another instrument, however remote", although it is also true that there are no hidden variables that "determine the results of individual measurements", because in the de Broglie–Bohm theory the results of measurements are determined not only by the hidden variables but also by the concrete experimental setup, as we saw in Sect. 5.1.4.

David Mermin summarized the situation described in this section in an amusing way:

> Contemporary physicists come in two varieties.
>
> Type 1 physicists are bothered by EPR and Bell's theorem.
>
> Type 2 (the majority) are not, but one has to distinguish two subvarieties.

Type 2a physicists explain why they are not bothered. Their explanations tend either to miss the point entirely (like Born's to Einstein)[60] or to contain physical assertions that can be shown to be false.

Type 2b are not bothered and refuse to explain why. Their position is unassailable. (There is a variant of type 2b who say that Bohr straightened out the whole business, but refuse to explain how.)

David Mermin [332]

However, the same David Mermin also wrote:

Bell's theorem establishes that the value assigned to an observable must depend on the complete experimental arrangement under which it is measured, even when two arrangements differ only far from the region in which the value is ascertained — a fact that Bohm theory[61] exemplifies, and that is now understood to be an unavoidable feature of any hidden-variables theory.

To those for whom nonlocality is anathema, Bell's Theorem finally spells the death of the hidden-variables program.

David Mermin [335, p. 814]

But Bell's theorem, combined with EPR, shows that nonlocality, whether we consider it anathema or not, is an unavoidable feature of the world.

Mermin adds in a footnote: "Many people contend that Bell's theorem demonstrates nonlocality independently of a hidden-variables program, but there is no general agreement about this." It is true that there is no agreement (to put it mildly), but Mermin's own remark on the two varieties of physicists seems to explain why this is so.[62]

To conclude, let us highlight some more positive reactions to Bell's result. As we mentioned at the end of Chap. 4, Henry Stapp, a particle theorist at Berkeley, wrote that "Bell's theorem is the most profound discovery of science" [458, p. 271]. David Mermin mentions an unnamed "distinguished Princeton physicist" who told him [332]: "Anybody who's not bothered by Bell's theorem has to have rocks in his head." And Feynman derived in 1982 an inequality similar to Bell's (no reference to Bell's work is made in his paper), from which he drew the following conclusion:

That's all. That's the difficulty. That's why quantum mechanics can't seem to be imitable by a local classical computer.[63] I've entertained myself always by squeezing the difficulty of quantum mechanics into a smaller and smaller place, so as to get more and more worried about this particular item. It seems to be almost ridiculous that you can squeeze it to a numerical question that one thing is bigger than another.

Richard Feynman [184, p. 485]

---

[60]See Sect. 7.2. (Note by J.B.).

[61]Which is what we call the de Broglie–Bohm theory. (Note by J.B.).

[62]What is even more surprising is that this comment comes at the end of a remarkably clear paper on the no hidden variables theorems [335].

[63]Which means what we call locality here. (Note by J.B.).

7.6   The Non-reception of de Broglie's and Bohm's Ideas

7.6.1   Reactions to de Broglie

Einstein, following Planck, introduced "quanta of light" (photons) and considered them as particles, thus seeing that "waves are particles" [160]. De Broglie, and then Schrödinger, associated waves with particles. So it may seem a bit odd that the people who supposedly never "understood" or "accepted" the wave–particle duality, according to the mainstream view of the history of quantum mechanics, actually discovered it.

In his 1924–1927 notes to the French *Académie des Sciences* [109–111], and in his PhD thesis [108], which pre-date all other major works on quantum physics, de Broglie not only associated waves with particles, but also introduced the idea that the motion of particles may be guided by waves.[64] The French physicist Paul Langevin sent a copy of de Broglie's thesis to Einstein, who replied that de Broglie had "lifted a corner of the great veil" [162] and Einstein circulated the thesis in Germany.

What were de Broglie's ideas? Throughout his life, both before 1927 and after 1952, de Broglie tried to develop what he called the theory of "the double solution", according to which particles would be associated with waves in two different ways: on the one hand, there would be what is now called the wave function, solution of the Schrödinger equation (which de Broglie did not write, but Schrödinger said that his ideas were influenced by de Broglie's early work [24, p. 124]); and, on top of that solution, there would also exist a "singular solution" that would be associated with each particle. In formulas, if $\psi(x, y, z, t) = R(x, y, z, t)e^{iS(x,y,z,t)}$ is the solution of the Schrödinger equation (for one particle), then de Broglie was looking for another solution, of the form $u(x, y, z, t) = a(x, y, z, t)e^{iS(x,y,z,t)}$, with the same phase $S(x, y, z, t)$ as in the solution of the Schrödinger equation, but a different amplitude $a(x, y, z, t)$, the latter being singular at a "moving singularity". This meant that it would be (in one dimension, to simplify notation) a function[65] like $1/|x - x_0(t)|$, which is smooth everywhere, but equal to infinity for $x = x_0(t)$; besides, $x_0(t)$ would satisfy a deterministic equation, which we will discuss below.

The idea would then be that the particle would be located at the singularity of the function $u(x, y, z, t)$. The singularity would move deterministically and the function u, unlike the solution of the Schrödinger equation, would not spread itself more and

---

[64]This idea had also been proposed by Einstein as early as 1909, when he wrote [161]: "Anyway, this conception seems to me the most natural: that the manifestation of light's electromagnetic waves is constrained at singularity points, like the manifestation of electrostatic fields in the theory of the electron. It cannot be ruled out that, in such a theory, the entire energy of the electromagnetic field could be viewed as localized in these singularities, just like the old theory of action-at-a-distance. I imagine to myself, each such singular point surrounded by a field that has essentially the same character as a plane wave, and whose amplitude decreases with the distance between the singular points."

[65]This is just an example of what such a function could look like; the actual functions in de Broglie's approach were more complicated. See, e.g., [117] for a discussion of de Broglie's ideas on the double solution.

more as time went by.[66] De Broglie was convinced that only such a "second" solution (the first one being the usual solution of Schrödinger's equation) could give a physical meaning to the theory. The motivation was that the solution of Schrödinger's equation was defined on the space of all the particle configurations of whatever system was being considered. This did not sound very physical, and it is of course the source of nonlocality. For many particles, de Broglie wanted to have a singular solution for each particle.

De Broglie's problem was that his program could never be carried out mathematically, by him or by his collaborators. Originally, de Broglie wanted the function $u(x, y, z, t)$ to satisfy a linear equation.[67] Later, together with his collaborator, the French physicist Jean-Pierre Vigier, he tried to consider nonlinear equations, which he thought were related to similar notions in general relativity. What de Broglie did notice, however, was that the moving singularity should propagate with a velocity proportional to the derivative of the phase $S(x, y, z, t)$, according to Eq. (5.1.1.3). Remember that this phase is common to both the solution of the Schrödinger equation and his putative singular solution. So even if one forgets the singular solution, one nevertheless obtains an equation for the motion of the moving singularity.

Near the end of his 1927 paper [112], he wrote that, if "one does not want to invoke the principle of the double solution", one can "adopt the following point of view":

> [...] one will assume the existence, as distinct realities, of the material point and of the continuous wave represented by the function, and one will take it as a postulate that the motion of the point is determined as a function of the phase of the wave [...]. One then conceives the continuous wave as guiding the motion of the particle. It is a pilot wave.
>
> Louis de Broglie [112], quoted in [24, p. 65] and reprinted in [116, p. 52]

In August 1927, Pauli reacted relatively favorably to that paper in a letter to Bohr:

> [...] even if this paper by de Broglie is off the mark (and I hope that actually), still it is very rich in ideas and very sharp, and on a much higher level than the childish papers by Schrödinger, who even today still thinks he may [...] abolish material points.[68]
>
> Wolfgang Pauli [374], quoted in [24, p. 55]

De Broglie was invited to present a report at the 5th Solvay conference in Brussels, in October 1927. There, he decided not to discuss the "double solution", because he knew that the mathematical problems were too hard, and he presented his "pilot wave" idea, but now for a non-relativistic system of N particles guided by a wave function in configuration space that determines the particle velocities according to the law of motion (5.1.1.3).

---

[66]See (2.A.2.24) in Appendix 2.A.2 for an example of such spreading.

[67]For the definition of a "linear" differential equation, see Appendix 2.A.

[68]This refers to the fact that Schrödinger hoped to have a theory where the wave function would produce a density of charge, similar to (6.1.2.1), but with the charge instead of the mass. See [14] for a modern discussion. (Note by J.B.).

Moreover, de Broglie did derive the Born rule by proving the equivariance formula (see Appendix 5.C), which implies the Born rule for probabilities, as we saw in Chap. 5. He also introduced the quantum potential, defined in (5.1.1.6).

As de Broglie said in his report to the Solvay Conference [24, p. 346], in order to justify the introduction of real particles, "it seems a little paradoxical to construct a configuration space with the coordinates of points that do not exist". But he also remarks that, if "the propagation of a wave in space has a clear physical meaning, it is not the same as the propagation of a wave in the abstract configuration space". That is what motivated de Broglie to consider "the corpuscles as 'exterior' to the wave Ψ, their motion being determined only by the propagation of the wave" [24, p. 355]. As we saw in Sect. 7.1.2, Einstein thought that [24, p. 441] "Mr de Broglie is right to search in this direction", i.e., to try to localise the particle during its propagation.

Another participant to the 1927 Solvay Conference, the French physicist Léon Brillouin, was even more positive during the discussion following de Broglie's talk:

Mr. Born can doubt the real existence of the trajectories calculated by Mr. de Broglie, and assert that one will never be able to observe them, but he cannot prove to us that these trajectories do not exist. There is no contradiction between the point of view of Mr. de Broglie and that of the other authors [...]

Léon Brillouin [24, p. 365]

But the general reaction was negative. In particular, Pauli raised an objection having to do with the account, within the de Broglie theory, of certain scattering phenomena. There is some discussion in the literature as to whether de Broglie answered Pauli's objection more or less correctly (see [24, Chap. 10] and [272] for opposing viewpoints), but de Broglie referred to that objection much later,[69] which suggests that he was not sure of the validity of his own answer. A complete answer to Pauli's objection was given in 1952 by Bohm, who should be credited for explaining how the de Broglie–Bohm theory accounts for all the quantum mechanical predictions.

In any case, de Broglie left the Solvay Conference discouraged. Not only because of Pauli's criticisms, or because of the generally negative reactions, but also because he could not take the wave function *defined on configuration space* seriously as a physical quantity, which is essential to understanding the de Broglie–Bohm theory. Already in his 1927 paper [112], written before the Solvay Conference, de Broglie had written:

Physically, there can be no question of a propagation in a configuration space whose existence is purely abstract: the wave picture of our system must include N waves propagating in real space and not a single wave propagating in the configuration space.

Louis de Broglie [116, p. 48]

---

[69]In [114], reprinted in [116, p. 66], he refers to "the defenders of the present interpretation, notably M. Pauli", by saying: "I was very much impressed by their arguments".

For de Broglie, the wave function had a purely probabilistic meaning, hence a subjective and non-physical meaning, and only the existence of another solution, or rather N solutions for N particles, more or less localized in space, would have given a meaning to the whole scheme. Since he could not solve the mathematical problems linked with this other solution, he gave up his alternative and followed the Copenhagen orthodoxy for the next twenty five years. That is, until he received a manuscript from a "young American physicist", David Bohm. He reacted by, of course, mentioning his priority, but also by restating his old objections towards his own theory [116, p. 67].

De Broglie stressed the obvious nonlocality of his and Bohm's theory in the EPR situation, about which he wrote:

> If the wave Ψ is a "physical reality", how can one understand that a measure done in one region of space could affect this physical reality in other regions of space that can be very far from the first one?
>
> Louis de Broglie [116, pp. 67–68]

One can of course criticize de Broglie for giving in to the orthodoxy after 1927, but to understand why he did so, one must first understand his attachment to the idea of the double solution (to which he came back after learning about Bohm's work in 1952) and therefore his lack of appreciation for his own theory, at least the one presented at the Solvay Conference.[70]

It would be a mistake to say, as Bohm and Hiley did in 1993, many years after the original Bohm paper:

> The idea of a 'pilot wave' that guides the movement of the electron was first suggested by de Broglie in 1927, but only in connection with the one-body system.
>
> David Bohm and Basil Hiley [70, p. 38]

Indeed, de Broglie did write the "guiding" equation of motion for N particles, but it is true that he never took that theory seriously for more than one particle because of its "abstract" and nonlocal nature.

What is more puzzling, however, is why people like Schrödinger and Einstein, who did not agree with the orthodoxy, either in 1927 or later on, did not pursue de Broglie's ideas. This is probably due once again to the seemingly unphysical nature of the quantum state defined on configuration space. Besides, Einstein had developed his own "hidden variables" theory, a sort of guiding wave similar to de Broglie, but with different equations, and he had realized that it did not work, in particular because it was nonlocal.[71]

---

[70]In [113], de Broglie gives other, more technical, reasons that led him to abandon his own theory. See [24, pp. 229–230] and [473, lecture 7] for a discussion.

[71]See [271] for a discussion of that theory, which Einstein did not publish.

Bell did put his finger on the root of the problem, but much later of course:

No one can understand this theory [the de Broglie–Bohm theory] until he is willing to think of
ψ as a real objective field rather than a 'probability amplitude'. Even though it propagates
not in 3-space but in 3N -space.

John Bell [49, p. 128] (italics in the original)

And this is something that nobody before Bohm was willing to think; certainly not de
Broglie, as we have seen, nor Einstein and Schrödinger, who did not like "telepathy"
or "magic".

De Broglie is a tragic figure in the history of physics. He was undoubtedly a
precursor and he did influence his successors in an indirect way. A case in point is
Schrödinger, who wrote [439, p. 46]: "My method was inspired by L. de Broglie and
by brief, yet infinitely far-seeing remarks of Einstein." However, his 1927 theory is to
a large extent forgotten, and he has been rather neglected in his own country, despite
his Nobel Prize and his having been the "perpetual secretary" for the mathematical
and physical sciences section of the French *Académie des Sciences* from 1942 until
1975.

It should be emphasized that de Broglie's theory had nothing to do with "solving the measurement problem", unlike almost all the subsequent discussions about
quantum mechanics. It was meant to be a physical theory, albeit a provisional one,
waiting for a fully fledged "double solution" that never came. Of course, de Broglie
did not fully understand his own theory, and neither did he really believe in it. But
who can blame him? Even the standard quantum measurement algorithm was not
fully developed then[72] and nobody could accept the most revolutionary feature of the
de Broglie theory, namely its nonlocality. De Broglie wrote 25 years before Bohm
and 37 years before Bell, and even Einstein could not propose in 1927 a hidden
variables theory on the same level of consistency as de Broglie's.

It is not clear what the "double solution" would have added if it could have
been developed: the motion would still have been determined by the usual guiding
equation (5.1.1.3) and would therefore have been just as nonlocal as what we call
the de Broglie–Bohm theory. The other solution, the singular one, would add a
psychological comfort for those who like classical pictures and it would have given
an illusion of locality, but it would not change anything regarding the statistical
predictions of the theory or the individual behavior of the particles.[73] It would have
been the theory's ether and could have been forgotten with no loss. It is sad that de
Broglie spent his entire life looking for this particular ghost.

---

[72]It is interesting to note that the report given at the Solvay Conference, which is supposed to
represent the "orthodox" view, namely that of Born and Heisenberg, is far from coinciding, when it
comes to the description of measurements, in particular the collapse rule, with what is now found
in textbooks. See [24, Chap. 6] and [23].

[73]One may remark that Newton and Maxwell also sought more "intuitive" or "representable"
versions of their own theories, but what remains of their work are basically the equations of motion
of particles and fields.

The tragedy of de Broglie is that he did "lift a corner of the great veil" and saw far beyond anybody else at the time; so far, indeed, that he could not clearly make things out himself.

7.6.2 Reactions to Bohm

In 1952, David Bohm published two papers in *Physical Review* which give in complete detail what we call the de Broglie–Bohm theory and others call Bohmian mechanics or Bohm's theory.[74] In Bohm's case, the little history of science got entangled with the big History of political events, in particular those of the Cold War. Bohm had been a member of the American Communist Party (CPUSA) for a few months in 1942–1943, at a time when the United States and the Soviet Union were allied in World War II. He left the party because he found the meetings boring [386, p. 58], but kept a lifelong interest in "Marxist" philosophy, and later in Hegelianism, and in particular the notion of dialectics. But Bohm had also been working for his PhD at the Radiation Laboratory of the University of Berkeley under the supervision of Robert Oppenheimer, the head of the Manhattan Project, and that, combined with his communist sympathies, led later to Bohm's troubles.

In 1946, after receiving his PhD, he was hired by Princeton University, where he taught, in particular, quantum mechanics, and did research on plasma physics. In May 1949, he was called by the House Committee on Un-American Activities (HUAC) to testify against some of his former colleagues who were suspected of pro-Soviet espionage. He refused to testify (as encouraged by Einstein, who warned him that he might be jailed), and, to justify himself, he invoked the Fifth Amendment of the United States Constitution, against self-incrimination.[75] Because of that, he was indicted in 1950 for "contempt of Congress", a charge of which he was eventually cleared in May 1951. However, Princeton University, upon hearing of the indictment, suspended Bohm from all his teaching duties and forbade him to set foot on campus (although his salary was still paid).

Even though Bohm was cleared of all charges, the university refused to renew his contract; the President of Princeton University even claimed that this was due to scientific and not political reasons, a transparent lie, and he was unable to find employment anywhere else in the United States. He left for Brazil, where the University of São Paulo offered him a job. He was then deprived of his United States passport, which he could get back only if he returned to the United States, and had to obtain Brazilian citizenship[76] in order to travel, which allowed him to move to

---

[74]For a history of the reactions to Bohm's theory, see Cushing [102, Chap. 9], and for a biography of Bohm, see the work of David Peat [386], that we follow here.

[75]That amendment says, among other things, that: "No person shall be compelled in any criminal case to be a witness against himself."

[76]In order to obtain it, Bohm had to renounce his American citizenship, which he was only able to retrieve decades later, in 1986.

Technion in Israel, in 1955. In 1957, he went to England, where he was eventually made Professor of Theoretical Physics at the University of London's Birkbeck College. He was elected Fellow of the Royal Society only in 1990, at the age of 72, and died two years later.

In 1951, Bohm published *Quantum Theory*, a book based on his courses at Princeton, which is often considered as one of the best "orthodox" textbooks on quantum mechanics [61], and where he emphasized the need to understand quantum mechanics, beyond the mathematical formalism. In this book, Bohm reformulates the EPR argument in term of spin variables, the formulation used in Chap. 4, but he sides with Bohr in using this argument against the idea of "hidden variables". However, according to Murray Gell-Mann, he discussed his book with Einstein and came back saying [205, p. 170]: "He [Einstein] talked me out of it. I'm back where I was before I wrote the book."

Bohm remarks that two Soviet physicists, Blokhintsev and Terletskii, had started a criticism of the interpretation of Bohr and Heisenberg, and viewed quantum mechanics as a statistical theory, but without proposing any theory for the behavior of individual systems.[77] Then, Bohm, "partly as a result of the stimulus of discussion with Dr. Einstein" [68, p. 110], wrote his two famous papers [62] in which he presented the complete "de Broglie–Bohm" theory.[78]

The first person to express a negative reaction to Bohm's papers, months before their publication, was, of all people, de Broglie. As we saw in the previous section, de Broglie not only established his priority, but also criticized Bohm's theory. Bohm replied through a letter sent to *Physical Review* [63]. Bohm's reaction was that de Broglie had not:

[…] really read my article, but simply reiterated Pauli's criticisms, which led him to abandon the theory, but did not point out my conclusion that these objections are not valid. He's going to look a little silly. That's what he gets from rushing into print 5 months before my article came out.

David Bohm [64]

Another somewhat surprising negative reaction to Bohm came from Einstein, who wrote to Born:

Have you noticed that Bohm believes (as de Broglie did, by the way 25 years ago) that he is able to interpret the quantum theory in deterministic terms? That way seems too cheap to me.

Albert Einstein, letter to Max Born, 12 May 1952 [79, p. 192]

---

[77] See Bohm [68, p. 110]. This influence is also mentioned by Max Jammer [281, p. 279] and by James T. Cushing [102, p. 152].

[78] The equations (5.1.1.5) of the theory had already been rediscovered by Rosen in 1945 [417], but Rosen did not develop the theory very much, and nor did he see it as a solution to the problems of quantum mechanics.

And writing about his contribution[79] to the Born *Festschrift* [172], Einstein explained to Born:

> I have written a little nursery song about physics, which has startled Bohm and de Broglie a little. It is meant to demonstrate the indispensability of your statistical interpretation of quantum mechanics, which Schrödinger too has recently tried to avoid.

Albert Einstein, letter to Max Born, 12 May 1952 [79, p. 199]

The reason for Einstein's hostility toward de Broglie and Bohm (which again shows that his main worry about quantum mechanics was not determinism) is not known, but can perhaps be understood if one remembers that, for Einstein, the main problem was nonlocality and that, of course, the de Broglie–Bohm theory did nothing to resolve that question. But Einstein was writing this 12 years before Bell's theorem.

We discussed many technical objections to the de Broglie–Bohm theory in Chap. 5. We will turn now towards the emotional reactions and the a priori objections to Bohm, which have often been quite spectacular. David Peat, who was both a friend and a biographer of Bohm, interviewed Max Dresden, a physicist who had read Bohm's papers and who visited Oppenheimer's group at the Princeton Institute for Advanced Study, where he enquired about Bohm, and even gave a seminar explaining his work. Here are the reactions of certain people at the Institute for Advanced Study in Princeton, according to Max Dresden[80]:

> "We consider it juvenile deviationism," Oppenheimer replied. No, no one had actually read the paper — "we don't waste our time." [...]

> Reactions to the theory were based less on scientific grounds than on accusations that Bohm was a fellow traveller, a Trotskyite, and a traitor.[81] It was suggested that Dresden himself was stupid to take Bohm's ideas seriously. [...] the overall reaction was that the scientific community should "pay no attention to Bohm's work". As Dresden recalled, Abraham Pais also used the term "juvenile deviationism". Another physicist said Bohm was a "public nuisance". Oppenheimer went so far as to suggest that "if we cannot disprove Bohm, then we must agree to ignore him."

David Peat [386, p. 133]

Hints of the Princeton reaction reached Bohm in exile in Brazil. He wrote to his friend Miriam Yevick [386, pp. 133–134]: "As for Pais and the rest of the 'Princetitute', what those little farts think is of no consequence to me. In the past 6 years, almost no work at all has come out of that place. [...] I am convinced that I am on the right track."[82]

---

[79]In which Einstein gave the example of the measurement of the velocity of a particle in a box, discussed in Sect. 5.1.4 and Appendix 5.D.

[80]According to Peat, Dresden also made those remarks on the floor at the American Physical Society meeting in Washington in May 1989; see [386, p. 340, Note 51].

[81]Given that the expression "fellow traveller" referred to people close to the Communist Party, Bohm could hardly be both a fellow traveller and a Trotskyite, since the latter were violently anti-communist. (Note by J.B.).

[82]Here Bohm refers of course to the Princeton Institute for Advanced Study.

Let us now see the reactions to Bohm's theory of the most famous people associated with the Copenhagen interpretation, Pauli, Heisenberg, and Bohr. In 1951, Pauli wrote to Bohm:

> I just received your long letter of 20 November, and I also have studied more thoroughly the details of your paper. I do not see any longer the possibility of any logical contradiction as long as your results agree completely with those of the usual wave mechanics and as long as no means is given to measure the values of your hidden parameters both in the measuring apparatus and in the observe [sic] system. As far as the whole matter stands now, your 'extra wave-mechanical predictions' are still a check, which cannot be cashed.

> Wolfgang Pauli [377], quoted in [386, p. 127]

But he also wrote, in 1951, to his friend Markus Fierz ([378], quoted in [386, p. 128]): "Bohm keeps writing me letters 'such as might have come from a sectarian cleric trying to convert me particularly to de Broglie's old theory of the pilot wave.' In the last analysis Bohm's whole approach is 'foolish simplicity', which 'is of course beyond all help'."

In the *Festschrift* devoted to de Broglie's sixtieth birthday (with contributions, among other people, from Einstein, Heisenberg, Pauli, Rosenfeld, and Schrödinger) [206], Pauli makes three objections to the revival of de Broglie's theory by Bohm: one is that it breaks the symmetry between position and momentum, an objection that we discussed in Sect. 5.3.3. Another objection is that, in a deterministic theory, one should be able to consider arbitrary probability distributions and not only the "quantum equilibrium" distribution. That was discussed in Sect. 5.1.7. Finally, Pauli observes that Bohm avoids being refuted by von Neumann's theorem because "measurements of observables", in his theory, do not generally measure a property of the system alone, but depend also on the way the measuring device is set up. Pauli finds that aspect of the theory "arbitrary" and even "metaphysical". But there, Bohm was in fact closer to Bohr than Pauli, since Bohr recognized the "active role" of those measuring devices.

What about Heisenberg's reaction? He wrote a rather detailed defense of the orthodoxy in *Physics and Philosophy* [259], where he discusses Bohm's ideas.[83] One objection that Heisenberg raised against Bohm was that the wave function could not be "real", because it was defined on configuration space [259, p. 116]. But of course, if even the wave function is not real, while supposedly constituting a "complete description" of physical systems, then nothing is real, and that surely cannot be what Heisenberg intended. Indeed, another objection he made is that, in certain states, as we have seen, the particle does not move (in the de Broglie–Bohm theory), yet if one measures their velocities one finds a nonzero result [259, p. 116].[84] This shows

---

[83]For earlier objections by Heisenberg to the idea of hidden variables, in particular in letters to Einstein and in an unpublished response to EPR, see Bacciagaluppi and Crull [25]. One of Heisenberg's objections was based on his belief that the introduction of hidden variables would destroy interference effects. But that would only be true if those hidden variables would in effect collapse the quantum state, whereas, in the de Broglie–Bohm theory, they do nothing of the sort and only complete the usual quantum description.

[84]This was explained within the de Broglie–Bohm theory in Sect. 5.1.4 and Appendix 5.D.

that Heisenberg was "guilty" of "naive realism with respect to operators" and did not really agree with Bohr about the active role of the "measurement" apparatus.

Heisenberg also falls back on the familiar idea that Bohm's theory is just like the Copenhagen interpretation, but expressed in a different language, which "from a strictly positivistic standpoint" amounts to an "exact repetition" and not a counter-proposal to the Copenhagen interpretation [259, p. 115]. This does not square with the fact that Heisenberg was also arguing that quantum mechanics had put an end to the "ontology of materialism", namely "to the idea of an objective real world whose smallest parts exist objectively in the same sense as stones or trees exist, independently of whether or not we observe them" [259, p. 115]. One cannot have it both ways: saying that Bohm's theory reintroduces an "objective ontology" and, at the same time, saying that it is just the same theory (expressed in a different language) as one that denies the possibility of such an ontology.

Heisenberg's objections may be contradictory, but he certainly expressed his dismissal of Bohm's idea in a contemptuous way. Referring without citation to Bohr, he said:

> When such strange hopes[85] were expressed, Bohr used to say that they were similar in structure to the sentence 'We may hope that it will later turn out that sometimes $2 \times 2 = 5$, for this would be of great advantage for our finances.'

> Werner Heisenberg [259, p. 117]

Another example of the "rhetoric of inevitability", to use Mara Beller's apt phrase [52, Chap. 9].

Concerning the reaction to Bohm from Bohr himself, David Peat quotes letters written to him by the philosopher of science Paul Feyerabend:

> The first Feyerabend heard of Bohm's new theory was during a seminar given by Niels Bohr. Following the lecture he asked Bohr to clarify certain points. The Danish physicist's reaction was, "Have you read Bohm"? As Feyerabend put it, "It seemed that, for him, the sky was falling in. ...Bohr was neither dismissive, nor shaken. He was amazed."

> In the midst of explaining to Feyerabend why Bohm's paper so disturbed him, Bohr was called away. The discussion continued without him for two more hours. Some of those present argued that the objections to Bohm's theory were not at all conclusive. As Feyerabend put it, the orthodox Copenhagen supporters tried to reply "in the Bohrian fashion". When this attempt was not successful, they said, "But von Neumann has proved ...", and that put an end to the discussion. Feyerabend noted, however, that Bohr himself did not use von Neumann's supposed proof as a crutch in that fashion.

> David Peat [386, p. 129]

However, the most vehement critique of Bohm was probably Léon Rosenfeld, who wrote to Bohm:

---

[85]Here, Heisenberg attributes to Bohm the hope that in the future the quantum theory may be proved false, but this was not the main point raised by Bohm, who stressed that his theory provided at least a coherent and empirically adequate alternative to the orthodoxy. (Note by J.B.).

I shall certainly not enter into any controversy with you or anybody else on the subject of complementarity, for the simple reason that there is not the slightest controversial point about it. [...] It is just because we have undergone this process of purification through error that we feel so sure of our results. [...] there is no truth in your suspicion that we may just be talking ourselves into complementarity by a kind of magical incantation. I am inclined to return that it is just among your Parisian admirers that I notice some disquieting signs of primitive mentality.[86]

Léon Rosenfeld, letter to Bohm, 20 May 1952, quoted in [386, p. 130]

In his contribution to a *Festschrift* for de Broglie's sixtieth birthday in 1952, Rosenfeld remarked:

[I]t is understandable that the pioneer who advances in an unknown territory does not find the best way at the outset; it is less understandable that a tourist loses his way again after this territory has been drawn and mapped in the twentieth century.

Léon Rosenfeld [419, pp. 55–56], quoted in [279]

Bohm responded to Rosenfeld that the term "tourist" could just as well be applied to Rosenfeld; Bohr would then be the pioneer who did "not find the best way at the outset". Thus, according to Bohm, it "worked both ways" ([65], quoted in [279, p. 23]).

Rosenfeld also considered that the mere idea of "interpreting a formalism" was a false problem and:

[...] appears as a short-lived decay product of the mechanistic philosophy of the nineteenth century. [...] every feature of it [the quantum theory] has been forced upon us as the only way to avoid the ambiguities which would essentially affect any attempt at an analysis in classical terms of typical quantum phenomena.

Léon Rosenfeld [421, p. 495]

He also suggested that the introduction of "hidden variables" would be "empty talk". These statements are remarkable both for their dismissiveness and for their naive empiricism: facts never determine by themselves a mathematical formalism and a formalism alone never includes its own "interpretation". Rosenfeld also thought that he was "merely translating in ordinary language the famous reasoning of von Neumann who showed long ago the formal incompatibility of quantum mechanics and determinism" [419, p. 56].

While Bohm was marginalized, partly for political reasons (see [364]), but also because of his heretical views on quantum mechanics, Rosenfeld was in a position of power, which, according to the historian of science A.S. Jacobsen, he did not always use very generously with his opponents[87]:

---

[86]He was probably referring to people around de Broglie, such as Jean-Pierre Vigier. (Note by J.B.).

[87]Another example, also noted by Jacobsen, is the way Rosenfeld treated a young Brazilian physicist, Klaus S. Tausk (see [394]).

In addition, [Rosenfeld] served as consultant or referee in matters of epistemology of physics and the like at several well-reputed publishing houses and at the influential journal *Nature*. In this capacity he used his influence effectively, and several books and papers, among them some by Frenkel,[88] Bohm, and de Broglie, were rejected on this account.

Anja Skaar Jacobsen [279, p. 23]

7.7   Quantum Mechanics, "Philosophy", and Politics

We have seen that Bohm was a victim of the witch hunts of the Cold War and that his work was derided rather than refuted. But things were much worse than that, because the discussions about quantum mechanics became entangled during that period with philosophical and political issues that were irrelevant to science, but had a great psychological impact. This was true on *both sides* of the Cold War divide. We shall discuss first the "Marxist" side, then the "Western" side.

Let us start by noting that Bohm and Rosenfeld, although polar opposites on the interpretation of quantum mechanics, were both philosophically "Marxists". We use the word "philosophically" here, because, although Bohm was briefly communist and Pauli jokingly wrote a letter to Rosenfeld starting with "Dear ($\sqrt{\text{Trotsky} \times \text{Bohr}} =$ Rosenfeld)" [279, p. 4], neither of them was very much engaged in political parties as such (even though they were both on the left). So their interest was mainly in Marxist philosophy. However, what this refers to is complicated and Marxist views are far from being as unified as they have tried to present them.

Marxist philosophy claims to be materialist. Classical eighteenth century materialism was a combination of realism (in the sense used in Chap. 3), an empirical approach to the world (rejection of metaphysics and a priori knowledge), and the idea that the mind is in some sense an emanation of the body. While not directly rejecting any of these ideas, Marxists have added two important adjectives: their materialism is supposed to be "historical" and "dialectical".

Because of the first adjective, Marxists tend to view all ideas as being produced by "social structures" or by the "ruling classes". To the extent that this applies to the natural sciences, it cuts science from its link to the natural world, since, in the end, it is society that imposes the frame of reference though which the scientists "view" the world.[89] As early as 1935, for example, the Soviet historian of science Boris Hessen, suggested that the "mechanical world view" of Newtonian physics was linked to the rise of the bourgeoisie [264]. This approach is a form of what we called, at the end of Sect. 3.2.2, the socialization of idealism, since it is not the individual subject or a

---

[88]Iakov Frenkel was a Soviet physicist who had submitted a paper critical of complementarity to *Nature* that Rosenfeld managed to have withdrawn by the translator, Pauline Yates [195, p. 38]. (Note by J.B.).

[89]This image of science bears some resemblance to Kuhn's ideas about paradigms, discussed in Sect. 3.2.2, except that Kuhn's paradigms are not connected to socio-economic structures and social classes in the same way that "dominant ideas" are connected to them in the Marxist view of science.

universal "mind" that imposes its view on the world, but society or social classes. One can also view contemporary social constructivism as a radicalized form of historical materialism (but without the emphasis on social classes).

The second adjective, "dialectical", is hard to define. Marxists have tried to combine eighteenth century materialism with the Hegelian idea of dialectics, not an obvious task, since Hegel was anything but a materialist or a scientifically minded philosopher. Marx indeed said [318]: "With [Hegel, the dialectics] is standing on its head. It must be turned right side up again, if you would discover the rational kernel within the mystical shell." But what this means is not clear.

Sometimes dialectics is presented as a set of "laws of thought", for example, trying to find a synthesis between opposite ideas, sometimes as a "law of society", showing how human societies develop historically from one "mode of production" to another, through "contradictions", and sometimes, especially in Engels' *Dialectics of Nature* [174][90] as "a law of Nature", showing that the latter is subjected, just as human societies are, to perpetual change and, supposedly, to progress. It also refers at times to holistic ways of thinking and to the alleged interconnectedness of everything in the universe. As Bertrand Russell observed, Hegel's philosophy "divides the friends of analysis from its enemies" [428, p. 714].

Marxists have tended to lambast non-dialectical materialists as being "mechanical", or "metaphysical". Yet, it is fair to say that what Marxists called "mechanical materialism", i.e., eighteenth century materialism, is in fact closer to the views of most practicing scientists (who do not use that expression of course), except, again, some of those involved in quantum philosophy.

Rosenfeld and Bohm were both exceptions in that sense. Indeed, the part of Marxism that they both liked was the "dialectical" one. But this word meant very different things for them: for Rosenfeld, it meant the need to abandon determinism (which was supposed to be a leftover from "mechanical materialism") and, of course, complementarity was viewed as a typically dialectical notion (indeed both notions shared the property of being ill-defined and thus able to fit any reader's favorite interpretation).[91] For Bohm, it was the holistic aspect of quantum mechanics, due to its nonlocal character, that sounded dialectical.

---

[90]This book was just a collection of personal notes that Engels never published. They were only published later in the Soviet Union, and sometimes worshipped by Communists.

[91]It is interesting to note that Max Born wrote a detailed and quite lucid critique of "dialectical materialism" in response to Rosenfeld's defense of that doctrine, in which Rosenfeld related it to Bohr's views on quantum mechanics [420]. Rosenfeld was of course, trying to prove that the Copenhagen interpretation of quantum mechanics was compatible with Marxism (to counter accusations of idealism, discussed below, leveled by some Marxists against that interpretation). Born shows that there is no basis in logic for the "law of dialectics", according to which thesis+antithesis leads, through the process of *aufhebung*, to a synthesis. He also shows that this "law" does not describe the history of ideas in physics, except, according to him, in one case: not surprisingly, the "wave–particle" duality. See [194] for Born's text and [193] for a discussion.

This can be viewed as an example of what Steven Weinberg calls the "unreasonable ineffectiveness of philosophy" in the sciences[92] [504, p. 169], if by "philosophy" one means the reliance on vague and a priori notions[93] which seem to provide a solid foundation for one's views, while in reality each author interprets those notions according to what he or she already believes for other, non-philosophical, reasons. Using this "method" one can arrive at radically different conclusions, as Rosenfeld and Bohm show, starting from the same "philosophy". On the other hand, one must admit that even this kind of "philosophy" can be inspirational, as it was in Bohm's case. The mistake, however, is to support one or another scientific view because of one's commitment to such a philosophy.

Of course, the association of Bohm with Marxism did not help his scientific reputation, at least in predominantly anti-Marxist circles in the West, and his later rapprochement with a figure like the Indian thinker Jiddu Krishnamurti only made matters worse.[94]

Rosenfeld, on the other hand, was fighting on two fronts, and as he said himself, seemed "to quarrel with everybody" [420, p. 482], because he was also critical of the idealism of some of the orthodox people, particularly Heisenberg.

Indeed, although prominent Soviet physicists, such as Lev Landau, were staunch defenders of the Copenhagen views, there was also a current in Soviet science and philosophy that considered those ideas "idealist".[95] In 1908, Lenin had published a book called *Materialism and Empirio-criticism*, where the word "materialism" refers more or less to what we call realism in Chap. 3 and "empirio-criticism" was a version of idealism that was fashionable at that time [309]. The book was a radical critique of idealism, which Lenin saw as a danger, because it encouraged various forms of subjectivism among some Marxists.

During the Soviet period, but also among non-Soviet Marxists, Lenin's book was referred to by people who did not like what they considered idealist tendencies in science or philosophy, and that included most of logical positivism, discussed below, or the Copenhagen interpretation of quantum mechanics.[96] It should be stressed that Lenin's views were criticized by other Marxists, for example by the Dutch astronomer Anton Pannekoek [373], who was closer to historical materialism, and therefore to what we call a socialized form of idealism. It is of course the idealism implicit in that

---

[92]This expression was used in contrast to what Wigner called the "unreasonable effectiveness" of mathematics in the natural sciences [513].

[93]This characterization would not of course apply to most of analytical philosophy.

[94]See [386] for a detailed description of the relationship between Bohm and Krishnamurti.

[95]The historian of science Alexei Kojevnikov has written detailed analyses of the strengths of Soviet science and the debates within it, including in relation to quantum mechanics and Bohm's ideas; see, e.g., [293–296]. Most prominent Soviet physicists were actually pro-Copenhagen, but were not harassed by the authorities for that reason.

[96]But not by Rosenfeld himself, who regarded Lenin's book as a return to "mechanistic materialism" [420, p. 482].

version of "historical materialism" that Lenin opposed, without, however, admitting that it had roots in the works of Marx.[97]

When Marxism fell out of fashion, its association with the critique of the Copenhagen school only made any such critique even more suspect, especially in places where Marxism had been influential, like France and, of course, the Soviet block. So for a while, discussing problems of quantum mechanics with scientists from those countries became especially difficult.

All this illustrates the enormous confusion created by mixing up science, philosophy, and politics, as happens all the time, not to mention the further confusion brought in when religion gets involved, something one should make the utmost effort to avoid. The problem of orthodox quantum mechanics is not that it disagrees with "materialism", dialectical or otherwise, but that it is an unclear theory that is just not "professional" enough, as Bell puts it:

> I think there are *professional* problems [with quantum mechanics]. That is to say, I'm a professional theoretical physicist and I would like to make a clean theory. And when I look at quantum mechanics I see that it's a dirty theory. The formulations of quantum mechanics that you find in the books involve dividing the world into an observer and an observed, and you are not told where that division comes [...]. So you have a theory which is fundamentally ambiguous [...]

> John Bell [107, pp. 53–54] (italics in the original)

But there were also associations between philosophy, politics, and quantum mechanics on the other side, so to speak. First of all, the historian of science Don Howard remarks that, among the "Copenhagen" people of the 1930s, it was Pascual Jordan, a member of the Nazi party, who was the closest to logical positivism and the Vienna circle (whose members were in general rather left wing)[98]:

> No other thinker was as central as Jordan, both to the core community of quantum physicists associated with Copenhagen and to the core community of philosophers of science associated with Vienna. Nowhere else in the literature of the 1930s will one find as extensive and technically adept a presentation of the case for an essential link between quantum mechanics and positivist epistemology. Never mind the fact that Bohr, himself, never endorsed such a linkage. That this was not Bohr's own understanding of the philosophical significance of quantum theory is of minimal relevance to our understanding how it was that the widespread, popular association of quantum mechanics and positivism was established.

> There is, second, the fact that the only way to make Jordan's odd mixing of quantum mechanics, positivism, and vitalism at all coherent is to embed the whole in the political context

---

[97] As Russell in particular observed in [427, Chap. 18], there was also a tendency in Marx towards a pragmatic view of truth, where truth is confused with usefulness, hence contradicting the notion of truth as corresponding to facts independent of human desires and interests. This can thus be regarded as another strand of idealism.

[98] The Vienna Circle was inspired by the work of the physicist and philosopher Ernst Mach and developed after World War I a current of thought called logical positivism (discussed below). The members of the Vienna Circle included Rudolf Carnap, Philipp Frank, Hans Hahn, Ernest Nagel, Otto Neurath, and, in a similar circle in Berlin, Carl Hempel and Hans Reichenbach. There was some proximity, at least in the minds of the logical positivists, between them and Albert Einstein, Kurt Gödel, Bertrand Russell, and Ludwig Wittgenstein.

of Germany in the mid-1930s. For it is Jordan's politically driven opposition to materialism that ties all of the pieces together. And therein lies a great irony. For it is Jordan, a member of the Nazi party, who in this way secured the popular association of quantum mechanics with a positivism that otherwise bore almost exclusively a left-liberal, even socialist political stamp.

Don Howard [277]

Actually, a member of the Vienna circle, Philipp Frank, was worried that Jordan was using the Copenhagen interpretation to support a "reactionary and barbaric *Weltanshauung*" and even wrote to Bohr to warn him about this (see Heilbron [255, p. 221]).

The association between philosophy, politics, and quantum mechanics also continued in the post-World War II period and on the "Western" side of the Cold War divide. In the United States, the dominant philosophy, at least in relation to the sciences in that period, was logical positivism. That current of thought, developed mostly in the Vienna circle, tried to purge philosophy of "metaphysics" and to base knowledge on secure foundations. This was certainly an admirable goal. The secure foundations would be logic and mathematics on the one hand, and direct experience, or sense data, on the other, hence the name logical positivism. But by proving that, given any system of axioms, there exist true arithmetical propositions that cannot be deduced from that system, Gödel showed that the logical part of the foundations was going to be more complicated than one might naively have thought.

However, this was not the main problem of logical positivism, which was rather related to the notion of "sense data". Certainly science relies on observations based on our senses, but as discussed in Chap. 3, the notion of observation is not simple: does it refer to our own subjective experience or to the perception of objects existing outside of us? If one adopts the first meaning, one easily falls into solipsism, but if one follows the second meaning, one loses "certainty" since our senses can always deceive us.

A related problem was connected with the rejection of "metaphysics". Indeed, what is metaphysics?[99] If it is the belief in entities whose existence cannot be "verified by experience", one has again to say precisely what "experience" means. It is true that the existence of angels cannot be verified, but how does one verify the existence of atoms? The problem is that, if one wants to have an ontology that includes atoms but not angels, one has to have a somewhat sophisticated view of what it means to be "verified by experience", and this is not easy to formulate explicitly, even if there is a huge difference between the evidence for the existence of atoms and the total absence of evidence for the existence of angels. This is because, even in the case of atoms, the evidence is indirect.[100]

---

[99]Not surprisingly, Rosenfeld, although not a logical positivist, also branded the rejection of complementarity as "metaphysical" [420, p. 472].

[100]Similarly, there is a big difference between, say, Carnap's critique of Heidegger as a metaphysician [91] and the "elimination" of quantities in physics that are not directly observable.

There has been a temptation within the logical positivist tradition to brand as "metaphysical" everything that goes beyond our immediate experiences. Naturally, this makes the accusation of metaphysics rather easy to level. As Russell observed ironically in the 1950s:

> The accusation of metaphysics has become in philosophy something like being a security risk in the public service. [...] The only definition I have found that fits all cases is: 'a philosophical opinion not held by the present author'.
>
> Bertrand Russell [429, p. 164]

With the rise of Nazism, the logical positivists had to emigrate and became rather influential in the United States. Although they clearly had their hearts in the right place, trying to be empirical and scientific and rejecting confused and a priori thinking, there was a certain youthful enthusiasm and naiveté in their endeavor. And when it came to quantum mechanics, even though prominent people like Bohr rejected the accusation of positivism, as we saw in Sect. 3.3, the positivist's emphasis on "observations" as being the only thing one could speak about "meaningfully" or "scientifically", with all the ambiguities of this term, helped to strengthen the dogmatism of the "Copenhagen" school.

But that philosophical baggage was not the main obstacle to an open discussion of the problems of quantum mechanics. Indeed there was in fact very little open-mindedness on these issues in the West, but this was mostly for socio-psychological and even political reasons. The physicist John Clauser, who made important contributions, both theoretical and experimental,[101] to the first tests of Bell's inequalities, describes in vivid terms the situation when he was a student (which is similar to my own experience, but in a very different place).[102] After characterizing the adhesion to the Bohr–von Neumann views (Bohr had defeated Einstein and von Neumann had shown with his no hidden variables theorem that there was no alternative to ordinary quantum mechanics) as a sort of religion, he wrote:

> During the post-war years, the United States quickly became embroiled both in the cold war and in an internal anti-communist frenzy. Driven by Sen. Joe McCarthy, stigmas then became into vogue. [...] Unfortunately, acceptance of stigmas by the populace appears to foster the malignant creation of additional stigmas, specially in an environment that is dominated by religious fanaticism. Thus, in keeping with the times, a very powerful secondary stigma began to develop within the physics community towards anyone who sacrilegiously was critical of quantum theory's fundamentals. The stigma long outlived the McCarthy era and persisted well into the 1970 and 1980s. Sadly, it effectively kept buried most of the untidiness left behind by quantum theory's founders, and physicists went on about their business in other areas. The net impact of this stigma was that any physicist who openly criticized or

---

[101] See [95] for a version of Bell's inequality, called the Clauser–Horne–Shimony–Holt inequality, that is more amenable than the original inequality to experimental tests. See [192] for early experimental tests and [229] for a discussion of the relation between Bell and Clauser–Horne–Shimony–Holt inequalities, and for references to more recent experimental tests.

[102] See also Gisin's book [214] for several examples of how, even not so long ago, people who raised questions about the foundations of quantum mechanics were silenced.

even seriously questioned those foundations (or predictions) was immediately branded as a "quack". Quacks naturally found it difficult to find decent jobs within the profession.

John F. Clauser [97. pp. 71–72]

Clauser mentions Einstein, Schrödinger, and de Broglie as well known critics who could hardly be dismissed as "quacks". But:

Instead, gossip among physicists branded these men "senile". This is not a joke. On many occasions, I was personally told as a student that these men had become senile, and that clearly their opinions could no longer be trusted in this regard. This gossip was repeated to me by a large number of well-known physicists from many different prestigious institutions. Given this branding, their leadership role in charting the course of progress in physics was thereby severely limited. Under the stigma's unspoken "rules", the worst sin that one might commit was to follow Einstein's teaching and to search for an explanation of quantum mechanics in terms of hidden variables, as Bohm and de Broglie did.

John F. Clauser [97, p. 72]

As an example of the consequence of this stigma, Clauser mentions a policy of the American Physical Society, editing *The Physical Review* and *Physical Review Letters* according to which referees should reject any paper on foundations of quantum mechanics that was not "mathematically based *and* gave new quantitative experimental predictions." As Clauser observes, under those rules, "Bohr's response to EPR certainly could not have been published" [97, p. 72], and the same is true for Bohm's articles (the rules were formulated in the 1970s). Clauser continues:

Religious zeal among physicists prompted an associated powerful proselytism of students. As part of the "common wisdom" taught in typical undergraduate and graduate physics curricula, students were told simply that Bohr was right and Einstein was wrong. That was the end of the story and the end of the discussion. Of course, it was the end, because the concluding chapters of the story were not yet written. Bohm's and de Broglie's alternative […] were neither thought, nor even cited. Any student who questioned the theory's foundations or, God forbid, considered studying the associated problems as a legitimate pursuit in physics was sternly advised that he would ruin his career by doing so. I was given this advice as a student on many occasions by many famous physicists on my faculty at Columbia and Dick Holt's faculty at Harvard gave him similar advice.

John F. Clauser [97, pp. 72–73]

Then Clauser discusses Bohm, de Broglie, and Bell. For Bohm, he writes: "Given the era, no one dared trust his opinion since he was openly communist." De Broglie was more prominent, because of his position in the *Académie des Sciences*, but "his publications went largely unread, since, of course, everyone already 'knew' that he was senile." As for Bell, Clauser explains that he was very careful: Bell avoided attending an awards ceremony for a prize given to him by Charles Brandon, a founder of *Federal Express*, in order not to appear linked to a non-scientific foundation. He never openly discussed his study of foundations of quantum mechanics with his colleagues at CERN and thus led a sort of double life.[103] Finally, when Alain Aspect, who was preparing his own experiments [21], came to visit him, "Bell's first question to him was 'Do you have a permanent position?'" [97, pp. 73–74].

Clauser also speaks of "evangelical theoreticians" who were not even willing to consider experimental tests of Bell's inequalities, because they were so convinced that quantum mechanics must be correct in all situations (and maybe also because they did not appreciate how radical the conclusions of the EPR–Bell argument were). Even the great Richard Feynman, when Clauser told him of his project to test Bell's inequalities, "immediately threw me out of his office saying 'Well, when you have found an error in quantum-theory experimental predictions, come back then, and we can discuss *your* problem with it'" [97, p. 71].

As a student, Clauser himself "had difficulty understanding the Copenhagen interpretation". And he "found Bohm's and de Broglie's work refreshing, since they do give real physical-space models of what is happening" [97, p. 78]. When he carried out his experiments, he "believed that hidden variables may indeed exist". By that time, the McCarthy era was "in the distant past" and:

> Instead the Vietnam war dominated the political agenda of my generation. Being a young student living in this era of revolutionary thinking, I naturally wanted to "shake the world".

John F. Clauser [97, p. 80]

Thus the testing of Bell's inequalities may have been a collateral benefit of opposition to the Vietnam war. But the upshot of the experiment was not what Clauser expected, and did not prove quantum mechanics wrong, but rather established the reality of nonlocality, something that probably "shakes the world" even more (provided that the world is willing to listen!).

The story recounted by Clauser illustrates the fact, which has been frequently observed, that in formally free societies, unpopular views and heterodox opinions can be silenced as effectively and sometimes more effectively than in dictatorships. Instead of putting people in jails or camps, it is enough, in order to impose the dominant views, to stigmatize the free thinkers, to censor unorthodox publications, and to refuse to give jobs to those who do not toe the line. Of course, such methods are less unpleasant for the victims than the more brutal ones inflicted by dictatorships, but that does not mean that they are less effective. Given the relative comfort of modern life in developed societies, who will risk marginalization and a ruined career simply for expressing publicly his or her thoughts?

Yet, as John Stuart Mill famously said, in any case, a great harm is done:

> But the peculiar evil of silencing the expression of an opinion is that it is robbing the human race; posterity as well as the existing generation; those who dissent from the opinion, still more than those who hold it. If the opinion is right, they are deprived of the opportunity of

---

[103] At least until a conference held in 1976 in Erice, Sicily, entitled *Thinkshops on Physics: Experimental Quantum Mechanics*, and in which Alain Aspect, John Bell, John Clauser, Philippe Eberhard, Michael Horne, Bernard d'Espagnat, and Abner Shimony participated, among others.

exchanging error for truth: if wrong, they lose, what is almost as great a benefit, the clearer perception and livelier impression of truth, produced by its collision with error.

John S. Mill [341, p. 76]

7.8 Conclusions

At the end of the previous chapter, we asked: why is the de Broglie–Bohm theory so little appreciated if it possesses all the qualities that its defenders (including the author of this book) attribute to it? It is difficult to give a complete answer to that complicated question, but the sketch of the history of quantum mechanics given here provides some elements of an answer.

It is obvious that one cannot refute an idea or a theory if one does not fully understand it. We have tried to show that Einstein and the EPR argument were not understood at the time they were formulated and even much later. This implied that Bell's proof of nonlocality was not grasped either, so that even now, many physicists are convinced that one has a choice: either abandon locality or drop determinism, realism, or hidden variables, while in reality there is no such choice, as we discussed in Chap. 4 and Sect. 7.5 of this chapter.

But the misunderstanding of Bell has had an enormous impact on the lack of appreciation of de Broglie–Bohm: either people think that this theory is impossible because Bell or others have shown that hidden variables are forbidden (without being precise about which kind of hidden variables are "forbidden") or because it is nonlocal, hence physically unacceptable. But of course, if Bell has shown, as he did, that the world *is* nonlocal, then the nonlocality of the de Broglie–Bohm theory is a quality, not a defect.

Moreover, the fact that this theory introduces "hidden variables" that are not hidden and that it accounts naturally for the results of those specific interactions that we call measurements, despite the no hidden variables theorems, is again a great quality and not a defect of de Broglie–Bohm. The reason why Einstein found Bohm's theory "too cheap" and why de Broglie never really believed in his own theory was perhaps because of the manifest nonlocality of this theory, but at least they could not know, before Bell, that this was a quality of the theory.

One must also take into account the influence of the version of "logical positivism" that we described above. That version is almost universally rejected by philosophers of science nowadays (in part, because of the imprecision of the word "observable"), but it survives in the backs of the minds of many physicists.

Given all this, it may not be too surprising to read in an encyclopedia, which is in principle unbiased:

Attempts have been made by Broglie, David Bohm, and others to construct theories based on hidden variables, but the theories are very complicated and contrived. For example, the electron would definitely have to go through only one slit in the two-slit experiment. To explain that interference occurs only when the other slit is open, it is necessary to postulate a special force on the electron which exists only when that slit is open. Such artificial additions make hidden variable theories unattractive, and there is little support for them among physicists.

Encyclopedia Britannica [457], quoted in [228].

After reading Born's summary of von Neumann's "proof" of the impossibility of hidden variables in his lectures on *Natural Philosophy of Cause and Chance*,[104] Bell says that he "relegated" questions about quantum mechanics to the back of his mind and "got on with more practical things". However, he adds:

> But in 1952 I saw the impossible done. It was in papers by David Bohm. Bohm showed explicitly how parameters could indeed be introduced, into nonrelativistic wave mechanics, with the help of which the indeterministic description could be transformed into a deterministic one. More importantly, in my opinion, the subjectivity of the orthodox version, the necessary reference to the 'observer', could be eliminated.
>
> Moreover, the essential idea was one that had been advanced already by de Broglie in 1927, in his 'pilot wave' picture.
>
> But why then had Born not told me of this 'pilot wave'? If only to point out what was wrong with it? Why did von Neumann not consider it? More extraordinarily, why did people go on producing 'impossibility' proofs, after 1952, and as recently as 1978?[105] When even Pauli [379], Rosenfeld [420], and Heisenberg [258], could produce no more devastating criticism of Bohm's version than to brand it as 'metaphysical' and 'ideological'? Why is the pilot wave picture ignored in text books? Should it not be taught, not as the only way, but as an antidote to the prevailing complacency? To show us that vagueness, subjectivity, and indeterminism are not forced on us by experimental facts, but by deliberate theoretical choice?

John Bell [49, p. 160]

Maybe to answer those questions, we should turn to a writer[106]:

> I know that most men, including those at ease with problems of the highest complexity, can seldom accept even the simplest and most obvious truth if it be such as would oblige them to admit the falsity of conclusions which they have delighted in explaining to colleagues, which they have proudly taught to others, and which they have woven, thread by thread, into the fabric of their lives.

Leo Tolstoy [472, p. 143]

---

[104]See [76, p. 109] and Sect. 7.4.

[105]Here Bell refers to a series of no hidden variables theorems due to Jauch and Piron, Misra, Kochen and Specker, Gudder, Jost, and Woolf. See [49, pp. 167–168] for the references. (Note by J.B.).

[106]The quote here is adapted from the published translation. The source for this version is usually [188].

Chapter 8
Quantum Mechanics and Our "Culture"

The concept of "truth" as something dependent upon facts largely outside human control has been one of the ways in which philosophy hitherto has inculcated the necessary element of humility. When this check upon pride is removed, a further step is taken on the road towards a certain kind of madness — the intoxication of power [...] to which modern men, whether philosophers or not, are prone. I am persuaded that this intoxication is the greatest danger of our time, and that any philosophy which, however unintentionally, contributes to it is increasing the danger of vast social disaster.

Bertrand Russell [428, p. 782]

A more emphatic title to this chapter could have been *The Harm Done to Western (now World) Culture by Certain Interpretations of Quantum Mechanics*. But that would have been misleading and it is easy to exaggerate along those lines. It is true that there has been an enormous amount of pseudo-science and nonsense, from "quantum healing" to telepathy, or from various forms of mysticism to postmodernist cultural studies, produced by people invoking quantum mechanics. However, the Copenhagen-like interpretations of quantum mechanics cannot take all the blame for that. Indeed, most areas of science are victims of either pseudo-scientific exploitation or misrepresentation, even those that are fairly simple to understand, like the theory of evolution, or others that are not so simple to understand but whose significance is perfectly clear, like the special theory of relativity.

Exploitation and misrepresentation of science are unavoidable and most people who do so with quantum mechanics do not really understand what is special about it, do not know its mathematical formalism, and do not grasp the reasons behind the weird-sounding statements about nonlocality or the central role of the observer that they refer to. But that does not mean that there is nothing wrong with the way most physicists have presented quantum mechanics.

© Springer International Publishing Switzerland 2016
J. Bricmont, *Making Sense of Quantum Mechanics*,
DOI 10.1007/978-3-319-25889-8_8

8.1   The Trouble with Quantum Mechanics

The historian of science Mara Beller has put her finger on some of specific problems linked to the presentation of quantum mechanics in her *Physics Today* article entitled *The Sokal hoax: at whom are we laughing?* [51].[1] One problem is the respect for authority, and in particular, the authority of Bohr[2]:

> Like the deconstructionist Jacques Derrida, whom Steven Weinberg attacked in his 1996 New York Review of Books article on Sokal's hoax [505], Bohr was notorious for the obscurity of his writing. Yet physicists relate to Derrida's and Bohr's obscurities in fundamentally different ways: to Derrida's with contempt, to Bohr's with awe. Bohr's obscurity is attributed, time and again, to a "depth and subtlety" that mere mortals are not equipped to comprehend. [...]
>
> Carl von Weizsäcker's testimony is a striking example of the overpowering, almost disabling, impact of Bohr's authority. After meeting with Bohr, von Weizsäcker asked himself [497]: "What had Bohr meant? What must I understand to be able to tell what he meant and why he was right? I tortured myself on endless solitary walks." Note that von Weizsäcker did not ask, "Was Bohr right?" or "To what extent, or on what issue, was Bohr right?" or "on what issues was Bohr right?", but, quite incredibly, he wondered what must one assume and in what way must one argue in order to render Bohr right?
>
> Mara Beller [51]

After quoting some of the statements by Bohr, Pauli, and Born that we discussed in Chap. 1, Beller remarks:

> Astonishing statements, hardly distinguishable from those satirized by Sokal, abound in the writings of Bohr, Heisenberg, Pauli, Born, and Jordan. And they are not just casual, incidental remarks. Bohr intended his philosophy of complementarity to be an overarching epistemological principle — applicable to physics, biology, psychology, anthropology. He expected complementarity to be a substitute for the lost religion. He believed that complementarity should be taught to children in elementary schools. Pauli argued that "the most important task of our time" was the elaboration of a new quantum concept of reality that would unify science and religion. Born stated that quantum philosophy would help humanity cope with the political reality of the era after World War II. Heisenberg expressed the hope that the results of quantum physics "will exert their influence upon the wider fields of the world of ideas [just as] the changes at the end of the Renaissance transformed the cultural life of the succeeding epochs."
>
> Mara Beller [51]

---

[1] The article could be interpreted as a critique of Sokal's hoax: the American physicist Alan Sokal managed to have an article full of postmodernist nonsense published in the rather fashionable journal *Social Text* [451, 455]. Beller rightly points out in her article that many of the founders of quantum mechanics have expressed views that are comparable to those satirized by Sokal. But one should stress that in our jointly authored book, *Fashionable Nonsense* [453], published after the hoax, we explicitly avoided criticizing non-scientists for the nonsense they might have said about quantum mechanics (otherwise the book would have been much longer), precisely because we were aware of the confusion spread by physicists on that topic. But the same was not true for fields such as topology, set theory, logic, or the theory of relativity, and about which we did criticize the "abuses".

[2] Another example of Bohr worship is provided by Wheeler, who compared his wisdom "to that of Confucius and Buddha, Jesus and Pericles, Erasmus and Lincoln [511, p. 226]" [195, p. 82].

And Beller adds, concerning Stanley Aronowitz, a founding editor of *Social Text*, the journal in which Sokal published his hoax:

> Aronowitz had been relying on the assertions of the inevitable and final overthrow of determinism, endlessly repeated by the most honored heroes of 20th-century physics. How can Aronowitz or other non-physicists resist the authority of such past eminences, unless the physicists of our time publicly declare that the Copenhagen orthodoxy is no longer obligatory? Such a public declaration could have diminished greatly the explosive proliferation of the postmodernist academic nonsense, so appalling to Sokal and Weinberg.

Mara Beller [51]

All this is fair enough, but one can still wonder why people who do not understand the complexities of quantum mechanics and who usually do not worship science, rely on it in order to promote their philosophical or political views. In fact, there is an apparent paradox: people who tend to be distrustful of scientists, postmodernists for example, often *love* the Copenhagen interpretation of quantum mechanics, probably because they see in it the antithesis of everything they dislike about science: determinism, materialism, objectivity, etc.

And that is what is specific about quantum mechanics: it is not just that the science has been abused by ignoramuses, but that the scientists themselves, at least a good fraction of them, have contributed to that "abuse". And they have done it in many ways: by claiming that one has put back an unspecified observer at the center of science (and, then, who could be blamed for identifying this observer with the human subject?); or by claiming that intelligibility is unattainable in the microscopic realm, and that this is so *ad vitam eternam*. They behaved like people celebrating defeats as if they were victories.

But this was, in general, more due to *hubris* than mysticism: it would have been more honest, but more humble too, to admit that quantum theory did have shortcomings and to look for a more detailed theory. The real trouble with the dominant discourse about quantum mechanics is that the *hubris* of the scientists, claiming that their understanding of quantum mechanics was somehow an ultimate understanding, provided ammunition to the mystics and to all the defenders of anti-scientific world views.

It is the scientific spirit, more than philosophy, which has been the main defender of the notion of objective truth in the modern world and which has thereby, to quote Russell, "inculcated the necessary element of humility" in human thought, as opposed to the wishful thinking of political, religious, and pseudo-scientific ideologies. If science itself gives up, or seems to give up that notion, then indeed a "check upon pride is removed", with unforeseeable consequences.

However, since every coin has two sides, we have to qualify that judgment.

8.2   A Plea for "Copenhagen"

The historian of science Paul Forman has offered a controversial but interesting conjecture, now called "Forman's thesis", which tries to account for the rejection of determinism and causality in quantum mechanics on the basis of the cultural climate of the post World War I Weimar Republic [189]. Forman's idea is that the hope of a German victory in the war was based in part on the ingenuity of German science, and that hope was still alive at the beginning of the summer of 1918. When everything collapsed, including the Empire, in the fall of 1918, the shock was unprecedented and its effect on the period after the Great War produced a general anti-scientific climate during the Weimar republic. Forman sees the anti-causal approach in quantum mechanics as an adaptation of scientists to this hostile cultural climate.

Such claims have a ring of plausibility, but they are difficult to evaluate because the facts that were observed and that led to quantum mechanics were truly strange, and the means that one had to successfully compute certain measurable quantities, such as atomic spectra, whether they relied on Heisenberg's matrix mechanics[3] or on Schrödinger's equation, were terribly abstract and counterintuitive. How was one supposed to think of quantities like position and momentum being represented by *matrices*? Or to interpret a wave defined and "propagating" on configuration space? As we saw in Sect. 7.6.1, even de Broglie, who came closer than anybody else to a more complete theory, was unable to swallow this idea.

When such a revolution happens, when all the familiar concepts are proven inadequate and the traditional certainties are radically unsettled, it is only natural to fall back on what seems to be the most secure basis of our knowledge: experimental data and the mathematical formalism. "Copenhagen" seemed to provide exactly such a fallback position.

However, it is not clear that there ever was a unified and coherent "Copenhagen" doctrine. Bell describes as follows the position of what he calls the "Three Musketeers of the Copenhagen interpretation", Bohr, Pauli, and Heisenberg: he saw Bohr as insisting on a sharp division between classical apparatus and the quantum system, while being "extraordinarily insensitive" as to where exactly that division should be. But Bohr was convinced that he had solved that problem and in so doing "not only contributed to atomic physics, but to epistemology, to philosophy, to humanity in general." Bell continues: "I like to speak of two Bohrs: one is a very pragmatic fellow who insists that the apparatus is classical, and the other is a very arrogant man who makes enormous claims for what he has done."

Concerning Pauli, Bell notes that while Bohr rejected the idea that the human mind "was somehow an important element in quantum mechanics", "Pauli was attracted to

---

[3]Heisenberg initially solved quantum mechanical problems using matrices, like those we encountered in the appendices of Chap. 2, but sometimes with an infinite number of rows and columns. That method was further developed by Born, Heisenberg and Jordan [74, 75]. We now think of those matrices as the representation of operators in certain bases. The equivalence between Heisenberg's approach and Schrödinger's based on his equation was proven by Schrödinger soon after he had introduced that equation [439].

that idea, and at the end of his life he became increasingly religious. He felt that it was wrong to separate science from religion; it was wrong to separate psychology from physics. He felt that the *real* Copenhagen interpretation did insist that the mind was something that you could not avoid referring to in formulating quantum mechanics. Pauli thought, as far as I can judge, that the division between system and apparatus was ultimately between mind and matter."

On Heisenberg, on the other hand, Bell is quite brief [56, pp. 52–53][4]: "It is perfectly obscure to me *what* Heisenberg thought." One could go on and analyze what von Neumann, Wigner, Wheeler, and others thought, but it is obvious, with insight, that nothing clear and definite could ever be called the *true* Copenhagen interpretation of quantum mechanics. And that is fundamentally because the problems of quantum mechanics lie within the physical theory itself and the various versions of the "Copenhagen interpretation" were just attempts to apply philosophical solutions to a conceptual but *physical* difficulty.

Moreover, it is not clear that what is now taught in physics courses would be recognized as being *the* Copenhagen interpretation by the people who already defined it as the orthodoxy at the Solvay Conference in 1927. Indeed, the mathematical formalism, including what is now the standard theory of measurement, was only developed by von Neumann in 1932.[5] When EPR raised the dilemma posed by entangled states (nonlocality or incompleteness) they were ignored or misunderstood by almost everybody, except Schrödinger. And obviously, the founding fathers could not anticipate either Bohm's theory or Bell's theorem.

So one could introduce the following "guilty plea" for "Copenhagen": yes, they vastly exaggerated the unavoidability of their own views, refused to listen to dissident voices such as de Broglie, Einstein, and Schrödinger, and claimed that quantum mechanics led to a revolution in human thought far greater than anything a scientific theory is able to do (e.g., having direct consequences for politics or religion). *But* they did face unprecedented difficulties and extraordinary novelties, and they *did* discover the most successful scientific theory ever found.

Besides, by freeing people's minds of conceptual questions (since Bohr had supposedly solved all those problems), the quantum formalism could be applied unhindered in atomic physics, nuclear physics, solid state physics, and chemistry. This has led to what has probably been the most productive period in the history of science and has laid the foundations for all the technologies of the contemporary world.

8.3   But What About Now?

With the rise of Nazism and to an even greater extent after World War II, the center of physics shifted from *Mitteleuropa* to the United States. The shift was not only geographic but also cultural; while most of the founding fathers were trained in

---

[4]For a sympathetic exegesis of Heisenberg's changing views, see Camilleri [90].

[5]The discussion of measurement in the report by Born and Heisenberg at the 1927 Solvay Conference is quite different from the modern one [23, 24].

philosophy (for better or for worse, since a good part of that philosophy was idealist!), and paid attention to the foundational problems, the dominant "philosophy" after the war was pragmatism. Its spirit is summarized in the well-known slogan: "Shut up and calculate!"[6]

However, it is remarkable that the two people who did challenge the Copenhagen monocracy after the war were both Americans: David Bohm and Hugh Everett (both at Princeton). But to say that they were not typical of the American academia is quite an understatement. We saw in Sects. 7.6 and 7.7 the trouble that Bohm had to go through, and Everett's ideas were also, in general, poorly received: he went to work as an advisor to the military even before finishing his PhD thesis and never came back to physics [87]. Everett's ideas would eventually become somewhat more popular, but only much later.[7]

Probably nobody was more critical of the "pragmatic" attitude than David Bohm, who despised what he called "resultlets". As Bohm's biographer F. David Peat says, in a reply to the question: "Why did so many scientists—why do so many scientists even now—seem to have so much trouble accepting or respecting [Bohm's] ideas?":

Well, I suppose in some cases it's because people like small little bits of work — 'resultlets,' as David called them, not results but 'resultlets'. When Dave did his work he really dealt with ideas, with concepts, and in very broad brush strokes; whereas the fashion in physics today is that it should all be hyper-mathematical, and he always mistrusted mathematics. Mathematics to him was a good tool, but it was a tool and no more. The thing with mathematics, even the most beautiful and elegant mathematics, is that somewhere in there a lot of assumptions have been hidden, and when we speak together, using ordinary language, it's a little bit easier to discover what those assumptions are. Mathematics tends to conceal a lot. He was also suspicious of other aspects of the way physics was being done — for example, all this reliance in particle physics on breaking things apart rather than seeing them in an all-embracing fashion. You see, Dave felt there had been a major revolution in this century in quantum mechanics and relativity, but that our thinking hadn't really caught up with it. In the old order you could fragment things, you could define everything on a Cartesian grid of space and time. Now we needed an entirely new order, and the implicate order, which is inherently infinite, was one of the approaches he was working on. But of course, that's asking too much of physicists. They like to see things small and finite, and Dave was too much of a global thinker, I think, for many of them — except the very good ones, who were sympathetic to Dave because they realized that something new was called for.

David Peat [387]

But that view is too unilateral: the "pragmatic" attitude allowed physicists to develop quantum field theory, including quantum electromagnetism, which led to the most accurately corroborated empirical predictions ever made in science, and later to the spectacularly successful standard model of elementary particles. And the penchant of Bohm for "deep thoughts" did not prevent him from having mystical tendencies at time.

---

[6]See Mermin [339] for the origin of this expression, often attributed to Richard Feynman.

[7]See the collection of papers [132], published in 1973, for the beginning of the renewed interest in the many-worlds interpretation.

But what is good at a given time may not be good for all times. As we saw in Sect. 7.7, the spirit of the Cold War led to an extraordinary closing of the physicist's mind when it came to the understanding of Bell's result, namely the most surprising and, in a sense, the most interesting consequence of quantum mechanics itself.[8] That closing of the physicist's mind is particularly harmful when it comes to physics education. Too many generations of students have been trained to "shut up and calculate"; but this is very discouraging for those who study physics in order to understand how the world works. Valia Allori, who started out as a physicist and became a philosopher of science, has forcefully expressed that sentiment[9]:

> My interest has always been to understand what the world is like. This is the main reason that I majored in physics: if physics is the study of nature, then to understand nature one should learn physics first. But my hopes were disappointed by what is (or at least seems to be) commonly accepted in many physics departments all over the world: after quantum mechanics, we should give up the idea that physics provides us with a picture of reality. At first, I believed this was really the case and I was so disappointed that I decided to forget about my "romantic" dream.

> Valia Allori

However, Allori adds:

> At some point, [...] I realized that some of the things I took for granted were not so obviously true, and I started to regain hope that quantum mechanics was not really the "end of physics" as I meant it. Therefore, I decided to go to graduate school in physics to figure out what the situation really was. While taking my PhD in the foundations of quantum mechanics, I understood that what physicists thought was an unavoidable truth was instead a blunt mistake: quantum mechanics does not force us to give up anything, and certainly not the possibility to investigate reality through physics.

> Valia Allori

What allowed Allori to regain hope was that she learned about the de Broglie–Bohm theory, and then contributed to it [11–15].

Concerning physics education, without necessarily studying "foundations" as deeply as Allori did, there are certain things that students should be told:

1 That there is no simple way to understand the quantum formalism, no obvious interpretation of it on which everyone agrees, and that there is certainly something rotten in the kingdom of Copenhagen.
2 They should also know that often repeated sentences about quantum mechanics "not dealing with the world but with our knowledge of it" do not make any sense, and that the cat paradox shows that establishing a link between the usual quantum formalism and the world is, to say the least, problematic.

---

[8] Another thesis defended by Paul Forman is that the massive scale of defense-related funding of science during World War II and the Cold War led to a shift in physics from basic to applied research [190, 191]. This may also have encouraged the "pragmatism" of those who were still doing fundamental research.

[9] The quotes here come from what she describes as her "personal background" on her website: http://www.niu.edu/~vallori/background.html.

3 One might go further and explain the meaning of EPR–Bell, namely nonlocality, and also the no hidden variables theorems. This would alert students against a naive "statistical" view of quantum mechanics.
4 One might also mention (without necessarily explaining it in introductory courses) that there is a completely coherent view of quantum theory in which no question is left unanswered (if the reader has got this far, it is probably unnecessary to say what we are referring to).

What sometimes discourages students from trying to go further in their understanding of quantum mechanics is that there seem to be a zillion different "interpretations". But that is not quite true. As we explained in Chap. 5, in the de Broglie–Bohm theory, everything falls into place; it is not *just* that one can associate trajectories with particles without running into contradictions. That theory also naturally accounts for all of the following:

1 The measurement formalism, including the collapse rule.
2 The no hidden variables theorems, which are explained by the contextuality of measurements and the active role of the measuring devices.
3 The apparent randomness of quantum mechanics, which follows, in a fully deterministic theory, from rather natural assumptions about initial conditions.
4 The unavoidable nonlocality of any theory reproducing the quantum predictions.

No alternative to the de Broglie–Bohm theory even comes close to achieving all that (see Chap. 6).

8.4 Understanding Quantum Mechanics: An Unfinished Story

Throughout the nineteenth century there has been a constant reappraisal of the foundations of mathematics: how to define limits? Real numbers? Sets? Which geometries are natural? In physics, following the great revolutions of the nineteenth century, including thermodynamics and statistical physics which introduced probabilistic notions at a rather fundamental level, and electromagnetism which introduced waves propagating in an unknown medium, the ether, later replaced by an even more mysterious vacuum, the beginning of the twentieth century dealt a final blow to all previous certainties: relativity showed, among other things, that the notion of simultaneity is relative to our state of motion, and quantum mechanics seemed to have made every possible intuitive image of the microscopic world impossible, and to have shown that Nature is intrinsically random.

It is understandable that such revolutions should produce a great deal of skepticism, and when that happens, there is a tendency to retreat towards what one hopes to be secure foundations. The same movement happened in many different fields in the twentieth century: logical positivism tried to reduce everything to sense-data plus logical constructions; behaviorism wanted to limit the study of the mind to a description of inputs and outputs, avoiding the introduction of mental states; formalism in mathematics viewed the latter as a purely formal game; the frequency interpretation of probability was trying to avoid making (Bayesian) rational inferences and was reducing probability to a mere theory of frequencies.

The Copenhagen interpretation of quantum mechanics, with its emphasis on observations was also part of that movement. But all those retreats carry a heavy cost: one may describe the world (or at least our "perception" of it) efficiently, but one refuses to understand it. And understanding is what makes science a distinctive and worthwhile enterprise.

Actually, all the attitudes described above have essentially disappeared in their respective fields, except, to some extent, the Copenhagen interpretation of quantum mechanics. Very few people adhere explicitly to that doctrine (assuming they have a clear idea of what it means), but it lingers in the textbook presentations of quantum mechanics, as well as in the discourses about "quantum information", where information is not information about anything else than itself.

Yet, despite all the claims coming from the greatest scientific authorities that it was impossible to do, claims apparently supported by theorems, de Broglie, Bohm, and Bell have shown us a way to understand the quantum formalism, not simply "how to use it" (which every physicist knows), but "what it means". This understanding is to some extent limited to non-relativistic quantum mechanics, but that does not arise from any intrinsic defect of the de Broglie–Bohm theory, but from the nonlocality of the world. A genuine merging of quantum mechanics and relativity is still an open problem, and might be called the great unrecognized frontier of physics.[10]

To paraphrase a famous dictum,[11] quantum physicists have been changing the world in various ways; it is time now to try to understand it.

---

[10]To repeat the problem in technical terms: how does one account for the collapse rule of quantum mechanics in relativistic terms when dealing with entangled states whose parts are spacelike separated?.

[11]Marx's eleventh *Thesis on Feuerbach*: The philosophers have only interpreted the world, in various ways; the point is to change it.

Glossary

We give here definitions of some of the main concepts used in the book and indicate where they are discussed.

Action at a Distance. This means that, by acting at some place A, one can instantaneously affect the physical situation at some place B, no matter how far A and B are from each other. See Chap. 4 and Sect. 5.2.1.

Beables. A term introduced by John Bell to refer to what exists, or what a given theory postulates as existing, as opposed to what is observable. In the de Broglie–Bohm theory, the beables are the positions of the particles and the quantum state. See Chap. 6 for a discussion of beables in other theories.

Bell Inequality. An inequality which is implied by certain assumptions about the existence of hidden variables and contradicted by quantum mechanical predictions (that have been verified). However, this statement can be misleading if one does not add that, under an assumption of locality, Einstein, Podolsky, and Rosen have shown that those hidden variables necessarily exist. So Bell's result, combined with the EPR argument is a nonlocality result rather than a mere no hidden variables theorem. This is explained in Chap. 4. For the misunderstandings of this result, see Sect. 7.5.

Born Rule. If we have a quantity A associated with vectors $|e_n\rangle$ and numbers λ_n according to

$$A|e_n\rangle = \lambda_n|e_n\rangle \, ,$$

and if it is measured in a state

$$|\text{state}\rangle = \sum_n c_n|e_n\rangle \, ,$$

© Springer International Publishing Switzerland 2016
J. Bricmont, *Making Sense of Quantum Mechanics*,
DOI 10.1007/978-3-319-25889-8

then the result will be λ_k with probability $|c_k|^2$. See Appendix 2.B for more details on Born's rule, and Sects. 5.1.4 and 5.1.5 for the justification of that rule in the de Broglie–Bohm theory.

Branches of the Quantum State. This refers to the different terms in sums like (2.5.1.1) or (5.1.6.2), where a microscopic system is entangled with a macroscopic one, after a "measurement".

Collapse (or Reduction) of the Quantum State. When one measures a quantity A associated with vectors $|e_n\rangle$ and numbers λ_n, if the pre-measurement state of the system is given by

$$|\text{state}\rangle = \sum_n c_n |e_n\rangle \,,$$

and if the observed result is λ_k, the post-measurement state becomes $|e_k\rangle$. See Sects. 2.2 and 2.3 and Appendix 2.B for the definition of this rule and Sect. 5.1.6 and Appendix 5.E for the corresponding notion ("effective collapse") in the de Broglie–Bohm theory.

Compatibility. Two quantities are compatible if one can measure them both together, that is, if the measurement of one quantity does not affect the measurement of the other. Mathematically, this means that the matrices or operators that represent those quantities commute. See Appendices 2.B and 2.C.2.

Complementarity. A concept central to Bohr's thinking, referring to the fact that there exist pairs of quantities that one cannot measure simultaneously (those that are mutually incompatible). Indeed, because of the collapse rule, measuring one quantity will, in general, affect the result of the measurement of the other quantity. See Sect. 2.1 and Appendix 2.C.

Contextuality. This refers to the fact that, in general, measurements do not measure a pre-existing property of a particle (see Sect. 2.5). In the de Broglie–Bohm theory, this is explained by the fact that the measurement of any "observable" other than position depends on the details of the experimental arrangement used to "measure" it. See Sects. 5.1.4 and 5.1.5.

Copenhagen Interpretation. This refers to a set of ideas, whose main defenders were Bohr, Born, Heisenberg, and Pauli, along with Jordan, Rosenfeld, Landau, von Neumann, Wigner, and Wheeler. These ideas emphasize the role of observations in the very formulation of the quantum mechanical laws. For some authors, the "observer" is an inanimate macroscopic object, for others a conscious subject, but in any case, it is introduced as a *deus ex machina* in the quantum formalism. This formulation lingers in most textbooks or popular presentations of quantum mechanics and has given rise to many philosophical commentaries. We discuss it in Sects. 1.2 and 2.5 and Chap. 8.

Decoherence. This refers to the fact that, after an interaction with a macroscopic system, the supports of the wave functions corresponding to different possible results of the interaction no longer overlap in configuration space. See Sect. 2.5 for the definition of decoherence and Sect. 5.1.6 and Appendix 5.E for the role of decoherence in the de Broglie–Bohm theory.

Determinism. A dynamical system is deterministic if, once the state of the system at a given time is specified, the dynamics determines a unique state for all later times.

Empirical Distribution. Consider a real-valued random variable ω with a probability density $p(x)$: for $B \subset \mathbb{R}$, $P(\omega \in B) = \int_B p(x)dx$. Let $\Omega = (\omega_i)_{i=1}^N$ be N independent variables with identical distributions whose density is $p(x)$. We define the empirical distribution of these N variables, for any subset $B \subset \mathbb{R}$ by

$$\rho_N(\Omega, B) = \frac{\left|\{i = 1, \ldots, N | \omega_i \in B\}\right|}{N} = \frac{\sum_{i=1}^N \mathbb{1}_B(\omega_i)}{N},$$

where $\mathbb{1}_B$ is the indicator function of the set B. One can show that $\lim_{N \to \infty} \rho_N(\Omega, B) = \int_B p(x)dx$, with probability 1. So $p(x)$ is the empirical density distribution of the variables $\Omega = (\omega_i)_{i=1}^\infty$. See Appendix 3.A for precise statements and for a similar definition for discrete random variables.

Entanglement. This refers to a situation comprising two quantum systems such as the one described by (4.4.2.1), in which the state is a not a product of states, one for each system, but a sum of such products. States of the form (2.5.1.1) or (2.D.8) are also entangled.

EPR (and EPRB). This refers to Einstein, Podolsky, and Rosen who showed that, in certain situations, assuming that the world is local implies that quantum mechanics is incomplete, in the sense that certain "hidden variables" must have definite values independently of any measurement. This is because conservation laws that hold in quantum mechanics allow us to predict with certainty the result of certain measurements done on one quantum mechanical system by doing another measurement on a similar system far away from the first. If there are no actions at a distance, reasoned Einstein, Podolsky, and Rosen, then the distant measurement cannot affect the first system and the latter must have had the predictable properties all along, independently of any measurement. The original argument involved positions and momenta, but it was reformulated in terms of spin by Bohm (hence the acronym EPRB). This is discussed in Chap. 4. For misunderstandings of this result, see Sect. 7.1.4.

Equivariance. If we assume, for a system of N variables and some initial time 0, that the positions of a set of particles have an empirical density distribution $\rho(\mathbf{x}, 0) = |\Psi(\mathbf{x}, 0)|^2$, where $\Psi(\mathbf{x}, 0)$ is the quantum state of a single particle in $\mathbb{R}^3$ at time 0, then, at any later time $t > 0$, we will have

$$\rho(\mathbf{x}, t) = |\Psi(\mathbf{x}, t)|^2 \,,$$

where $\Psi(\mathbf{x}, t)$ is the solution to Schrödinger's equation (5.1.1.1), $\rho(\mathbf{x}, t)$ is the empirical density distribution defined in (5.1.3.1), and $\phi^t(\mathbf{x})$ in (5.1.3.1) comes from the deterministic de Broglie–Bohm dynamics through (5.1.1.4). See Sects. 5.1.3 and 5.1.4 and Appendix 5.C for the definition, use, and proof of equivariance in the de Broglie–Bohm theory.

FAPP. An acronym introduced by John Bell, meaning "for all practical purposes". It is used to mean that the ordinary quantum formalism is perfectly adequate in order to predict results of measurements or for technological applications. The "only" thing the de Broglie–Bohm theory does is to explain what the formalism means and why it works so well.

Heisenberg Uncertainty Principle. A lower bound on the variance of the statistics of measurements of positions and momenta of a particle, or more generally on the variance of the statistics of measurements of a pair of "incompatible observables", viz., pairs of observables such that the measurement of one of them may affect the result of a measurement of the other. See Appendix 2.C for a precise definition and Sect. 5.1.8 for the status of that "principle" in the de Broglie–Bohm theory.

Hidden Variables. Any variables that would characterize the state of an individual system beyond the quantum state. In the de Broglie–Bohm theory, the hidden variables are the positions of the particles (see Sect. 5.1.1). Assuming that there are hidden variables other than the positions of the particles that determine the result of a measurement usually runs into contradictions because of the no hidden variables theorems (see Sect. 2.5).

Interference. This refers to the possibility for two wave functions whose supports are disjoint to evolve into a situation where their supports overlap again. Examples of such a phenomenon include the Mach–Zehnder interferometer (see Sect. 2.2) and the double-slit experiment (see Appendix 2.E). See also Sect. 5.2.1.

Law of Large Numbers. This says that the empirical distribution $\rho_N(\Omega, B)$ converges, as $N \to \infty$, to the probability distribution $\int_B p(x)dx$, with probability one, for suitable sets B. See Appendix 3.A for precise statements.

Measurement. The most misleading term in traditional discussions about quantum mechanics, since it gives the impression that some pre-existing property of the quantum system is being revealed by the "measurement". This is discussed in Sect. 2.5, where this naive view of measurements is shown to be untenable because of the no hidden variables theorem. In the de Broglie–Bohm theory, measurements are viewed as interactions between a microscopic system and a piece of apparatus, interactions whose statistical results are given by the usual quantum formalism (see Sect. 5.1.4).

No Hidden Variables Theorems. A set of theorems that show that one cannot assign values to various sets of observables that would pre-exist their "measurement" and that the latter would simply reveal. In Sect. 2.5 and Appendix 2.F, we state and prove the main no hidden variables theorem used in this book. Bell's result in Chap. 4 is also a no hidden variables theorem. We mention another such theorem in Sect. 5.3.4 and use yet another version of those theorems in Sect. 6.3 and Appendix 6.C. However, the best known no hidden variables theorem is the one due to John von Neumann, and that theorem is quite misleading (see Sect. 7.4).

Nonlocality. A theory is nonlocal if it allows actions at a distance. See Chap. 4 and Sect. 5.2.1.

Observable. Any quantity that can be "measured". Sometimes observables correspond to quantities that have a classical analogue, such as position, momentum, angular momentum, or energy; but there are other observables, such as the spin, that do not have a classical analogue. Mathematically, observables are associated with self-adjoint matrices or operators, but essentially with a family of states and numbers, so that the results of the measurement of an observable are described by Born's rule and the collapse of the quantum state. See Sect. 2.5 and Appendix 2.B for the definition of observables and Sects. 5.1.4 and 5.1.6 and Appendix 5.E for the status of observables in the de Broglie–Bohm theory.

Positivism. This refers to logical positivism, or logical empiricism, a philosophical school of thought originally centered around the post-World War I Vienna Circle. The aim was to base all our knowledge on direct observations and logical deductions. See Sect. 7.7 and Chap. 8.

Probabilities, Probability Densities, and Conditional Probabilities. The probability distribution P of a random variable taking continuous values, e.g., in $\mathbb{R}$, is often given in terms of an integrable function $p(z)$ such that, for $B \subset \mathbb{R}$, $P(x \in B) = \int_B p(z)dz$. Then $p(z)$ is the probability density of P. The conditional probability of an event A, given an event B, is given by the usual formula $P(A, B)/P(B)$. For continuous random variables, one can also associate a conditional probability density with a conditional probability. We use conditional probabilities in Sect. 5.1.7.

Quantum Equilibrium. A system composed of many copies of identical subsystems, each of them having an effective wave function $\psi(x, t)$, is in quantum equilibrium if the empirical density distribution $\rho(x, t)$ of the positions of the system is given by $\rho(x, t) = |\psi(x, t)|^2$. See Sect. 5.1.7.

Quantum State. Either a wave function $\Psi(x, t)$ or a linear combination of products of a spatial part, viz., the wave function, and a spin part, e.g., a product of the form $\Psi(x, t)|1 \uparrow\rangle$ or a linear combination of such products, such as $\Psi^\uparrow(x, t)|1 \uparrow\rangle + \Psi^\downarrow(x, t)|1 \downarrow\rangle$. See Sect. 2.3.

Randomness. One must distinguish between intrinsic or genuine randomness and effective randomness. The first notion refers to what defines a stochastic process, for example a Markov chain or the spontaneous collapse theories of Sect. 6.2. The second notion refers to the fact that a deterministic system may appear random, or be effectively random, at least for certain dynamics and initial conditions. See Sects. 3.4.2 and 3.4.5 for general examples of effective randomness and Sects. 5.1.3 and 5.1.7 for the effective randomness of the de Broglie–Bohm theory.

Schrödinger's Equation. This governs the time evolution of the quantum state. Other equations, such as the Dirac or Pauli equations, play a similar role for particles with spin. See Appendix 2.A.2 for the definition of Schrödinger's equation, and see Sect. 5.1.2 and Appendix 5.D for its role in the de Broglie–Bohm theory.

Solipsism. The idea that everything outside my consciousness is an illusion, or that I am constantly dreaming. This is a radical version of idealism, but idealists cannot easily avoid falling into this position, as we discuss in Sect. 3.1.

Spin. This is often considered as the quintessentially quantum property of a particle, since it has no classical analogue. We discuss it phenomenologically in Sects. 2.1 and 2.2. From the point of view of the de Broglie–Bohm theory, the existence of spin is expressed by the fact that the quantum state is a linear combination of products of a wave function and a spin state. The value of the spin is not an intrinsic property of a particle but the result of interactions between that particle and Stern–Gerlach setups. See Sects. 5.1.4 and 5.1.5.

State (or Complete State). In the de Broglie–Bohm theory, the complete state of a system with N variables at time t is a pair $\big(\Psi(t), \mathbf{X}(t)\big)$, where $\Psi(t) = \Psi(x_1, \ldots, x_N, t)$ is the usual quantum state and $\mathbf{X}(t) = \big(X_1(t), \ldots, X_N(t)\big)$ are the actual positions of the particles. Capital letters denote the positions of the particles and lower case letters the arguments of $\Psi(x_1, \ldots, x_N, t)$. See Sect. 5.1.1.

Typicality. A sequence of realizations of N independent identically distributed random variables $\mathbf{x} = (x_1, \ldots, x_N)$ is typical if the corresponding empirical density distribution is close to the probability distribution of those variables. See Appendix 3.A for a precise definition; by the law of large numbers, most sequences of realizations of N independent identically distributed random variables are typical for $N \to \infty$. See Sect. 5.1.7 and Appendix 5.C for the use of the notion of typicality in the de Broglie–Bohm theory.

Underdetermination. Theories are always underdetermined by data, or in other words, there always exist different theories that can "account for" any given set of facts. In philosophy of science, this is called the Duhem–Quine thesis. It is discussed in Sect. 3.2.1. Most of these theories are ad hoc and have no predictive power. But more genuine cases of underdetermination also exist, for example, in the de Broglie–Bohm theory. See Sect. 5.4.1.

Wave Function. The spatial part of the quantum state, $\Psi(x, t)$, whose meaning in the usual quantum mechanical formalism is that $|\Psi(x, t)|^2$ is the probability density of the particle being found at x upon measurement at time t. In the de Broglie–Bohm theory, $|\Psi(x, t)|^2$ is the probability density of the particle actually being at x at time t. See Sects. 2.3 and 5.1.3.

References

1. A. Ádám, L. Jánossy, P. Varga, Coincidence between photons contained in coherent light. Acta Phys. Hung. **4**, 301 (1955)
2. Y. Aharonov, D. Bohm, Significance of electromagnetic potentials in the quantum theory. Phys. Rev. **115**, 485–491 (1959)
3. Y. Aharonov, D.Z. Albert, L. Vaidman, How the result of a measurement of a component of the spin of a spin-1/2 particle can turn out to be 100. Phys. Rev. Lett. **60**, 1351–1354 (1988)
4. Y. Aharonov, J. Anandan, L. Vaidman, Meaning of the wave function. Phys. Rev. A **47**, 4616–4626 (1993)
5. D. Albert, B. Loewer, Interpreting the many-worlds interpretation. Synthese **77**, 195–213 (1988)
9. D. Albert, B. Loewer, Two no-collapse interpretations of quantum mechanics. Noûs **12**, 121–138 (1989)
7. D. Albert, B. Loewer, The measurement problem: some "solutions". Synthese **86**, 87–98 (1991)
8. D. Albert, *Quantum Mechanics and Experience* (Harvard University Press, Cambridge, 1992)
9. D. Albert, Bohm's alternative to quantum mechanics. Sci. Am. **270**, 32–39 (May 1994)
10. M.G. Alford, Ghostly action at a distance: a non-technical explanation of the Bell inequality, preprint arXiv:1506.02179v1 (2015)
11. V. Allori, D. Dürr, S. Goldstein, N. Zanghì, Seven steps towards the classical world. J. Opt. B 4, 482–488 (2002) (Reprinted in [153, Chap. 5])
12. V. Allori, Fundamental Physical Theories: Mathematical Structures Grounded on a Primitive Ontology. Ph.D. Thesis, Department of Philosophy, Rutgers University (2007)
13. V. Allori, S. Goldstein, R. Tumulka, N. Zanghì, On the common structure of Bohmian mechanics and the Ghirardi-Rimini-Weber theory. Br. J. Philos. Sci. **59**, 353–389 (2008)
14. V. Allori, S. Goldstein, R. Tumulka, N. Zanghì, Many-worlds and Schrödinger's first quantum theory. Br. J. Philos. Sci. **62**, 1–27 (2011)
15. V. Allori, S. Goldstein, R. Tumulka, N. Zanghì, Predictions and primitive ontology in quantum foundations: a study of examples. Br. J. Philos. Sci. **65**, 323–352 (2014)
16. H. Almanspacher, The hidden side of Wolfgang Pauli. Extraordinary encounter with depth psychology. J. Conscious. Stud. **3**, 112–126 (1996)
17. J. Anandan, H.R. Brown, On the reality of space-time geometry and the wavefunction. Found. Phys. **25**, 349–360 (1995)
18. V.I. Arnold, *Ordinary Differential Equations* (MIT Press, Cambridge, 1973)
19. A. Aspect, J. Dalibard, G. Roger, Experimental tests of realistic local theories via Bell's theorem. Phys. Rev. Lett. **47**, 463–464 (1981)

© Springer International Publishing Switzerland 2016
J. Bricmont, *Making Sense of Quantum Mechanics*,
DOI 10.1007/978-3-319-25889-8

20. A. Aspect, J. Dalibard, G. Roger, Experimental realization of Einstein-Podolsky-Rosen-Bohm Gedanken experiment: a new violation of Bell's inequalities. Phys. Rev. Lett. **49**, 91–94 (1982)
21. A. Aspect, J. Dalibard, G. Roger, Experimental test of Bell's inequalities using time-varying analysers. Phys. Rev. Lett. **49**, 1804–1807 (1982)
22. G. Auletta, *Foundations and Interpretation of Quantum Mechanics. With a Foreword by Giorgio Parisi* (World Scientific, Singapore, 2001)
23. G. Bacciagaluppi, The statistical interpretation according to Born and Heisenberg, in *HQ-1: Conference on the History of Quantum Physics*, (vol. II), ed. by C. Joas, C. Lehner, J. Renn. Max-Planck-Institut für Wissenschaftsgeschichte, Berlin, preprint series, vol. 350 (2008) pp. 269–288
24. G. Bacciagaluppi, A. Valentini, *Quantum Mechanics at the Crossroads. Reconsidering the 1927 Solvay Conference* (Cambridge University Press, Cambridge, 2009)
25. G. Bacciagaluppi, E. Crull, Heisenberg (and Schrödinger, and Pauli) on hidden variables. Stud. Hist. Philos. Sci. Part B: Stud. Hist. Philos. Modern Phys. **40**, 374–382 (2009)
26. G. Bacciagaluppi, J. Ismael, Review of the emergent multiverse by David Wallace [501]. Philos. Sci. **82**, 129–148 (2015)
27. L.E. Ballentine, The statistical interpretation of quantum mechanics. Rev. Modern Phys. **42**, 358–380 (1970)
28. J.A. Barrett, The persistence of memory: surreal trajectories in Bohm's theory. Philos. Sci. **67**, 680–703 (2000)
29. A. Bassi, G.C. Ghirardi, Can the decoherent histories description of reality be considered satisfactory? Phys. Lett. A **257**, 247–263 (1999)
30. A. Bassi, G.C. Ghirardi, Decoherent histories and realism. J. Stat. Phys. **98**, 457–494 (2000)
31. A. Bassi, G.C. Ghirardi, About the notion of truth in the decoherent histories approach: a reply to Griffiths. Phys. Lett. A **265**, 153–155 (2000)
32. A. Bassi, G.C. Ghirardi, Dynamical reduction models. Phys. Rep. **379**, 257–427 (2003)
33. D. Bedingham, D. Dürr, G.C. Ghirardi, S. Goldstein, R. Tumulka, N. Zanghì, Matter density and relativistic models of wave function collapse. J. Stat. Phys. **154**, 623–631 (2014)
34. F.J. Belinfante, *A Survey of Hidden-Variables Theories* (Pergamon Press, Oxford, 1973)
35. J.S. Bell, On the Einstein-Podolsky-Rosen paradox. Physics **1**, 195–200 (1964) (Reprinted as Chap. 2 in [49])
36. J.S. Bell, On the problem of hidden variables in quantum mechanics. Rev. Modern Phys. **38**, 447–452 (1966) (Reprinted as Chap. 1 in [49])
37. J.S. Bell, The theory of local beables. Epistemol. Lett. (1976) (Reprinted as Chap. 7 in [49])
38. J.S. Bell, de Broglie-Bohm, delayed-choice double-slit experiment, and density matrix. Int. J. Quantum Chem.: Quantum Chem. Symp. **14**, 155–159 (1980) (Reprinted as Chap. 14 in [49])
39. J.S. Bell, Quantum mechanics for cosmologists, in *Quantum Gravity 2*, ed. by C. Isham, R. Penrose, D. Sciama (Oxford University Press, New York, 1981), pp. 611–637 (Reprinted as Chap. 15 in [49])
40. J.S. Bell, Bertlmann's socks and the nature of reality. Journal de Physique **42** C2, 41–61 (1981) (Reprinted as Chap. 16 in [49])
41. J.S. Bell, On the impossible pilot wave. Found. Phys. **12**, 989–999 (1982) (Reprinted as Chap. 17 in [49])
42. J.S. Bell, Beables for quantum field theory. CERN-TH 4035/84 (1984) (Reprinted as Chap. 19 of [49])
43. J.S. Bell, Six possible worlds of quantum mechanics, in *Proceedings of the Nobel Symposium 65: Possible Worlds in Arts and Science*, Stockholm, August 11–15 (1986) (Reprinted as Chap. 20 of [49])
44. J.S. Bell, Are there quantum jumps? in *Schrödinger. Centenary Celebration of a Polymath*, ed. by C.W. Kilmister (Cambridge University Press, Cambridge, 1987), pp. 41–52 (Reprinted as Chap. 22 of [49])
45. J. Bell, John Bell, Physicist, interview by Charles Mann and Robert Crease, Omni Magazine, May 1988, pp. 85–92, 121

46. J.S. Bell, Against measurement. Phys. World **3**, 33–40 (1990) (Reprinted as Chap. 23 in [49])
47. J.S. Bell, Unpublished interview with the philosopher Renée Weber (1990)
48. J.S. Bell, La nouvelle cuisine, in *Between Science and Technology*, ed. by A. Sarlemijn, P. Kroes (Elsevier Science Publishers, Amsterdam, 1990) (Reprinted as Chap. 24 in [49])
49. J.S. Bell, *Speakable and Unspeakable in Quantum Mechanics. Collected Papers on Quantum Philosophy*, 2nd edn, with an introduction by Alain Aspect (Cambridge University Press, Cambridge, 2004) (1st edn 1993, The page numbers refer to the first edition)
50. M. Beller, The rhetoric of antirealism and the Copenhagen spirit. Philos. Sci. **63**, 183–204 (1996)
51. M. Beller, The Sokal hoax: at whom are we laughing? Phys. Today **51**, 29–34 (1998)
52. M. Beller, *Quantum Dialogue: The Making of a Revolution* (University of Chicago Press, Chicago, 1999)
53. G. Berkeley, *On the Principles of Human Knowledge*, in *Complete Works*, ed. by Fraser, vol. 1 (Oxford University Press, Oxford, 1901)
54. K. Berndl, D. Dürr, S. Goldstein, G. Peruzzi, N. Zanghì, On the global existence of Bohmian mechanics. Commun. Math. Phys. **173**, 647–673 (1995)
55. K. Berndl, M. Daumer, D. Dürr, S. Goldstein, N. Zanghì, A survey of Bohmian mechanics. Il Nuovo Cimento **110B**, 737–750 (1995)
56. J. Bernstein, *Quantum Profiles* (Princeton University Press, Princeton, 1991)
57. R.A. Bertlmann, A. Zeilinger (eds.), *Quantum [Un]speakables From Bell to Quantum Information* (Springer, Berlin, 2002)
58. G. Birkhoff, J. von Neumann, The logic of quantum mechanics. Ann. Math. **37**, 823–843 (1936)
59. D.I. Blokhintsev, *The Philosophy of Quantum Mechanics* (Reidel Dordrecht, 1968). Original Russian edition (1965)
60. P. Boghossian, *Fear of Knowledge: Against Relativism and Constructivism* (Oxford University Press, Oxford, 2006)
61. D. Bohm, *Quantum Theory* (Dover Publications, New York, 1989) First edition (Prentice Hall, Englewood Cliffs, 1951)
62. D. Bohm, A suggested interpretation of the quantum theory in terms of "hidden variables", Parts 1 and 2. Phys. Rev. **89**, 166–193 (1952) (Reprinted in [510] pp. 369–390)
63. D. Bohm, Reply to a criticism of the causal re-interpretation of quantum theory. Phys. Rev. **87**, 389–390 (1952)
64. D. Bohm, Letter to Miriam Yevick [n.d.] quoted in [386], p. 128
65. D. Bohm, Letter to Léon Rosenfeld [n.d.], *Rosenfeld Papers, Correspondance générale, Volume de Broglie* (The Niels Bohr Archive, Copenhagen, 1952)
66. D. Bohm, *A discussion of certain remarks by Einstein on Born's probability interpretation of the ψ-function, in Scientific Papers Presented to Max Born* (Hafner, New York, 1953), pp. 13–19
67. D. Bohm, Proof that probability density approaches $|\psi|^2$ in causal interpretation of quantum theory. Phys. Rev. **89**, 458–466 (1953)
68. D. Bohm, *Causality and Chance in Modern Physics* (Harper, New York, 1957)
69. D. Bohm, C. Dewdney, B.J. Hiley, A quantum potential approach to the Wheeler delayed-choice experiment. Nature **315**, 294–297 (1985)
70. D. Bohm, B.J. Hiley, *The Undivided Universe* (Routledge, London, 1993)
71. N. Bohr, The quantum of action and the description of nature, in *Atomic Theory and the Description of Nature*, ed. by N. Bohr (Cambridge University Press, Cambridge, 1934) new edition 2011. Originally published in German: Wirkungsquantum und Naturbeschreibung, Die Naturwissenschaften **17**, 483–486 (1929)
72. N. Bohr, Can quantum mechanical description of reality be considered complete? Phys. Rev. **48**, 696–702 (1935)
73. N. Bohr, Discussion with Einstein on epistemological problems in atomic physics (in, [436], pp. 201–241)

74. M. Born, P. Jordan, Zur Quantenmechanik, I. Zeitschrift für Physik **34**, 858–888 (1925) (English transation in [491])
75. M. Born, W. Heisenberg, P. Jordan, Zur Quantenmechanik, II. Zeitschrift für Physik **35**, 557–615 (1926) (English transation in [491])
76. M. Born, *Natural Philosophy of Cause and Chance* (Clarendon, Oxford, 1949)
77. M. Born, Continuity, determinism and reality. Dan. Mat. Fys. Medd. **30**, 1–26 (1955)
78. M. Born, *Physics in My Generation. A Collection of Papers* (Pergamon, London, 1956)
79. M. Born (ed.), *The Born-Einstein Letters* (Macmillan, London, 1971)
80. J. Bricmont, Science of chaos, or chaos in science? Physicalia Mag. **17**, 159–208 (1995). P.R. Gross, N. Levitt, M.W. Lewis (eds.), *The Flight from Science and Reason Annals of the New York Academy of Sciences*, **775**, 131–175 (1996)
81. J. Bricmont, What is the meaning of the wave function? in *Fundamental Interactions: From Symmetries to Black Holes*, ed. by J.-M. Frère, M. Henneaux, A. Sevrin, Ph. Spindel, pp. 53–67, Conference held on the occasion of the "Eméritat" of François Englert, Université Libre de Bruxelles, Brussels (1999)
82. J. Bricmont, Sociology and epistemology, in *After Postmodernism. An Introduction to Critical Realism*, ed. by J. Lopez, G. Porter (Athlone Press, London, 2001)
83. J. Bricmont, Bayes, Boltzmann and Bohm: Probabilities in Physics (in, [84], pp. 3–21)
84. J. Bricmont, D. Dürr, M.C. Galavotti, G. Ghirardi, F. Petrucione, N. Zanghì (eds.), *Chance in Physics. Foundations and Perspective* (Springer, Berlin, 2001)
85. P.W. Bridgman, Review of Louis de Broglie's book. *Non-Linear Wave Mechanics: A Causal Interpretation.* Sci. Am. **203**, 206 (October 1960)
86. H.R. Brown, D. Wallace, Solving the measurement problem: de Broglie-Bohm loses out to Everett. Found. Phys. **35**, 517–540 (2005)
87. P. Byrne, The many-worlds of Hugh Everett. Sci. Am. **297**, 98–105 (December 2007)
88. A. Cabello, Bibliographic guide to the foundations of quantum mechanics and quantum information. arXiv:quant-ph/0012089
89. C. Callender, R. Weingard, Trouble in paradise? Problems for Bohm's theory. Monist **80**, 24–43 (1997)
90. K. Camilieri, *Heisenberg and the Interpretation of Quantum Mechanics. The Physicist as Philosopher* (Cambridge University Press, Cambridge, 2009)
91. R. Carnap, Überwindung der Metaphysik durch logische Analyse der Sprache. Erkenntnis **2**, 220–241 (1931). English translation: The elimination of metaphysics through logical analysis of language, in *Logical Positivism*, ed. by A.J. Ayer (The Free Press, Glencoe, Ill, 1959), pp. 60–81
92. S. Carroll, *The Particle at the End of the Universe: How the Hunt for the Higgs Boson Leads Us to the Edge of a New World* (Oneworld Publications, London, 2012)
93. H.B.G. Casimir, *Koninklijke Nederlandse Akademie van Wetenschappen 1808–1958* (Noord-Hollandsche Uitgeversmij, Amsterdam, 1958), pp. 243–251
94. C. Caves, C. Fuchs, R. Schack, Quantum probabilities as Bayesian probabilities. Phys. Rev. A **65**, 022305 (2002)
95. J.F. Clauser, M.A. Horne, A. Shimony, R.A. Holt, Proposed experiment to test local hidden-variable theories. Phys. Rev. Lett. **23**, 880–884 (1969) (Reprinted in [510], pp. 409–413)
96. J.F. Clauser, Von Neumann's informal hidden-variable argument. Am. J. Phys. **39**, 1095–1099 (1971)
97. J.F. Clauser, Early history of Bell's theorem, in *Quantum [Un]speakables: From Bell to Quantum Information*, ed. by R.A. Bertlmann, A. Zeilinger (Springer, Berlin, 2002)
98. R. Clifton, Complementarity between position and momentum as a consequence of Kochen-Specker arguments. Phys. Lett. A **271**, 1–7 (2000)
99. S. Colin, W. Struyve, A Dirac sea pilot-wave model for quantum field theory. J. Phys. A **40**, 7309–7342 (2007)
100. S. Colin, W. Struyve, Quantum non-equilibrium and relaxation to equilibrium for a class of de Broglie-Bohm-type theories. New J. Phys. **12**, 043008 (2010)

101. J.T. Cushing, E. MacMullin (eds.), *Philosophical Consequences of Quantum Theory. Reflections on Bell's Theorem* (University of Notre Dame Press, Notre Dame, 1989)
102. J.T. Cushing, *Quantum Mechanics. Historical Contingency and the Copenhagen Hegemony* (University of Chicago Press, Chicago, 1994)
103. J.T. Cushing, A. Fine, S. Goldstein (eds.), *Bohmian Mechanics and Quantum Theory: An Appraisal* (Kluwer Academic Publishers, Dordrecht, 1996)
104. T. Damour, *Once Upon Einstein* (A.K. Peters, Wellesley, 2006)
105. M. Daumer, D. Dürr, S. Goldstein, N. Zanghì, Naive realism about operators. Erkenntnis **45**, 379–397 (1996)
106. M. Daumer, D. Dürr, S. Goldstein, T. Maudlin, R. Tumulka, N. Zanghì, The Message of the Quantum? (2006). arxiv.org/pdf/quant-ph/0604173.pdf
107. P.C.W. Davies, J.R. Brown (eds.), *The Ghost in the Atom: A Discussion of the Mysteries of Quantum Physics* (Cambridge University Press, Cambridge, 1993)
108. L. de Broglie, Recherches sur la théorie des quanta. Ph.D. Thesis, Université de Paris (1924)
109. L. de Broglie, Sur la dynamique du quantum de lumière et les interférences. C. R. Acad. Sci. Paris **179**, 1039–1041 (1924) (Reprinted in [116, pp. 23–25])
110. L. de Broglie, Sur la possibilité de relier les phénomènes d'interférence et de diffraction à la théorie des quantas de lumières. C. R. Acad. Sci. Paris **183**, 447 (1926) (Reprinted in [116, pp. 25–27])
111. L. de Broglie, La structure atomique de la matière et du rayonnement et la mécanique ondulatoire. C. R. Acad. Sci. Paris **184**, 273–274 (1927) (Reprinted in [116, pp. 27–29])
112. L. de Broglie, La mécanique ondulatoire et la structure atomique de la matière et du rayonnement. Le Journal de Physique et le Radium **8**, 225–241 (1927) (Reprinted, with additional comments, in [116, pp. 29–61])
113. L. de Broglie, *An Introduction to the Study of Wave Mechanics* (E.P. Dutton and company, New York, 1930)
114. L. de Broglie, Remarques sur la théorie de l'onde-pilote. C. R. Acad. Sci. Paris **233**, 641–644 (1951) (Reprinted in [116, pp. 65–69])
115. L. de Broglie, Sur la possibilité d'une interprétation causale et objective de la mécanique ondulatoire. C. R. Acad. Sci. Paris **234**, 265–268 (1952) (Reprinted in [116, pp. 69–72])
116. L. de Broglie, *La physique quantique restera-t-elle indéterministe?* (Gauthier-Villars, Paris, 1953)
117. L. de Broglie, *Une tentative d'interprétation causale et nonlinéaire de la mécanique ondulatoire* (Gauthier-Villars, Paris, 1956). English translation: *Non-linear Wave Mechanics: A Causal Interpretation* (Elsevier, Amsterdam, 1960)
118. L. de Broglie, L'interprétation de la mécanique ondulatoire. Le Journal de Physique et le Radium **20**, 963–979 (1959)
119. L. de Broglie, *The Current Interpretation of Wave Mechanics: A Critical Study* (Elsevier, Amsterdam, 1964)
120. R. Deltete, R. Guy, Einstein's opposition to the quantum theory. Am. J. Phys. **58**, 673–683 (1990)
121. E. Dennis, T. Norsen, Quantum theory: interpretation cannot be avoided. arxiv.org/abs/quant-ph/0408178
122. E. Deotto, G.C. Ghirardi, Bohmian mechanics revisited. Found. Phys. **28**, 1–30 (1998)
123. R. Descartes, *Meditations and Other Metaphysical Writings*, trans. Desmond M. Clarke (Penguin, London, 1998) original in Latin, 1641
124. B. d'Espagnat, The quantum theory and reality. Sci. Am. **241**, 158–181 (November 1979)
125. B. d'Espagnat, Quantum weirdness: what we call 'reality' is just a state of mind. The Guardian (20 March 2009)
126. M. Devitt, *Realism and Truth*, 2nd edn. (Princeton University Press, Princeton, 1991)
127. D. Deutsch, Comment on Lockwood. Br. J. Philos. Sci. **47**, 222–228 (1996)
128. D. Deutsch, Quantum theory of probability and decisions. Proc. Royal Soc. Lond. A **455**, 3129–3137 (1999)

129. D. Deutsch, A. Ekert, R. Lupacchini, Machines, logic and quantum physics. Bull. Symbol. Logic **6**, 265–283 (2000)
130. C. Dewdney, L. Hardy, E.J. Squires, How late measurements of quantum trajectories can fool a detector. Phys. Lett. A **184**, 6–11 (1993)
131. B. DeWitt, Quantum mechanics and reality. Phys. Today **23**, 30–35 (1970) (Reprinted in [132, pp. 155–165])
132. B. DeWitt, R.N. Graham (eds.), *The Many-Worlds Interpretation of Quantum Mechanics* (Princeton University Press, Princeton, 1973)
133. D. Diderot, *Letter on the Blind*, in *Diderot's Early Philosophical Works*, trans. and ed. by M. Jourdain (The Open Court Pub. Co., Chicago, 1916)
134. L. Diósi, Gravitation and quantum-mechanical localization of macro-objects. Phys. Lett. A **105**, 199–202 (1984)
135. L. Diósi, Models for universal reduction of macroscopic quantum. Phys. Rev. A **40**, 1165–1174 (1989)
136. L. Diósi, Gravity-related spontaneous wave function collapse in bulk matter. New J. Phys. **16**, 105006 (2014)
137. P.A.M. Dirac, *The Principles of Quantum Mechanics* (Oxford University Press, Oxford, 1930)
138. P.A.M. Dirac, The evolution of the physicist's picture of nature. Sci. Am. **208**, 45–53 (May 1963) (Reprinted in Scientific American, June 25, 2010)
139. P. Duhem, *The Aim and Structure of Physical Theory*. Translated by P. Wiener (Princeton University Press, Princeton, 1954) French original: *La Théorie physique: son objet, sa structure*, 2e éd. revue et augmentée (Rivière, Paris 1914)
140. H. Dukas, B. Hoffmann (eds.), *Albert Einstein: The Human Side* (Princeton University Press, Princeton, 1979)
141. D. Dürr, S. Goldstein, N. Zanghì, Quantum equilibrium and the origin of absolute uncertainty. J. Stat. Phys. **67**, 843–907 (1992) (Reprinted in [153, Chap. 2])
142. D. Dürr, S. Goldstein, N. Zanghì, Quantum chaos, classical randomness and Bohmian mechanics. J. Stat. Phys. **68**, 259–270 (1992)
143. D. Dürr, W. Fusseder, S. Goldstein, N. Zanghì, Comment on "Surrealistic Bohm Trajectories". Z. Naturforsch **48a**, 1261–1262 (1993)
144. D. Dürr, S. Goldstein, N. Zanghì, Bohmian mechanics and the meaning of the wave function, contribution to experimental metaphysics, in: *Quantum Mechanical Studies for Abner Shimony*, ed. by R.S.Cohen, M. Horne, J. Stachel, vol. 1. (Boston Studies in the Philosophy of Science **193**, Kluwer, Dordrecht, 1997), pp. 25–38
145. D. Dürr, S. Goldstein, R. Tumulka, N. Zanghì, Trajectories and particle creation and annihilation in quantum field theory. J. Phys. A: Math. Gen. **36**, 4143–4149 (2003)
146. D. Dürr, S. Goldstein, R. Tumulka, N. Zanghì, Bohmian mechanics and quantum field theory. Phys. Rev. Lett. **93**, 1–4 (2004) (Reprinted in [153, Chap. 10])
147. D. Dürr, S. Goldstein, N. Zanghì: Quantum equilibrium and the role of operators as observables in quantum theory. J. Stat. Phys. **116**, 959–1055 (2004) (Reprinted in [153, Chap. 3])
148. D. Dürr, S. Goldstein, R. Tumulka, N. Zanghì, Quantum Hamiltonians and stochastic jumps. Commun. Math. Phys. **254**, 129–166 (2005)
149. D. Dürr, S. Goldstein, R. Tumulka, N. Zanghì, John Bell and Bell's theorem, in *Encyclopedia of Philosophy*, ed. by D.M. Borchert (Macmillan Reference, New York, 2005)
150. D. Dürr, S. Goldstein, R. Tumulka, N. Zanghì, Bell-type quantum field theories. J. Phys. A: Math. Gen. **38**, R1–R43 (2005)
151. D. Dürr, S. Goldstein, N. Zanghì, On the weak measurement of velocity in Bohmian mechanics. J. Stat. Phys. **13**, 1023–1032 (2009) (Reprinted in [153, Chap. 7])
152. D. Dürr, S. Teufel, *Bohmian Mechanics. The Physics and Mathematics of Quantum Theory* (Springer, Berlin, 2009)
153. D. Dürr, S. Goldstein, N. Zanghì, *Quantum Physics Without Quantum Philosophy* (Springer, Berlin, 2012)
154. D. Dürr, S. Goldstein, T. Norsen, W. Struyve, N. Zanghì, Can Bohmian mechanics be made relativistic? Proc. R. Soc. A **470**, 20130699 (2014)

155. H. Dym, H. McKean, *Fourier Series and Integrals* (Academic Press, New York, 1972)
156. J. Earman, *A Primer on Determinism* (Reidel Publishing Co., Dordrecht, 1986)
157. The queerness of quanta, The Economist (January 7, 1989), pp. 73–76
158. P.H. Eberhard, Bell's theorem and the different concepts of locality. Il Nuovo Cimento **46B**, 392–419 (1978)
159. P. Ehrenfest, Letter to Bohr, 9 July 1931, in *Bohr Scientific Correspondence, Archive for History of Quantum Physics*. Quoted in [274, p. 98]
160. A. Einstein, Über einen die Erzeugung und Verwandlung des Lichtes betreffenden heuristischen Gesichtspunkt. Annalen der Physik **17**, 132–148 (1905). English translation: On a heuristic point of view concerning the production and transformation of light, in *The Collected Papers of Albert Einstein. Volume 2: The Swiss Years: Writings, 1900–1909*, ed. by J. Stachel, D.C. Cassidy, J. Renn, R. Schulmann (Princeton University Press, Princeton, 1990), pp. 149–169
161. A. Einstein, Entwicklung unserer Anschauungen über das Wesen und die Konstitution der Strahlung. Physikalische Zeitschrift **10**, 817–825 (1909). English translation: On the development of our views concerning the nature and constitution of radiation, in *The Collected Papers of Albert Einstein. Volume 2: The Swiss Years: Writings, 1900–1909*, ed. by J. Stachel, D.C. Cassidy, J. Renn, R. Schulmann (Princeton University Press, Princeton, 1990), pp. 563–583
162. A. Einstein, Letter to Paul Langevin (December 16, 1924) (Einstein Archives, Jerusalem), 15–376
163. A. Einstein, Letter to Erwin Schrödinger (May 31, 1928) (in [407, p. 31])
164. A. Einstein, B. Podolsky, N. Rosen, Can quantum mechanical description of reality be considered complete? Phys. Rev. **47**, 777–780 (1935)
165. A. Einstein, Letter to Erwin Schrödinger (June 19, 1935) (in [186, p. 35])
166. A. Einstein, L. Infeld, The Evolution of Physics: From Early Concepts to Relativity and Quanta, ed. by C.P. Snow (Cambridge University Press, Cambridge, 1938)
167. A. Einstein, Letter to Erwin Schrödinger (August 9, 1939) (in [407, p. 36])
168. A. Einstein, Letter to Cornel Lanczos (March 21, 1942) (in [140, p. 68])
169. A. Einstein, Quantum mechanics and reality. Dialectica **2**, 320–324 (1948)
170. A. Einstein, Remarks concerning the essays brought together in this co-operative volume. In [436, pp. 665–688]
171. A. Einstein, Letter to Maurice Solovine (May 7, 1952) (translated by Don Howard, in [325])
172. A. Einstein, Elementary considerations on the interpretation of the foundations of quantum mechanics, in *Scientific Papers Presented to Max Born on his retirement from the Tait Chair on Natural Philosophy in the University of Edimburgh* (Oliver and Boyd, 1953), pp. 33–40
173. A. Elitzur, L. Vaidman, Quantum mechanical interaction-free measurements. Found. Phys. **23**, 987–997 (1993)
174. F. Engels, *Dialectics of Nature* (Progress Publisher, Moscow, 1976)
175. B-G. Englert, M.O. Scully, G. Sussman, H. Walther, Surrealistic Bohm trajectories. Z. Naturforsch. **47a**, 1175–1186 (1992)
176. B-G. Englert, M.O. Scully, G. Sussman, H. Walther, Reply to comment on "Surrealistic Bohm trajectories". Z. Naturforsch. **48a**, 1263–1264 (1993)
177. M. Esfeld, Essay review. Wigner's view of physical reality. Stud. Hist. Philos. Modern Phys. **30B**, 145–154 (1999)
178. L. Euler, Lettres à une princesse d'Allemagne. Lettre **97** (1791), in *Leonhardi Euleri Opéra Omnia*, série III, Volume 11, Turici, 1911 pp. 219–220. English translation: Refutation of the idealists. In: *Letters of Euler to a German Princess*, translated by H. Hunter, vol. 1 (Thoemmes Press, London, 1997) (originally published 1795) pp. 426–430
179. H. Everett, The Theory of the Universal Wavefunction. Ph.D. thesis, Department of Physics, Princeton University (1955) (Reprinted in [132], pp. 3–140)
180. H. Everett, 'Relative state' formulation of quantum mechanics. Rev. Modern Phys. **29**, 454–462 (1957) (Reprinted in [132, pp. 141–149])
181. J. Faye, Copenhagen Interpretation of Quantum Mechanics, in *The Stanford Encyclopedia of Philosophy* (Fall 2014 Edition), ed. by E.N. Zalta. http://www.plato.stanford.edu/archives/fall2014/entries/qm-copenhagen/

182. I. Fényes, Eine wahrscheinlichkeitstheoretische begründung und interpretation der Quanten-mechanik. Zeitschrift für Physik **132**, 81–106 (1952)

183. R. Feynman, R.B. Leighton, M. Sands, *The Feynman Lectures on Physics, vol. 1 Mainly Mechanics, Radiation, and Heat* (Addison-Wesley, Reading, 1966)

184. R. Feynman, R.B. Leighton, M. Sands, *The Feynman Lectures on Physics, vol. 3, Quantum Mechanics* (Addison-Wesley, Reading, 1966)

185. R. Feynman, *The Character of Physical Law (Messenger Lectures, 1964)* (MIT Press, Boston, 1967)

186. A. Fine, *The Shaky Game: Einstein Realism and the Quantum Theory* (University of Chicago Press, Chicago, 1986)

187. A. Fine, Do correlations need to be explained? (1989) (In [101, pp. 175–194])

188. J. Ford, Chaos: solving the unsolvable, predicting the unpredictable, in *Chaotic Dynamics and Fractals*, ed. by M.F. Barnsley, S.G. Demko (Academic Press, New York, 1986)

189. P. Forman, Weimar culture, causality, and quantum theory: adaptation by German physicists and mathematicians to a hostile environment. Hist. Stud. Phys. Sci. **3**, 1–115 (1971)

190. P. Forman, Kausalität, Anschaulichkeit, and Individualität, or How cultural values prescribed the character and lessons ascribed to quantum mechanics, in *Society and Knowledge. Contemporary Perspectives in the Sociology of Knowledge and Science* ed. by N. Stehr, V. Meja (Transaction Publishers, New Brunswick, 1984), pp. 333–347

191. P. Forman, Behind quantum electronics: National security as basis for physical research in the United States, 1940–1960. Hist. Stud. Phys. Biol. Sci. **18**, 149–229 (1987)

192. S.J. Freedman, J.F. Clauser, Experimental test of local hidden-variable theories. Phys. Rev. Lett. **28**, 938–941 (1972) (Reprinted in [510, pp. 414–421])

193. O. Freire, Science, philosophy and politics in the fifties-On Max Born's unpublished paper entitled "Dialectical Materialism and Modern Physics". Hist. Sci. **10**, 248–254 (2001)

194. O. Freire, C. Lehner, Dialectical materialism and modern physics, an unpublished text by Max Born. Notes and Records of the Royal Society. (2010). doi:10.1098/rsnr.2010.0012

195. O. Freire, *The Quantum Dissidents: Rebuilding the Foundations of Quantum Mechanics (1950–1990)* (Springer, Berlin, 2015)

196. C. Fuchs, A. Peres: Quantum theory needs no 'interpretation'. Phys. Today **53**, 70–71 (March 2000)

197. C. Fuchs, QBism: the perimeter of quantum Bayesianism (2010). arxiv.org/abs/1003.5209v1

198. C. Fuchs, R. Schack, Quantum-Bayesian coherence: the no-nonsense version. Rev. Modern Phys. **85**, 1693–1715 (2013)

199. C. Fuchs, N.D. Mermin, R. Schack, An introduction to QBism with an application to the locality of quantum mechanics. Am. J. Phys. **82**, 749–754 (2014)

200. M. Gell-Mann, What are the building blocks of matter?, in *The Nature of the Physical Universe: 1976 Nobel Conference*, ed. by D. Huff, O. Prewett (Wiley, New York, 1979), pp. 29–45

201. M. Gell-Mann, J.B. Hartle, Quantum mechanics in the light of quantum cosmology, in *Complexity, Entropy, and the Physics of Information*, ed. by W. Zurek (Addison-Wesley, Reading, 1990), pp. 425–458

202. M. Gell-Mann, J.B. Hartle, Alternative decohering histories in quantum mechanics, in *Proceedings of the 25th International Conference on High Energy Physics, Singapore, 1990*, ed. by K.K. Phua, Y. Yamaguchi (World Scientific, Singapore, 1991)

203. M. Gell-Mann, Speech at the annual meeting of the American Association for the Advancement of Science, Chicago (1992) (Reported in [450, pp. 177–178])

204. M. Gell-Mann, J.B. Hartle, Classical equations for quantum systems. Phys. Rev. D **47**, 3345–3382 (1993)

205. M. Gell-Mann, *The Quark and the Jaguar* (Little Brown and Co., London, 1994)

206. A. George (ed.), *Louis de Broglie, physicien et penseur* (Albin Michel, Paris, 1953)

207. M. Ghins, Can common sense realism be extended to theoretical physics? Logic J. IGPL **13**, 95–111 (2005)

208. G.C. Ghirardi, A. Rimini, T. Weber, A general argument against superluminal transmission through the quantum mechanical measurement process. Lettere Al Nuovo Cimento **27**, 293–298 (1980)

209. G.C. Ghirardi, A. Rimini, T. Weber, Unified dynamics for microscopic and macroscopic systems. Phys. Rev. D **34**, 470–491 (1986)

210. G.C. Ghirardi, R. Grassi, F. Benatti, Describing the macroscopic world: closing the circle within the dynamical reduction program. Found. Phys. **25**, 5–38 (1995)

211. G.C. Ghirardi, *Sneaking a Look at God's Cards-Unraveling the Mysteries of Quantum Mechanics* (Princeton University Press, Princeton, revised edition, 2007)

212. G.C. Ghirardi, Collapse Theories, in *The Stanford Encyclopedia of Philosophy* (Winter 2011 Edition), ed. by E.N. Zalta. http://www.plato.stanford.edu/entries/qm-collapse/

213. N. Gisin, Weinberg non-linear quantum mechanics and supraluminal communications. Phys. Lett. A **143**, 1–2 (1990)

214. N. Gisin, *Quantum Chance. Nonlocality. Teleportation and Other Quantum Marvels, Foreword by Alain Aspect* (Springer, Berlin, 2014)

215. A.M. Gleason, Measures on the closed subspaces of a Hilbert space. J. Math. Mech. **6**, 885–893 (1957)

216. J. Glimm, A. Jaffe, *Quantum Physics: A Functional Integral Point of View* (Springer, Berlin, 1987)

217. S. Goldstein, Stochastic mechanics and quantum theory. J. Stat. Phys. **47**, 645–667 (1987)

218. S. Goldstein, J.L. Lebowitz, Quantum mechanics: does the wave function provide a complete description of physical reality? in *The Physical Review: The First Hundred Years*, ed. by H.H. Stroke (AIP Press, Melville, 1994)

219. S. Goldstein, Nonlocality without inequalities for almost all entangled states for two particles. Phys. Rev. Lett. **72**, 1951 (1994)

220. S. Goldstein, Bohmian mechanics and the quantum revolution. Synthese **107**, 145–165 (1996)

221. S. Goldstein, Quantum philosophy: the flight from reason in science, in *The Flight from Science and Reason*, ed. by P. Gross, N. Levitt, M.W. Lewis (Annals of the New York Academy of Sciences **775**, 119–125 (1996)) (Reprinted in [153, Chap. 4])

222. S. Goldstein, Quantum theory without observers. Phys. Today **51**, Part I, March 1998, pp. 42–46; Part II, April 1998, pp. 38–42

223. S. Goldstein et al., Letters and replies in Physics Today **52**, 11 (February 1999)

224. S. Goldstein, S. Teufel, Quantum spacetime without observers: ontological clarity and the conceptual foundations of quantum gravity, in *Physics Meets Philosophy at the Planck Scale*, ed. by C. Callender, N. Huggett (Cambridge University Press, Cambridge, 2001)

225. S. Goldstein, Boltzmann's approach to statistical mechanics (In: [84, pp. 39–54])

226. S. Goldstein, J. Taylor, R. Tumulka, N. Zanghì, Are all particles real? Stud. Hist. Philos. Sci. Part B: Stud. Hist. Philos. Modern Phys. **36**, 103–112 (2005)

227. S. Goldstein, W. Struyve, On the uniqueness of quantum equilibrium in Bohmian mechanics. J. Stat. Phys. **128**, 1197–1209 (2007)

228. S. Goldstein, Bohmian mechanics and quantum information. Found. Phys. **40**, 335–355 (2010)

229. S. Goldstein, T. Norsen, D.V. Tausk, N. Zanghì, Bell's theorem. Scholarpedia **6**(10), 8378 (2011)

230. S. Goldstein, R. Tumulka, N. Zanghì, The quantum formalism and the GRW formalism. J. Stat. Phys. **149**, 142–201 (2012)

231. S. Goldstein, Bohmian mechanics, in *The Stanford Encyclopedia of Philosophy*, ed. by E.N. Zalta (Spring 2013 Edition). http://www.plato.stanford.edu/archives/spr2013/entries/qm-bohm/

232. S. Goldstein, W. Struyve, On quantum potential dynamics. J. Phys. A: Math. Theor. **48**, 025303 (2015)

233. S. Goldstein, *Bell on Bohm, to appear in Quantum Nonlocality and Reality*, ed. by S. Gao (Cambridge University Press, Cambridge, 2015)

234. M. Gondran, R. Hoblos, Trajectoires de de Broglie pour la particule dans la boîte. Annales de la fondation Louis de Broglie **30**, 273–281 (2005)

235. M. Gondran, A. Gondran, *Mécanique quantique. Et si Einstein et de Broglie avaient aussi raison?* (Éditions Matériologiques, Paris, 2014)

236. K. Gottfried, *Quantum Mechanics. vol. 1: Fundamentals* (Benjamin, New York, 1966)

237. G. Gouesbet, *Hidden Worlds in Quantum Physics* (Dover Publications, Mineola, 2013)

238. N. Graham, The measurement of relative frequency (In [132, pp. 229–252])

239. D.M. Greenberger, The neutron interferometer as a device for illustrating the strange behavior of quantum systems. Rev. Modern Phys. **55**, 875–905 (1983)

240. D.M. Greenberger, M.A. Horne, A. Shimony, A. Zeilinger, Bell's theorem without inequalities. Am. J. Phys. **58**, 1131–1143 (1990)

241. R.B. Griffiths, Consistent histories and the interpretation of quantum mechanics. J. Stat. Phys. **36**, 219–272 (1984)

242. R.B. Griffiths, Correlations in separated quantum systems: a consistent history analysis of the EPR problem. Am. J. Phys. **55**, 11–17 (1987)

243. R.B. Griffiths, A consistent interpretation of quantum mechanics using quantum trajectories. Phys. Rev. Lett. **70**, 2201–2204 (1993)

244. R.B. Griffiths, Consistent histories and quantum reasoning. Phys. Rev. A **54**, 2759–2774 (1996)

245. R.B. Griffiths, Choice of consistent family, and quantum incompatibility. Phys. Rev. A **57**, 1604–1618 (1998)

246. R.B. Griffiths, Consistent quantum realism: a reply to Bassi and Ghirardi. J. Stat. Phys. **99**, 1409–1425 (2000)

247. R.B. Griffiths, Consistent histories, quantum truth functionals, and hidden variables. Phys. Lett. A **265**, 12–19 (2000)

248. R.B. Griffiths, *Consistent Quantum Theory* (Cambridge University Press, Cambridge, 2002)

249. R.B. Griffiths, Consistent histories: questions and answers. http://quantum.phys.cmu.edu/CHS/quest.html

250. G. Grössing, Serious Matter: the John Bell Scandal. http://www.nonlinearstudies.at/Bell_E.php

251. S. Haack, *Defending Science—Within Reason. Between Scientism and Cynicism* (Prometheus Books, Amherst, 2003)

252. L. Hardy, Nonlocality for two particles without inequalities for almost all entangled states. Phys. Rev. Lett. **71**, 1665–1668 (1993)

253. J. Hartle, The quantum mechanics of closed systems, in *Directions in General Relativity*, ed. by B.-L. Hu, M.P. Ryan, C.V. Vishveshwars, Proceedings of the 1993 International Symposium, Maryland: Papers in Honor of Charles Misner, vol. 1 (Cambridge University Press, Cambridge, 1993), pp. 104–125

254. S. Hawking: Does god play dice? (1999). http://www.hawking.org.uk/doesgod-play-dice.html

255. J.L. Heilbron, The earliest missionaries of the Copenhagen spirit. Revue d'histoire des sciences **38**, 195–230 (1985)

256. W. Heisenberg, Über den anschaulichen Inhalt der quantentheoretischen Kinematik und Mechanik, Zeitschrift für Physik **43**, 172–198 (1927). English translation, The physical content of quantum kinematics and mechanics (in [510], pp. 62–84)

257. W. Heisenberg, *The Physical Principles of Quantum Theory*, translated by C. Eckart, F. Hoyt (Dover publications, New York, 1949). First edition: University of Chicago Press, Chicago (1930)

258. W. Heisenberg, The development of the interpretation of the quantum theory, in *Niels Bohr and the Development of Physics*, ed. by W. Pauli, L. Rosenfeld, V. Weisskopf (Pergamon Press, New York, 1955)

259. W. Heisenberg, *Physics and Philosophy. The Revolution in Modern Science* (Allen and Unwin, London, 1959). First edition: Harper and Row, New York, 1958

260. W. Heisenberg, *The Physicist's Conception of Nature*, translated by A.J. Pomerans (Harcourt Brace, New York, 1958)

261. W. Heisenberg, *Physics and Beyond. Encounters and Conversations*, translated by A.J. Pomerans (Harper and Row, New York, 1971)

262. B. Hensen, H. Bernien, A.E. Dréau, A. Reiserer, N. Kalb, M.S. Blok, J. Ruitenberg, R.F.L. Vermeulen, R.N. Schouten, C. Abellán, W. Amaya, V. Pruneri, M.W. Mitchell, M. Markham, D.J. Twitchen, D. Elkouss, S. Wehner, T.H. Taminiau, R. Hanson, Experimental loophole-free violation of a Bell inequality using entangled electron spins separated by 1.3 km. arxiv.org/abs/1508.05949

263. G. Hermann, Die naturphilosophischen grundlagen der Quantenmechanik. Abhandlungen der Friesschen Schule **6**, 69–152 (1935)

264. B. Hessen, The social and economic roots of Newton's Principia, in *Science at the Crossroads. Papers Presented to the International Congress of the History of Science and Technology, 1931, by the Delegates of the U.S.S.R.*, ed. by N.I. Bukharin (Franck Cass. and Co., London 1931), pp. 151–212. New translation, in *The Social and Economic Roots of the Scientific Revolution*, ed. by G. Freudenthal, P. McLaughlin (Springer, Berlin, 2009), pp. 41–101

265. B.J. Hiley, R.E. Callaghan, Delayed choice experiments and the Bohm approach. Phys. Scr. **74**, 336–348 (2006)

266. J. Hilgevoord, J. Uffink, The Uncertainty Principle, in *The Stanford Encyclopedia of Philosophy*, ed. by E.N. Zalta (Spring 2014 Edition). http://www.plato.stanford.edu/archives/spr2014/entries/qt-uncertainty/

267. M. Hirsch, S. Smale, R. Devaney, *Differential Equations, Dynamical Systems and an Introduction to Chaos*, 2nd edn. (Academic Press, San Diego, 2004)

268. P. Holland, *The Quantum Theory of Motion, An Account of the de Broglie-Bohm Causal Interpretation of Quantum Mechanics* (Cambridge University Press, Cambridge, 1993)

269. P. Holland, Uniqueness of paths in quantum mechanics. Phys. Rev. A **60**, 4326–4330 (1999)

270. P. Holland, Uniqueness of conserved current in quantum mechanics. Ann. der Phys. (Leipzig) **12**, 446–462 (2003)

271. P. Holland, What's wrong with Einstein's 1927 hidden-variable interpretation of quantum mechanics? Found. Phys. **35**, 177–196 (2005)

272. P. Holland, A quantum of history. Contemp. Phys. **52**, 355–358 (2011)

273. J. Horgan, Quantum philosophy. Sci. Am. **267**, 94–104 (July 1992)

274. D. Howard, "Nicht sein kann was nicht sein darf," Or the prehistory of EPR, 1909–1935, in *Sixty-Two Years of Uncertainty*, ed. by A.I. Miller (Plenum, New York, 1990), pp. 61–111

275. D. Howard, Who invented the "Copenhagen interpretation"? A study in mythology. Philos. Sci. **71**, 669–682 (2004)

276. D. Howard, Revisiting the Einstein-Bohr dialogue. Iyyun: The Jerusalem Philosophical Quaterly **56**. Special Issue dedicated to the memory of Mara Beller, 2007, pp. 57–90

277. D. Howard, Quantum mechanics in context: Pascual Jordan's 1936 Anschauliche Quantentheorie, in *Research and Pedagogy: A History of Quantum Physics through Its Textbooks*, ed. by M. Badino, J. Navarro, Chap. 11 (Edition Open Access, Berlin, 2013)

278. D. Hume, *An Enquiry Concerning Human Understanding* (Prometheus, Amherst, 1988) (original edition, 1748)

279. A.S. Jacobsen, Léon Rosenfeld's Marxist defense of complementarity. Hist. Stud. Phys. Biol. Sci. **37** (Supplement), 3–34 (2007)

280. V. Jacques, E. Wu, F. Grosshans, F. Treussart, P. Grangier, A. Aspect, J.-F. Roch, Experimental realization of Wheeler's delayed-choice Gedanken experiment. Science **315**, 966–968 (2007)

281. M. Jammer, *The Philosophy of Quantum Mechanics, The Interpretation of Quantum Mechanics in Historical Perspective* (Wiley, New York, 1974)

282. E.T. Jaynes, *Papers on Probability, Statistics and Statistical Physics*, ed. by R.D. Rosencrantz (Reidel, Dordrecht, 1983)

283. E.T. Jaynes, Clearing up mysteries-the original goal, in *Maximum Entropy and Bayesian Methods*, ed. by J. Skilling (Kluwer Academic Publishers, Dordrecht, 1989)

284. E.T. Jaynes, *Probability Theory: The Logic of Science* (Cambridge University Press, Cambridge, 2003)

285. P. Jordan, Quantenphysikalische Bemerkungen zur Biologie und Psychologie. Erkenntnis **4**, 215–252 (1934)
286. I. Kant, *Critique of Pure Reason*, translated by N.K. Smith (McMillan, London, 1950). First edition in German, *Kritik der Reinen Vernunft*, 1781
287. E.H. Kennard, Zur Quantenmechanik einfacher Bewegungstypen. Zeitschrift für Physik **44**, 326–352 (1927)
288. A. Kent, Against many-worlds interpretations. Int. J. Mod. Phys. A **5**, 1745–1762 (1990)
289. A. Kent, One world versus many: the inadequacy of Everettian accounts of evolution, probability, and scientific confirmation (in [435] Chap. 10)
290. D. Kleppner, R. Jackiw, One hundred years of quantum physics. Science **289**, 893–898 (2000)
291. S. Kochen, E.P. Specker, The problem of hidden variables in quantum mechanics. J. Math. Mech. **17**, 59–87 (1967)
292. S. Kocsis, B. Braverman, S. Ravets, M.J. Stevens, R.P. Mirin, L.K. Shalm, A.M. Steinberg, Observing the average trajectories of single photons in a two-slit interferometer. Science **332**, 1170–1173 (2011)
293. A. Kojevnikov, David Bohm and collective movement. Hist. Stud. Phys. Biol. Sci. **33**, 161–192 (2002)
294. A. Kojevnikov, *Stalin's Great Science: The Times and Adventures of Soviet Physicists* (Imperial College Press, London, 2004)
295. A. Kojevnikov, The phenomenon of Soviet science. Osiris **23**, 115–135 (2008)
296. A. Kojevnikov, Probability, Marxism, and quantum ensembles. Yearb. Eur. Cult. Sci. **6**, 211–235 (2012)
297. H. Krivine, A. Grossman, *De l'atome imaginé à l'atome découvert: Contre le relativisme* (De Boeck, Bruxelles, 2015)
298. T. Kuhn, *The Structure of Scientific Revolutions*, 2nd edn. (University of Chicago Press, Chicago, 1970)
299. T. Kuhn, J.L. Heilbron, Interview with Léon Rosenfeld, Sources for the history of quantum physics. www.amphilsoc.org/library/guides/ahqp/index.htm
300. P. Kwiat, H. Weinfurter, T. Herzog, A. Zeilinger, M.A. Kasevich, Interaction-free measurement. Phys. Rev. Lett. **74**, 4763–4766 (1995)
301. F. Laloë, *Do We Really Understand Quantum Mechanics?* (Cambridge University Press, Cambridge, 2012)
302. L. Landau, E. Lifshitz, *Quantum Mechanics, vol. 3, A Course of Theoretical Physics* (Pergamon Press, Oxford, 1965)
303. P.S. Laplace, A Philosophical Essay on Probabilities, translated by F.W. Truscott, F.L. Emory (Dover Publications, New York, 1951)
304. J. Laskar, A numerical experiment on the chaotic behaviour of the solar system. Nature **338**, 237–238 (1989)
305. L. Laudan, A confutation of convergent realism. Philos. Sci. **48**, 19–49 (1981)
306. K.V. Laurikainen, *Beyond the Atom* (Springer, Berlin, 1988)
307. J.L. Lebowitz, Microscopic origins of irreversible macroscopic behavior. Round Table on Irreversibility at STATPHYS20, Paris, July 22, 1998. Physica A **263**, 516–527 (1999)
308. La physique quantique a sauvé Wilkinson de la dépression, interview of Etienne Klein by Adrien Pécout, Le Monde (June 23, 2014)
309. V.I. Lenin, *Materialism and Empirio-criticism: Critical Comments on a Reactionary Philosophy* (University Press of the Pacific, Honolulu, 2002). First edition, in Russian, 1908
310. J. Lighthill, The recently recognized failure of predictability in Newtonian dynamics. Proc. R. Soc. Lond. A **407**, 35–50 (1986)
311. O. Lombardi, D. Dieks, Modal interpretations of quantum mechanics, in *The Stanford Encyclopedia of Philosophy*, ed. by E.N. Zalta (Spring 2014 Edition). https://www.plato.stanford.edu/archives/spr2014/entries/qm-modal/
312. F. London, E. Bauer, *La théorie de l'observation en mécanique quantique* (Hermann, Paris 1939). English translation: The theory of observation in quantum mechanics (in [510, pp. 217–259])

313. S.N. Lyle, Review of *Quantum Chance* by Nicolas Gisin (2014). http://www.stephenlyle.org/
314. E. Mach, *The Analysis of Sensations and the Relation of the Physical to the Psychical*, translated by C.M. Williams, S. Waterlow, introduction by T.S. Szasz (Dover, New York, 1959)
315. E. Madelung, Eine anschauliche Deutung der Gleichung von Schrödinger. Die Naturwissenschaften **14**, 1004 (1926)
316. E. Madelung, Quantentheorie in hydrodynamischer Form. Zeitschrift für Physik **40**, 322–326 (1927)
317. D. Marian, X. Oriols, N. Zanghi, Proposal for measuring the Bohmian velocity in mesoscopic systems using the weak value of the total electrical current (in Preparation)
318. K. Marx, *Capital: A Critique of Political Economy—The Process of Capitalist Production*, ed. by F. Engels (Cosimo Inc, New York, 2007)
319. T. Maudlin, *Quantum Nonlocality and Relativity* (Blackwell, Cambridge, 1st edn, 1994, 3rd edn, 2011)
320. T. Maudlin, Kuhn defanged: incommensurability and theory-choice, translated by Jean-Pierre Deschepper and Michel Ghins and published in French under the title: Kuhn édenté: incommensurabilité et choix entre théories. Rev. Philosophique de Louvain **94**, 428–446 (1996)
321. T. Maudlin, Space-time in the quantum world (in [103], pp. 285–307)
322. T. Maudlin, Descrying the world in the wave function. Monist **80**, 3–23 (1997)
323. T. Maudlin, Completeness, supervenience and ontology. J. Phys. A Math. Theor. **40**, 3151–3171 (2007)
324. T. Maudlin, *The Metaphysics Within Physics* (Oxford University Press, Oxford, 2007)
325. T. Maudlin, Can the world be only wavefunction? (in [435] Chap. 4)
326. T. Maudlin, *Philosophy of Physics: Space and Time* (Princeton University Press, Princeton, 2012)
327. T. Maudlin, What Bell did. J. Phys. A Math. Theor. **47**, 424010 (2014)
328. T. Maudlin, Critical Study. *The Emergent Multiverse: Quantum Theory according to the Everett Interpretation*, ed. by D. Wallace (Oxford University Press, 2012), Noûs **48**, 794–808 (2014)
329. C. McGinn, *Problems in Philosophy, The Limits of Enquiry* (Blackwell, Oxford, 1993)
330. E. McMullin, The Explanation of Distant Action: Historical Notes (in [101, pp. 272–302])
331. D. Mermin, Quantum mysteries for anyone. J. Philos. **78**, 397–408 (1981) (Reprinted in [101, pp. 49–59], and in [334, pp. 81–94])
332. D. Mermin, Is the moon there when nobody looks? Reality and the Quantum Theory. Phys. Today **38**, 38–47 (April, 1985)
333. D. Mermin et al., *Letters and replies*. Phys. Today **43**, 11–13 (December 1990)
334. D. Mermin, *Boojums All the Way Through* (Cambridge University Press, Cambridge, 1990)
335. D. Mermin, Hidden variables and the two theorems of John Bell. Rev. Mod. Phys. **65**, 803–815 (1993)
336. D. Mermin, The best version of Bell's theorem. Ann. N. Y. Acad. Sci. **775**, Fundam. Prob. Quantum Theory 616–623 (1995)
337. D. Mermin, The science of science: a physicist reads Barnes. Bloor and Henry. Soc. Stud. Sci. **28**, 603–623 (1998)
338. D. Mermin, Whose knowledge? (in [57], pp. 271–280)
339. D. Mermin, Could Feynman have said this? Phys. Today **57**, 10 (May 2004)
340. D. Mermin, QBism puts the scientist back into science. Nature **507**, 421–423 (2014)
341. J.S. Mill, *On Liberty*, ed. by G. Himmelfarb (Penguin Books, London, 1982) First edition, 1859
342. W. Moore, *Schrödinger: Life and Thought* (Cambridge University Press, Cambridge, 1989)
343. A. Mouchet, *L'étrange subtilité quantique. Quintessence de poussières*, 2nd edn. (Dunod, Paris, 2015)
344. W.C. Myrvold, On some early objections to Bohm's theory. Int. Stud. Philos. Sci. **17**, 7–24 (2003)
345. T. Nagel, What is it like to be a bat? Philos. Rev. **83**, 435–450 (1974). Reprinted in *Mortal Questions*, ed. by T. Nagel (Cambridge University Press, Cambridge, 1979)

346. T. Nagel, The sleep of reason. New Repub. **35** (October 12, 1998)
347. J. Nash, An Interesting Equation. http://www.sites.stat.psu.edu/~babu/nash/intereq.pdf
348. E. Nelson, *Dynamical Theories of Brownian Motion* (Princeton University Press, Princeton, 1967)
349. E. Nelson, *Quantum Fluctuations* (Princeton University Press, Princeton, 1985)
350. I. Newton, Letter to Richard Bentley, 25 February 1693, in *The Correspondence of Isaac Newton*, vol. 3, ed. by H.W. Turnbull (Cambridge University Press, Cambridge, 1959), pp. 253–254
351. I. Newton, *Philosophiae Naturalis Principia Mathematica, General Scholium* (1726). Third edition, translated by I.B. Cohen, A. Whitman (University of California Press, Oakland, 1999)
352. M.A. Nielsen, I.L. Chuang, *Quantum Computation and Quantum Information* (Cambridge University Press, Cambridge, 2000)
353. H. Nikolić, EPR before EPR: a 1930 Einstein-Bohr thought experiment revisited. Eur. J. Phys. **33**, 1089–1097 (2012)
354. C. Norris, *Quantum Theory and the Flight from Realism, Philosophical Responses to Quantum Mechanics* (Routledge, London, 2000)
355. T. Norsen, Einstein's boxes. Am. J. Phys. **73**, 164–176 (2005)
356. T. Norsen, EPR and Bell locality, in *Quantum Mechanics: Are there Quantum Jumps? and On the Present Status of Quantum Mechanics, AIP Conference Proceedings, 844*, ed. by A. Bassi, D. Dürr, T. Weber, N. Zanghi (American Institute of Physics, Melville, 2006), pp. 281–293
357. T. Norsen, Against 'realism'. Found. Phys. **37**, 311–340 (2007)
358. T. Norsen, J.S. Bell's concept of local causality. Am. J. Phys. **79**, 1261–1275 (2011)
359. T. Norsen, The pilot-wave perspective on quantum scattering and tunneling. Am. J. Phys. **81**, 258–266 (2013)
360. T. Norsen, The pilot-wave perspective on spin. Am. J. Phys. **82**, 337–348 (2014)
361. T. Norsen, Quantum solipsism and nonlocality. Int. J. Quantum Found. (electronic journal) (2014). http://www.ijqf.org/archives/1548
362. T. Norsen, W. Struyve, Weak measurement and (Bohmian) conditional wave functions. Ann. Phys. **350**, 166–178 (2014)
363. A. Olding, Religion as smorgasbord. Quadrant **42**, 73–75 (1998)
364. R. Olwell, Physical isolation and marginalization in physics—David Bohm's cold war exile. Isis **90**, 738–756 (1999)
365. R. Omnès, Logical reformulation of quantum mechanics. I. Foundations. J. Stat. Phys. **53**, 893–932 (1988). II. Interference and the Einstein-Podolsky-Rosen experiment. J. Stat. Phys. **53**, 933–955 (1988). III. Classical limit and irreversibility. J. Stat. Phys. **53**, 957–975 (1988). IV. Projectors in semiclassical physics. J. Stat. Phys. **57**, 357–382 (1989)
366. R. Omnès, Consistent interpretations of quantum mechanics. Rev. Mod. Phys. **64**, 339–382 (1992)
367. R. Omnès, *The Interpretation of Quantum Mechanics* (Princeton University Press, Princeton, 1994)
368. R. Omnès, *Understanding Quantum Mechanics* (Princeton University Press, Princeton, 1999)
369. A. Pais, Einstein and the quantum theory. Rev. Mod. Phys. **51**, 863–914 (1979)
370. A. Pais, *"Subtle is the Lord": the Science and the Life of Albert Einstein* (Oxford University Press, Oxford, 1982)
371. A. Pais, *Niels Bohr's Times: In Physics, Philosophy, and Polity* (Oxford University Press, Oxford, 1991)
372. A. Pais, *Einstein Lived Here* (Oxford University Press, Oxford, 1994)
373. A. Pannekoek, *Lenin as Philosopher: a Critical Examination of the Philosophical Basis of Leninism*, revised edition; edited, annotated, and with an introduction by Lance Byron Richey (Marquette University Press, Milwaukee, 2003)
374. W. Pauli, Letter to Niels Bohr, 6 August 1927, in *Wissenschaftlicher Briefwechsel mit Bohr, Einstein, Heisenberg u.a./Scientific Correspondence with Bohr, Einstein, Heisenberg a.o.: Band/Volume I History of Mathematics and Physical Sciences*, ed. by K. von Meyenn, A. Hermann, V.F. Weisskopf (Springer, Berlin, 2005), pp. 404–405

375. W. Pauli, Letter to Markus Fierz on 12 August 1948, from the Pauli Letter Collection at CERN, Geneva. Translation. In: K.V. Laurikainen, *Beyond the Atom* (Springer, Berlin, 1988)

376. W. Pauli, Editorial on the concept of complementarity. Dialectica **2**, 307–311 (1948)

377. W. Pauli, Letter to David Bohm (December 3, 1951) (in [386], p. 127)

378. W. Pauli, Letter to Markus Fierz, 1951, in *Wissenschaftlicher Briefwechsel mit Bohr, Einstein, Heisenberg u.a./Scientific Correspondence with Bohr, Einstein, Heisenberg a.o.: Band/Volume IV. History of Mathematics and Physical Sciences*, ed. by K. von Meyenn (Springer, Berlin, 2005), pp. 131–132

379. W. Pauli, Remarques sur le problème des paramètres cachés dans la mécanique quantique et sur la théorie de l'onde pilote (in [206], pp. 33–42)

380. W. Pauli, Letter to Max Born (March 31, 1954) (in [79], p. 221)

381. W. Pauli, Wave Mechanics. *Pauli Lectures on Physics*, vol. 5, C.P. Enz (ed.). Translated by S. Margulies, H.R. Lewis (MIT Press, Cambridge, 1973). Reprinted by Dover, New York, 2000

382. W. Pauli, *Writings on Physics and Philosophy*, ed. by C.P. Enz, K. von Meyenn, translated by R. Schlapp (Springer, Berlin, 1994)

383. P. Pearle, Reduction of the state-vector by a nonlinear Schrödinger equation. Phys. Rev. D **13**, 857–868 (1976)

384. P. Pearle, Toward explaining why events occur. Int. J. Theor. Phys. **18**, 489–518 (1979)

385. P. Pearle, Combining stochastic dynamical state-vector reduction with spontaneous localization. Phys. Rev. A **39**, 2277–2289 (1989)

386. D. Peat, *Infinite Potential: The Life and Times of David Bohm* (Basic Books, New York, 1997)

387. D. Peat, Look for truth—no matter where it takes you, interview of F. David Peat by Simeon Alev, in What is Enlightenment? Spring/Summer 1997, **6**, Number 1

388. R. Peierls, *Surprises in Theoretical Physics* (Princeton University Press, Princeton, 1979)

389. R. Penrose, *The Emperor's New Mind* (Oxford University Press, Oxford, 1989)

390. R. Penrose, On gravity's role in quantum state reduction. Gen. Relativ. Gravit. **28**, 581–600 (1996)

391. R. Penrose, *The Road to Reality: A Complete Guide to the Laws of the Universe* (Alfred A. Knopf, New York, 2005)

392. A. Peres, Incompatible results of quantum measurements. Phys. Lett. A **151**, 107–108 (1990)

393. A. Peres, Two simple proofs of the Kochen-Specker theorem. J. Phys. A Math. Gen. **24**, L175–L178 (1991)

394. O. Pessoa, Jr., O. Freire, Jr., A. De Greiff, The Tausk controversy on the foundations of quantum mechanics: physics, philosophy, and politics. Phys. Perspect. **10**, 138–162 (2008) (Reprinted in [195, pp. 175–195])

395. A. Petersen, The philosophy of Niels Bohr. Bull. At. Sci. **19**(7), 8–14 (1963)

396. A. Petersen, *Quantum Physics and the Philosophical Tradition* (MIT Press, Cambridge, 1968)

397. C. Philippidis, C. Dewdney, B.J. Hiley, Quantum interference and the quantum potential. Il Nuovo Cimento B **52**, 15–28 (1979)

398. T.J. Pinch, What does a proof do if it does not prove? A study of the social conditions and metaphysical divisions leading to David Bohm and John von Neumann failing to communicate in quantum physics, in *The Social Production of Scientific Knowledge*, ed. by E. Mendelsohn, P. Weingart, R. Whitly (Reidel, Boston, 1977), pp. 171–215

399. X.O. Pladevall, J. Mompart (eds.), *Applied Bohmian Mechanics: From Nanoscale Systems to Cosmology* (Pan Stanford Publishing, Singapore, 2012)

400. H. Poincaré, *New Methods of Celestial Mechanics*, ed. by D. Goroff (American Institute of Physics Press, 1992). French original: Les méthodes nouvelles de la mécanique céleste, 3 volumes, Gauthier-Villars, Paris, 1892–1899

401. H. Poincaré, *La Valeur de la Science* (Flammarion, Paris, 1905). English translation in [403]

402. H. Poincaré, *Science et Méthode* (Flammarion, Paris, 1909). English translation in [403]

403. H. Poincaré, *The Foundations of Science: Science and Hypothesis, The Value of Science, Science and Method*, translated by G.B. Halsted (The Science Press, New York, 1913)

404. J. Preskill, Lectures on quantum information and quantum computation. http://www.theory.caltech.edu/preskill/

405. I. Prigogine, I. Stengers, *Entre le temps et l'éternité* (Fayard, Paris, 1988)
406. I. Prigogine, *Les Lois du Chaos* (Flammarion, Paris, 1994)
407. K. Przibram (ed.), *Letters on Wave Mechanics, Schrödinger, Planck, Einstein, Lorentz*. Translated and with an Introduction by M.J. Klein (Philosophical Library, New York, 1967)
408. S. Psillos, *Scientific Realism, How Science Tracks Truth* (Routledge, London, 1999)
409. H. Putnam, *Reason, Truth, and History* (Cambridge University Press, Cambridge, 1981)
410. H. Putnam, A philosopher looks at quantum mechanics (Again). Br. J. Philos. Sci. **56**, 615–634 (2005)
411. W.V.O. Quine, Two dogmas of empiricism, in *From a Logical Point of View*, ed. by W.V.O. Quine. 2nd edn., revised (Harvard University Press, Cambridge, 1980), 1st edn., 1953
412. M. Reed, B. Simon, *Methods of Modern Mathematical Physics I: Functional Analysis* (Academic Press, New York, 1972)
413. M. Reed, B. Simon, *Methods of Modern Mathematical Physics II: Fourier Analysis, Self-Adjointness* (Academic Press, New York, 1975)
414. L.E. Reichl, *The Transition to Chaos in Conservative Classical Systems: Quantum Manifestations* (Springer, New York, 1992)
415. J.-F. Revel, *L'Histoire de la Philosophie Occidentale, De Thalès à Kant* (NiL éditions, Paris, 1994)
416. D.A. Rice, A geometric approach to nonlocality in the Bohm model of quantum mechanics. Am. J. Phys. **65**, 144–147 (1997)
417. N. Rosen, On waves and particles. J. Elisha Mitchell Sci. Soc. **61**, 67–73 (1945)
418. L. Rosenfeld, Letter to Bohm (May 20, 1952) (quoted in [386], p. 130)
419. L. Rosenfeld, L'évidence de la complémentarité, in *Louis de Broglie: Physicien et penseur*, ed. by André George (Albin Michel, Paris, 1953), pp. 43–65
420. L. Rosenfeld, Strife about complementarity (in [424], pp. 465–484) Revised version of [419]
421. L. Rosenfeld, Misunderstandings about the foundations of quantum theory, in *Observation and Interpretation*, ed. by S. Körner (Butterworths, London, 1957), pp. 41–61 (Reprinted in [424], pp. 495–502)
422. L. Rosenfeld, Niels Bohr's contribution to epistemology. Phys. Today **16** (November 1963) (Reprinted in [424], pp. 522–535)
423. L. Rosenfeld, Niels Bohr in the thirties, in *Niels Bohr: His life and Times as Seen by His Friends and Colleagues*, ed. by S. Rozental (North Holland, Amsterdam, 1964), pp. 114–136
424. L. Rosenfeld, Rosenfeld, *Selected Papers*, ed. by R.S. Cohen, J.J. Stachel (Reidel, Dordrecht, 1979)
425. B. Russell, *The Problems of Philosophy* (Williams and Norgate, London, 1912)
426. B. Russell, On the notion of cause with application to the free-will problem. Chap. 8 in B. Russell, *Our Knowledge of the External World: As a Field for Scientific Method in Philosophy* (The Open Court Pub. Co., Chicago, 1914)
427. B. Russell, *Freedom Versus Organization 1814–1914* (Norton and co., New York, 1934)
428. B. Russell, *A History of Western Philosophy and Its Connection with Political and Social Circumstances from the Earliest Times to the Present Day* (Allen and Unwin Ltd, London, 1945)
429. B. Russell, *My Philosophical Development* (Routledge, London, 1993), 1st edn., Unwin Hyman, 1959
430. B. Russell, *Autobiography* (Routledge Classics, New York, 2010). 1st edn., *The Autobiography of Bertrand Russell*, 3 vols., Allen, Unwin, London, 1951–1969
431. H. Sankey, *Rationality, Relativism and Incommensurability* (Ashgate Publishing Co., Aldershot, 1997)
432. T. Sauer, An Einstein manuscript on the EPR paradox for spin observables. Stud. Hist. Philos. Sci. Part B Stud. Hist. Philos. Mod. Phys. **38**, 879–887 (2007)
433. S. Saunders, Time, decoherence, and quantum mechanics. Synthese **102**, 235–66 (1995)
434. S. Saunders, D. Wallace, Branching and uncertainty. Brit. J. Philos. Sci. **59**, 293–305 (2008)
435. S. Saunders, J. Barrett, A. Kent, D. Wallace (eds.), *Many Worlds? Everett, Quantum Theory, and Reality* (Oxford University Press, Oxford, 2012)

436. P.A. Schilpp (ed.), *Albert Einstein: Philosopher-Scientist, The Library of Living Philosophers* (Evanston, Illinois, 1949)

437. E. Schrödinger, Quantisierung als Eigenwertproblem (Vierte Mitteilung). Annalen der Physik **81**, 109–139 (1926). English translation: Quantisation as a problem of proper values Part IV, in [440]

438. E. Schrödinger, Letter to Wilhem Wien (August 26, 1926). Wien Archiv, Deutsches Museum, Munich, quoted in [342, p. 228]

439. E. Schrödinger, Über das Verhältnis der Heisenberg Born Jordanischen Quantenmechanik zu der meinen. Annalen der Physik **79**, 734 (1926). English translation: On the relation between the quantum mechanics of Heisenberg, Born, and Jordan, and that of Schrödinger, in [440]

440. E. Schrödinger, Collected Papers on Wave Mechanics, translated by J.F. Shearer (Chelsea, New York, 1927)

441. E. Schrödinger, Die gegenwärtige Situation in der Quantenmechanik. Naturwissenschaften **23**, 807–812; 823–828; 844–849 (1935). English translation: The present situation in quantum mechanics, translated by J.-D. Trimmer, Proceedings of the American Philosophical Society **124**, 323–338 (1984) (Reprinted in [510, pp. 152–167])

442. E. Schrödinger, Discussion of probability relations between separated systems. Math. Proc. Camb. Philos. Soc. **31**, 555–563 (1935)

443. E. Schrödinger, Probability relations between separated systems. Math. Proc. Camb. Philos. Soc. **32**, 446–452 (1936)

444. E. Schrödinger, Letter to John Synge (November 9, 1959) (quoted in [342], pp. 472–473)

445. E. Schrödinger, Letter to Max Born (October 10, 1960). Staatsbibliothek Preussischer Kulturbesitz, Berlin, Born Nachlass 704, quoted in [342, p. 479]

446. J.T. Schwartz, The pernicious influence of mathematics on science, in *Logic, Methodology and Philosophy of Science*, ed. by E. Nagel, P. Suppes, A. Tarski (Stanford University Press, Stanford, 1962)

447. R. Shankar, *Principles of Quantum Mechanics* (Springer, Berlin, 1980)

448. A. Shimony, *The Search for a Naturalistic World View, vol. 1. Scientific Method and Epistemology, vol. 2, Natural Science and Metaphysics* (Cambridge University Press, Cambridge, 1993)

449. A. Shimony, Bell's Theorem, in *The Stanford Encyclopedia of Philosophy*, ed. by E.N. Zalta (Winter 2013 Edition). http://plato.stanford.edu/archives/win2013/entries/bell-theorem/

450. T. Siegfried, *The Bit and the Pendulum. From Quantum Computing to M Theory—The New Physics of Information* (Wiley, New York, 2000)

451. A. Sokal, Transgressing the boundaries: towards a transformative hermeneutics of quantum gravity. Soc. Text **46/47**, 217–252 (1996)

452. A. Sokal, *A Physicist Experiments with Cultural Studies* (Lingua Franca, May/June, 1996)

453. A. Sokal, J. Bricmont, *Fashionable Nonsense: Postmodern Intellectuals' Abuse of Science* (Picador, New York, 1998). British edition: *Intellectual Impostures* (Profile Books, London, 1998)

454. A. Sokal, J. Bricmont, Defense of a modest scientific realism, in *Knowledge and the World: Challenges Beyond the Science Wars*, ed. by M. Carrier, J. Roggenhofer, G. Küppers, Ph. Blanchard (Springer, Berlin, 2004)

455. A. Sokal, *Beyond the Hoax: Science, Philosophy, and Culture* (Oxford University Press, Oxford, 2008)

456. Electrons et photons: rapports et discussions du 5e Conseil de Physique tenu à Bruxelles du 24 au 29 octobre 1927, Instituts Solvay, Gauthier-Villars et cie, Paris, 1928. English translation in [24]

457. G.L. Squires, Quantum mechanics, in *Encyclopedia Britannica* (2014). http://www.britannica.com/science/quantum-mechanics-physics

458. H.P. Stapp, Bell's theorem and world process. Il Nuovo Cimento **29B**, 270 (1975)

459. H. Stapp, Bell's theorem and the foundations of quantum physics. Am. J. Phys. **53**, 306–317 (1985)

460. H. Stapp, Quantum nonlocality and the description of nature (in [104], pp. 154–174)

461. D. Stove, *Popper and After, Four Modern Irrationalists* (Pergamon Press, Oxford, 1982)
462. D. Stove, *The Plato Cult and Other Philosophical Follies* (Blackwell, Oxford, 1991)
463. W. Struyve, H. Westman, A new pilot-wave model for quantum field theory. AIP Conf. Proc. **844**, 321–339 (2006)
464. W. Struyve, Pilot-wave theory and quantum fields. Rep. Prog. Phys. **73**, 106001 (2010)
465. W. Struyve, Pilot-wave approaches to quantum field theory. J. Phys. Conf. Ser. **306**, 012047 (2011)
466. W. Struyve, Semi-classical approximations based on Bohmian mechanics (2015). arXiv:1507.04771
467. L. Susskind, A. Friedman, *Quantum Mechanics: The Theoretical Minimum* (Basic Civitas Books, New York, 2014)
468. J. Taylor, Interview with P.C.W. Davies and J.R. Brown (in [107], Chap. 7)
469. E.F. Taylor, J.A. Wheeler, *Spacetime Physics* (W.H. Freeman and Co, San Francisco, 1963)
470. M. Tegmark, J.A. Wheeler, 100 years of quantum mysteries. Sci. Am. **284**, 72–79 (February 2001)
471. S. Teufel, R. Tumulka, Simple proof for global existence of Bohmian trajectories. Commun. Math. Phys. **258**, 349–365 (2005)
472. L. Tolstoy, *What Is Art and Essays on Art*, translated by A. Maude (Oxford University Press, Oxford, 1930). Russian original, 1897
473. M. Towler, de Broglie–Bohm pilot-wave theory and the foundations of quantum mechanics, lectures. http://www.tcm.phy.cam.ac.uk/~mdt26/
474. R. Tumulka, Understanding Bohmian mechanics—a dialogue. Am. J. Phys. **72**, 1220–1226 (2004)
475. R. Tumulka, On spontaneous wave function collapse and quantum field theory. Proc. R. Soc. A **462**, 1897–1908 (2006)
476. R. Tumulka, A relativistic version of the Ghirardi-Rimini-Weber model. J. Stat. Phys. **125**, 825–844 (2006)
477. R. Tumulka, The "unromantic pictures" of quantum theory. J. Phys. A **40**, 3245–3273 (2007)
478. R. Tumulka, The assumptions of Bell's proof. Int. J. Quantum Found. (electronic journal) (2014). http://www.ijqf.org/groups-2/bells-theorem/forum/topic/the-assumptions-of-bells-proof/
479. L. Vaidman, On schizophrenic experiences of the neutron or why we should believe in the many-worlds interpretation of quantum theory. Int. Stud. Philos. Sci. **12**, 245–261 (1998)
480. L. Vaidman, Many-worlds interpretation of quantum mechanics, in *Stanford Encyclopedia of Philosophy*, ed. by E.N. Zalta (Fall 2008 Edition). http://plato.stanford.edu/archives/fall2008/entries/qm-manyworlds/
481. A. Valentini, Signal-locality, uncertainty and the subquantum H-theorem, I. Phys. Lett. A **156**, 5–11 (1991)
482. A. Valentini, Signal-locality, uncertainty and the subquantum H-theorem, II. Phys. Lett. A **156**, 1–8 (1991)
483. A. Valentini, On Galilean and Lorentz invariance in pilot-wave dynamics. Phys. Lett. A **228**, 215–222 (1997)
484. A. Valentini, Hidden variables, statistical mechanics and the early universe (In [84, pp. 165–181])
485. A. Valentini, Pilot-wave theory of fields, gravitation and cosmology (in [103, pp. 45–66])
486. A. Valentini, Signal-locality in hidden variable theories. Phys. Lett. A **297**, 273–278 (2002)
487. A. Valentini, H. Westman, Dynamical origin of quantum probabilities. Proc. Royal Soc. A **461**, 253–272 (2005)
488. A. Valentini, Inflationary cosmology as a probe of primordial quantum mechanics. Phys. Rev. D **82**, 063513 (2010)
489. A. Valentini, de Broglie–Bohm pilot-wave theory: many worlds in denial? (In [435, pp. 476–509])
490. A.W. van der Vaart, *Asymptotic Statistics* (Cambridge University Press, Cambridge, 1998)

491. B.L. van der Waerden, *Sources of Quantum Mechanics, edited with a historical introduction* (Dover Publications, New York, 1968)

492. B.C. van Fraasen, The charybdis of realism: epistemological implications of Bell's inequality. Synthese 52, 25–38 (1982) (Reprinted in [101, pp. 97–113])

493. B.C. van Fraassen, Discussion, in *Physics and Our View of the World*, ed. by J. Hilgevoord (Cambridge University Press, Cambridge, 1994)

494. N.G. van Kampen, Ten theorems about quantum mechanical measurements. Physica A **153**, 97–113 (1988)

495. J.P. Vigier, C. Dewdney, P.E. Holland, A. Kyprianidis, Causal particle trajectories and the interpretation of quantum mechanics, in *Quantum Implications: Essays in Honour of David Bohm*, ed. by B.J. Hiley, F.D. Peat (Routledge, London, 1987)

496. J. von Neumann, *Mathematical Foundations of Quantum Mechanics* (Princeton University Press, Princeton, 1955). First edition in German, *Mathematische Grundlagen der Quantenmechanik*, 1932

497. C.F. von Weizsäcker, A reminiscence from 1932, in *Niels Bohr: A Centenary Volume*, ed. by A.P. French, P.J. Kennedy (Harvard University Press, Cambridge, 1985), pp. 183–190

498. D. Wallace, Everett and structure. Stud. Hist. Philos. Modern Phys. **34**, 87–105 (2003)

499. D. Wallace, Epistemology quantized: circumstances in which we should come to believe in the Everett interpretation. Br. J. Philos. Sci. **57**, 655–689 (2006)

500. D. Wallace, Quantum probability from subjective likelihood: improving on Deutsch's proof of the probability rule. Stud. Hist. Philos. Modern Phys. **38**, 311–332 (2007)

501. D. Wallace, *The Emergent Multiverse: Quantum Theory According to the Everett Interpretation* (Oxford University Press, Oxford, 2012)

502. S. Weinberg, Testing quantum mechanics. Ann. Phys. **194**, 336–386 (1989)

503. S. Weinberg, Precision tests of quantum mechanics. Phys. Rev. Lett. **62**, 485–488 (1989)

504. S. Weinberg, *Dreams of a Final Theory: The Scientist's Search for the Ultimate Laws of Nature* (Vintage, New York, 1994). First edition: Pantheon, New York, 1992

505. S. Weinberg, Sokal's hoax. New York Rev. Books (August 8, 1996)

506. S. Weinberg, The revolution that didn't happen. New York Rev. Books (October 8, 1998)

507. J.A. Wheeler, The 'past' and the 'delayed-choice double-slit experiment', in *Mathematical Foundations of Quantum Theory*, ed. by A.R. Marlow (Academic Press, New York, 1978), pp. 9–48

508. J.A. Wheeler, From the Big Bang to the Big Crunch, interview by Mirjana R. Gearhart. Cosmic Search Mag. **1**(4) (1979)

509. J.A. Wheeler, Law without law (In [510, pp. 182–213])

510. J.A. Wheeler, W.H. Zurek (eds.), *Quantum Theory and Measurement* (Princeton University Press, Princeton, 1983)

511. J.A. Wheeler, Physics in Copenhagen in 1934 and 1935, in *Niels Bohr: A Centenary Volume*, ed. by A.P. French, P.J. Kennedy (Harvard University Press, Cambridge (Ma) 1985), pp. 221–226

512. D. Wick, *The Infamous Boundary: Seven Decades of Controversy in Quantum Physics* (Birkhauser, Boston, 1995)

513. E.P. Wigner, The unreasonable effectiveness of mathematics in the natural sciences. Commun. Pure Appl. Math. **13**, 1–14 (1960) (Reprinted in [515, pp. 22–235])

514. E.P. Wigner, Remarks on the mind-body question, in *The Scientist Speculates*, ed. by I.J. Gould (Heinemann, London, 1961), pp. 284–302 (Reprinted in [515, pp. 171–184] and in [510, pp. 168–181])

515. E.P. Wigner, *Symmetries and Reflections* (Indiana University Press, Bloomington, 1967)

516. E.P. Wigner, On hidden variables and quantum mechanical probabilities. Am. J. Phys. **38**, 1005–1009 (1970)

517. E.P. Wigner, Interpretation of quantum mechanics (In: [510, pp. 260–314])

518. A. Wilce, Quantum logic and probability theory, in *The Stanford Encyclopedia of Philosophy*, ed. by E.N. Zalta (Fall 2012 Edition). http://plato.stanford.edu/archives/fall2012/entries/qt-quantlog/

519. J. Wilkinson, E. Klein, J. Iliopoulos, *Rugby Quantique* (Les Presses de l'ENSTA, Paris, 2011)
520. M.N. Wise, Pascual Jordan: quantum mechanics, psychology, National Socialism, in *Quantum Mechanics: Science and Society*, ed. by P. Galison, M. Gordin, D. Kaise (Routledge, New York, 2001)
521. H.M. Wiseman, From Einstein's theorem to Bell's theorem: a history of quantum nonlocality. Contemp. Phys. **47**, 79–88 (2006)
522. H.M. Wiseman, Grounding Bohmian mechanics in weak values and Bayesianism. New J. Phys. **9**, 165 (2007)
523. H.M. Wiseman, Bell's theorem still reverberates. Nature **510**, 467–469 (2014)
524. R.E. Wyatt, *Quantum Dynamics with Trajectories* (Springer, New York, 2005)
525. H.D. Zeh, Why Bohm's quantum theory? Found. Phys. Lett. **12**, 197–200 (1999)
526. A. Zeilinger, The message of the quantum. Nature **438**, 743 (2005)

Author Index

A

Aharonov, Yakir, 135
Albert, David Z., 18, 21, 130, 183–184, 199–200, 212
Allori, Valia, 200, 207, 291
Aspect, Alain, 222, 260, 282

B

Bacciagaluppi, Guido, 234, 236, 253, 272
Ballentine, Leslie E., 34
Bassi, Angelo, 219–220
Bauer, Edmond, 9
Bell, John, 2–4, 8, 13, 15, 18–19, 24, 27, 31, 35, 40–41, 43, 62, 66–67, 77–78, 112, 115, 120–121, 123–126, 129, 137, 151, 153, 162, 166, 169, 171–173, 175–176, 184–185, 200, 207, 210–211, 214, 218, 223, 226, 232–235, 241, 244–246, 248, 254, 256–260, 262, 268, 278, 280–284, 288–289, 292–293, 295, 298
Beller, Mara, 5, 7, 93, 234, 273, 286–287
Berkeley, George, 17, 75, 89
Bernstein, Jeremy, 2, 115
Bertlmann, Reinhold, 259–260, 262
Blokhintsev, Dmitrii I., 34, 270
Bohm, David, 2, 8, 13, 16, 18–19, 62, 91, 121, 129, 131, 135, 137, 150, 153, 154, 172, 233–235, 241, 256, 258, 266–277, 281, 284, 290, 293, 297
Bohr, Niels, 4–8, 11–13, 16, 23–24, 35, 61, 65, 75, 93–94, 112, 148, 151, 206, 221, 233–247, 249, 252–253, 263, 265, 270, 272–276, 278–281, 286, 288–289, 296

Born, Max, 4, 6–8, 12, 117, 150–151, 185, 206, 234–235, 243, 247–248, 255, 263, 266, 268, 270–271, 276, 284, 286, 288, 296
Bridgman, Percy W., 256
Brillouin, Léon, 266
Brown, Harvey, 180, 211

C

Carnap, Rudolf, 278, 279
Carroll, Sean, 10–11, 199
Casimir, Hendrik, 117, 163
Clauser, John F., 112, 257, 280–282
Clifton, Robert, 43, 69
Cushing, James T., 234, 269–270

D

Damour, Thibault, 206
Daumer, Martin, 76
De Broglie, Louis, 4–5, 11, 16, 18–19, 113, 137, 211, 234, 236–237, 256, 264–271, 274–275, 281, 283–284, 288–289
D'Espagnat, Bernard, 10, 282
Deutsch, David, 146, 207, 211
Dewdney, Christopher, 18, 134
DeWitt, Bryce S., 201–202
Diderot, Denis, 77
Diósi, Lajos, 216
Dirac, Paul A.M., 2, 4, 14, 15, 40
Duhem, Pierre, 81, 82, 87, 125, 182, 215, 300
Dürr, Detlef, 18, 76, 129, 152, 160, 180

© Springer International Publishing Switzerland 2016
J. Bricmont, *Making Sense of Quantum Mechanics*,
DOI 10.1007/978-3-319-25889-8

Subject Index

A

Action-reaction, 179
Alice and Bob, 118, 123, 126, 223, 225–226
Angular momentum, 42, 173, 248, 299
Anthropomorphic, 118–119, 121, 123
Anticommute, anticommuting, 67, 178
Asymptotic position, 53, 160
Atomic theory, 84, 125, 202
Avogadro's number, 153, 195, 229

B

Basis (of vectors), 46–47, 49–50, 54–57, 61, 63, 192, 231
Bayesian, 101–105, 107, 159, 223, 293
Beable, 199–200, 210–211, 214, 295
Bell inequality, 121, 280, 282, 295
Born's rule, 55, 149, 158–159, 185, 203–207, 209, 223, 266, 295–296, 299
Branch, 155, 201–202, 206–207, 209, 211, 296
Butterfly (effect), 99

C

Cat (Schrödinger's), 11–12, 38–40, 44, 92, 153, 182, 197, 200, 201, 203, 208–210, 249–251, 291
Chaos, chaotic, 98, 100–101
Coin (tossing), 38–39, 41, 101–107, 114–115, 203, 205, 223–224, 226
Collapse (of the quantum state), 9, 30–32, 35–36, 39–44, 54–55, 58, 60, 93, 98, 114, 122, 124, 126, 145, 151–156, 161–162, 172–173, 177, 179, 195–197, 199–200, 203, 213–219, 222–

224, 228–229, 243, 247, 249–252, 268, 272, 292–293, 296, 299
Commutation relations, 62, 68, 178
Commutator, 60
Commute, commuting, 61–62, 66–68, 70–71, 178–179, 219, 231–232, 242, 254, 296
Compatible (quantities), 296
Complementarity, 8, 12, 23–24, 58, 60–62, 65, 246, 274–276, 279, 286, 296
Complementary, 23, 24, 236
Completeness (of quantum mechanics), 13, 150, 242, 253
Configuration space, 52, 133, 162, 203, 207–208, 211, 265–267, 272, 288, 297
Consciousness, 4, 7, 9–10, 39–40, 74–75, 89, 92, 183–184, 201, 224, 300
Conservation law, 111, 135, 173, 297
Contextual(ity), 67, 148, 151–152, 292, 296
Continuous matter density, 207–209, 212, 214–216, 228
Counting probability, 204–206, 209

D

Decoherence, 40, 117, 154–155, 195, 202, 218–219, 230–232, 250, 297
Decoherent histories, 155, 199, 216–222, 228–232, 260
Degenerate (eigenvalue), 61
Delayed choice, 26–27, 33, 89, 145
Demon (evil), 78
Demon (Laplacian), 94, 102
Dice, 102, 106, 226, 234–235
Dirac equation, 29, 300
Dirac delta function, 57
Dirac sea, 181

© Springer International Publishing Switzerland 2016
J. Bricmont, *Making Sense of Quantum Mechanics*,
DOI 10.1007/978-3-319-25889-8

Printed in the United States
By Bookmasters